STUDY GUIDE FOR STARR AND TAGGART'S

BIOLOGY
THE UNITY AND DIVERSITY OF LIFE

THIRD EDITION

JANE B. TAYLOR
Northern Virginia Commmunity College, Woodbridge

Wadsworth Publishing Company
Belmont, California
A Division of Wadsworth, Inc.

ISBN 0-534-02743-1

2 3 4 5 6 7 8 9 10---88 87 86 85 84

CONTENTS

PREFACE

This study guide is like a tutor; it increases the efficiency of your study periods; it condenses the major ideas of your text; it asks you to do a series of specific tasks to demonstrate your ability to recall key concepts and terms and relate them to life; it tests you on your understanding of the factual material and indicates what you might wish to reexamine or clarify; and it gives you a preliminary estimate of your next test score based on specific material. Most important, though, the study guide and text together help you to make informed decisions about matters that affect your own well-being and the well-being of your environment. In the years to come our survival will increasingly depend on administrative and managerial decisions based on an informed biological background.

HOW TO USE THIS STUDY GUIDE

After this preface you will find an outline that will show you how the study guide is organized and help you use it efficiently. Each chapter of the study guide begins with an outline of the topics discussed, in order to provide an overview of what follows. The content of each text chapter is then broken up into sections, which are labelled 1-I, 1-II, and so on. Each section has many parts. The *summary* stresses important concepts and indicates the text pages covered; it is followed immediately by a list of *key terms*. To help you memorize the terms so that you can improve your grades, flash cards (with a term or name on one side and its definition on the reverse side) are extremely useful and well worth the time and effort.

A series of learning *objectives* follows each key terms section. These are tasks that you should be able to accomplish if you have understood the assigned reading in the text. There are generally three levels of difficulty here: some objectives require that you memorize the terms and concepts; some objectives require that you understand the material; and others require that you apply your understanding of the terms and concepts to different situations.

So that you can immediately evaluate your mastery of the terms and concepts, each objectives section is followed by a series of *self-quiz questions,* which are answered in the back of the guide. Any wrong answers will show you which portions of the text you need to reexamine.

Each chapter ends with a short section entitled *Integrating and Applying Key Concepts*. These invite you to try your hand at applying the major concepts to situations in which there is not necessarily a single pat answer (so none is provided in the back of the guide). Your text will generally provide enough clues to get you started on an answer, but these sections are intended to stimulate your thought processes and provoke group discussions.

Finally, *crossword puzzles* are included following chapters 2, 8, 19, 29, 35, and 41, as another way to help you match terms with definitions. Answers to the puzzles are in the back of the study guide.

I would like to thank Jane Townsend and Jonas Weisel for their care in editing the manuscript.

Jane B. Taylor
Northern Virginia Community College,
Woodbridge Campus

The outline below indicates how each chapter in the guide is structured.

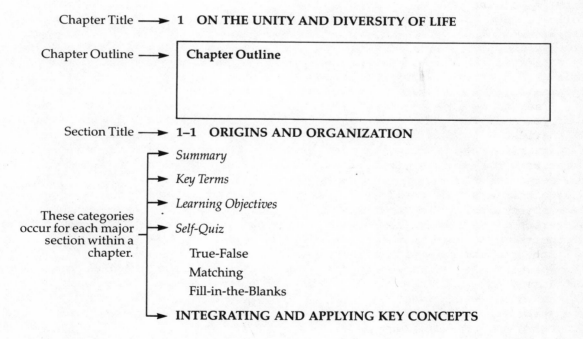

Chapter Title ⟶ **1 ON THE UNITY AND DIVERSITY OF LIFE**

Chapter Outline ⟶ **Chapter Outline**

Section Title ⟶ **1–1 ORIGINS AND ORGANIZATION**

These categories occur for each major section within a chapter.

→ *Summary*

→ *Key Terms*

→ *Learning Objectives*

→ *Self-Quiz*

True-False

Matching

Fill-in-the-Blanks

⟶ **INTEGRATING AND APPLYING KEY CONCEPTS**

1
ON THE UNITY AND DIVERSITY OF LIFE

ORIGINS AND ORGANIZATION

UNITY IN BASIC LIFE PROCESSES

  Metabolism
  Growth, Development, and
    Reproduction
  Homeostasis
  DNA: Storehouse of Constancy
    and Change

DIVERSITY IN FORM AND FUNCTION

  The Tropical Reef
  The Savanna
  A Definition of Diversity

ENERGY FLOW AND THE CYCLING
OF MATERIAL RESOURCES

PERSPECTIVE

1–I ORIGINS AND ORGANIZATIONS

Summary

Living organisms tend to have specific form, function, and behavior. They tend to have specific types of atoms and molecules, characteristic patterns of organization, and particular ways of obtaining and systematically using energy and materials. Organisms are constructed from one or more cells. They can generally adjust to many long-term and short-term environmental changes. Living forms can reproduce themselves and are committed to characteristic programs of growth and development. And organisms interact on several different levels of organization in nature.

Key Terms

NOTE: You should be able to define and explain all of the following terms. Under-lined terms are especially important and are often boldfaced or italicized in the text.

life (p. 2)
adaptive (p. 3)
cell (p. 4)
tissue (p. 4)
organ (p. 4)
bacterium (p. 4)
virus (p. 4)

molecule (p. 4)
potential (p. 4)
energy (p. 5)
atom (p. 5)
matter (p. 5)
population (p. 5)
multicelled (p. 5)

organism (p. 5)
community (p. 6)
ecosystem (p. 6)
biosphere (p. 6)
subatomic particle (p. 6)
organelle (p. 6)
organ system (p. 6)

Objectives

After reading the chapter and thinking about the contents, you should be able to:

1. Compare the basic structures of (a) a frog and a bacterium, and (b) a virus and a rock.

2. List some specific functions or examples of behavior carried out by:
   a frog
   a bacterium
   a virus
   a rock

3. Define and contrast energy and matter. Explain how each is related to a living organism and to work.

4. Arrange in order from smallest to largest the levels of organization that occur in nature. Define each as you list it.

Self-Quiz Questions

True-False If false, explain why.

____ (1) A bacterium can reproduce itself by itself if it has the appropriate raw materials, but a virus cannot.

____ (2) A rock falling three meters and smashing a glass vase is an example of work being done.

____ (3) Food storage, reproduction, and locomotion are all examples of work that cells do.

Matching Choose the one most appropriate answer to match each of the following terms.

(4) __F__ atom

(5) __C__ cell

(6) __E__ community

(7) __G__ ecosystem

(8) __D__ molecule

(9) __B__ organelle

(10) __H__ population

(11) __A__ subatomic particles

(12) __I__ tissue

A. proton, neutron, or electron

B. a well-defined structure within a cell that carries out a particular function

C. smallest unit of life

D. two or more atoms bonded together

E. all of the populations interacting in a given area

F. smallest unit of a pure substance that has the properties of that substance

G. a community interacting with its nonliving environment

H. a group of individuals of the same species in a particular place at a particular time

I. a group of cells that work together to carry out a particular function

Metabolism   6

Growth, Development, and Reproduction   6-7

Summary

All forms of life show metabolic activity; they extract and transform energy from
their environment and use it for manipulating materials in ways that ensure their own
maintenance, growth, and reproduction. Each living organism is a continuum of patterns
that unfold during its life cycle. The patterns unfold in the same way for all of its
kind, and they correspond to specific aspects of the environment.

Key Terms

energy transformation (p. 6)   metabolism (p. 6)   development (p. 6)
maintenance (p. 6)   reproduction (p. 6)   larva (p. 7)
ATP (p. 6)   growth (p. 6)   pupa (p. 7)

Objectives

1. Explain what is meant by *energy transformation* and give an example.

2. State which energy sources you tap each day and the forms into which you transform
    that energy for both long-term storage and immediate use.

3. State which energy sources are tapped by green plants and the forms in which
    plants store energy.

4. Explain what rocks would be able to do (that they cannot do now) if they carried
    out metabolism.

5. List in order the stages of the moth life cycle; then state why you think the
    larval form does not eat the same food as the adult.

Self-Quiz Questions

    True-False If false, explain why.

____ (1) Converting light to heat is an example of an energy transformation.

____ (2) Trapping light energy in the chemical bonds of sugar is an example of an
          energy transformation.

____ (3) All of the energy stored in food molecules on earth came from our sun.

____ (4) To stay alive, an organism must obtain energy from someplace else.

____ (5) Green plants absorb sugar, water, and minerals from the soil. The sugar
          absorbed in this manner is used to do cellular work.

____ (6) Larvae that consume foods not eaten by the adults can coexist in the same re-
          gion without competing with the adults for the same food supply.

____ (7) The moth adult is equipped with jaws that enable it to obtain enough energy
          to fly.

Fill-in-the-Blanks

In humans (8) _____ serves as the main long-term storehouse of energy, and (9) _____ is used as a more immediate energy source to do metabolic work. (10) _____ is the capacity for acquiring and using energy for stockpiling, tearing down, building up, and eliminating materials in controlled ways. The four stages in the moth life cycle, in sequence, are (11) _____, (12) _____, (13) _____, and (14) _____.

1–III Homeostasis

DNA: Storehouse of Constancy and Change

Summary

All forms of life depend on homeostatic controls. These maintain internal conditions within some tolerable range even when external conditions change. They also govern new kinds of adjustments in the internal state as the life cycle unfolds. DNA is a storehouse of patterns for all heritable traits in living things. Mutations introduce variations in the patterns, and the environment—both internal and external—determines whether or not any changes in the patterns permit the organism to survive.

Key Terms

internal environment (p. 8)	inheritance (p. 8)	trait (p. 8)
homeostasis (p. 8)	hereditary (p. 8)	DNA (p. 8)
external environment (p. 8)	information (p. 8)	mutations (p. 8)
dynamic homeostasis (p. 8)	variations (p. 8)	

Objectives

1. List some examples of homeostatic activities in living organisms.

2. Define *inheritance* and indicate what is passed from parent to offspring that issues the instructions resulting in particular developmental and maintenance patterns.

3. Explain how variations can occur in offspring and why variations are important for the long-term survival of a population.

4. Imagine a community of different populations in which mutations never occur. Predict what will probably occur in these populations over thousands of years as the environment changes.

Self-Quiz Questions

True-False If false, explain why.

_____ (1) Instructions that result in particular developmental and maintenance patterns being followed are encoded in the ATP molecule.

_____ (2) From an evolutionary point of view, the populations that are likely to survive the longest have the fewest variations.

_____ (3) A mutation is never advantageous to the organism in which it occurs.

_____ (4) One of the major groups of bacteria lacks homeostatic controls.

_____ (5) Sweating is part of a homeostatic control system that keeps body temperature more or less constant in many mammals.

Fill-in-the-Blanks

(6) _____ is the capacity to maintain internal conditions within some tolerable range, even when external conditions vary. (7) _____ are changes that occur in the kind, structure, sequence, or number of DNA's component parts.

1-IV DIVERSITY IN FORM AND FUNCTION 9-15

Summary

The array of different organisms on earth is the total of variations that have proved adaptive over time in obtaining available resources, in tolerating diseases and toxic substances, and in escaping from predators and parasites.

Key Terms

diversity (p. 9)   blue-green algae (p. 12)   scavenger (p. 12)
reef (p. 10)   sea anemone (p. 12)   available resources (p. 13)
coral (p. 12)   predator (p. 12)   function (p. 13)
sponge (p. 10)   sea star (p. 12)   behavior (p. 13)
green algae (p. 12)   savanna (p. 12)   natural selection (p. 13)
red algae (p. 12)   ungulate (p. 12)

Objectives

1. Explain why there are so many different kinds of fish and coral on a tropical reef.

2. Explain why there are so many hoofed, herbivorous mammals on the African savanna.

Self-Quiz Questions

True-False If false, explain why.

____ (1) There are more different species of organisms associated with a tropical coral reef than there are living off the coast of Greenland because the mutation rate is much higher in the coral reef ecosystem.

____ (2) There are more different species of organisms associated with the savanna than with the arctic tundra because there is a greater variety and abundance of food in the savanna.

____ (3) Organisms that are well adapted to prevailing ecological conditions obtain a larger share of the resources and thus tend to survive longer and reproduce more often than less well-adapted organisms.

Matching Choose the one most appropriate answer for each.

(4) ____ coral

(5) ____ diversity

(6) ____ predator

(7) ____ savanna

(8) ____ scavenger

(9) ____ sea anemone

(10) ____ sponge

(11) ____ ungulate

A. tiny marine organisms with tentacles and cylindrical bodies encased in calcareous skeletons

B. small marine organism with weapon-studded tentacles and cylindrical body

C. has pores opened toward the oncoming food-laden currents

D. hoofed, plant-eating mammal

E. all of the variations in form and behavior that have accumulated in different types of organisms

F. grassland of tropical or subtropical regions punctuated with scattered trees and shrubs

G. feeds on dead animal flesh or other decaying matter

H. an organism that lives by preying on other organisms

1-V ENERGY FLOW AND THE CYCLING OF MATERIAL RESOURCES

16-17

PERSPECTIVE

18-19

Summary

Almost all existing forms of life depend directly or indirectly on one another for materials and energy. There is a one-way flow of energy from the sun, through producer organisms, on through consumers and decomposers.

Key Terms

geologic record (p. 16)
carbon (p. 16)
compound (p. 16)
nutrient (p. 17)
ecology (p. 17)
energy flow (p. 17)
raw materials (p. 17)

unity (p. 18)
evolution (p. 18)
natural selection (p. 18)
principle (p. 18)
Darwin (p. 18)
Wallace (p. 18)
consumer (p. 19)

herbivore (p. 19)
carnivore (p. 19)
decomposer (p. 19)
producer (p. 19)
cycling of materials (p. 19)
microorganisms (p. 19)

Objectives

1. Explain how huge, ponderous elephants depend on and benefit from the efforts of tiny dung beetles.

2. Diagram how raw materials, producers, consumers, and decomposers are interrelated in the flow of energy and the cycling of materials through an ecosystem.

3. Consult Figure 1.12 and explain why energy cannot be totally recycled even though matter can.

Self-Quiz Questions

True-False If false, explain why.

____ (1) Dung beetles act as consumers that help decomposers to recycle some of the nitrogen and minerals in elephant fecal material.

____ (2) A carnivore is a consumer, but a herbivore is a producer.

____ (3) Each time an energy transformation occurs, some energy is lost, usually as heat.

____ (4) The theory of evolution by natural selection explains how there can be so much diversity among the earth's organisms, but it does not explain why organisms share so many of the same characteristics.

Matching Choose the one most appropriate answer for each.

(5) ____ consumer

(6) ____ Darwin

(7) ____ decomposer

(8) ____ producer

(9) ____ Wallace

A. traps solar energy in its tissues

B. formulated the principle of evolution by natural selection

C. organism that eats other organisms

D. organism that breaks down the tissues of other organisms and releases the raw materials contained therein

Fill-in-the-Blanks

Solar energy is first trapped in the chemical bonds of substances in the tissues of

(10) _____, then passed to plant-eating organisms called (11) _____, the

tissues of which are in part consumed by flesh-eating organisms called (12) _____.

These last are in turn consumed by other flesh-eaters, which, when they die, are

broken down by (13) _____ such as bacteria and fungi into (14) _____

such as CO_2, O_2, nitrogen, and minerals.

INTEGRATING AND APPLYING KEY CONCEPTS

(1) Humans have the ability to maintain body temperature very close to 37°C.

    (a) What conditions would tend to make the body temperature drop?

    (b) What measures do you think your body takes to raise body temperature when it drops?

    (c) What conditions would cause body temperature to rise?

    (d) And what measures do you think your body takes to lower body temperature when it rises?

(2) (a) What is required for an animal to become well-adapted to a particular environment?

    (b) Do you consider yourself to be biologically well-adapted to your environment?

    (c) Are there any ways that you could improve your ability to adjust to desert life?

    (d) Arctic life?

    (e) Life in outer space?

2
METHODS AND ORGANIZING CONCEPTS IN BIOLOGY

Summary

The scientific method of approaching questions is a commitment to systematic observation and testing. A variety of different tools and experimental designs is used to record observations, test hypotheses, and draw conclusions. Most experiments compare a control group (used as a standard by which to judge results) with the experimental group, and only one variable at a time is tested quantitatively. Discipline, objectivity, suspended judgment, testing, and repeatability of experimental results are the bases of scientific principles. Occasionally, accidents and intuition contribute to the experimental process.

Key Terms

evolution (p. 20)
scientific (p. 20)
generalization (p. 21)
inductive reasoning (p. 21)
deductive reasoning (p. 21)
"scientific method" (p. 21)
hypothesis (p. 21)
variables (p. 21)
suspended judgment (p. 21)

testable (p. 22)
independent
  variables (p. 23)
dependent variables (p. 23)
controlled
  variables (p. 23)
experimental group (p. 23)
control group (p. 23)
randomization (p. 24)

sampling error (p. 24)
significant (p. 24)
theory (p. 24)
principle (p. 24)
"law" (p. 25)
Darwin (p. 25)
Wallace (p. 25)
discipline (p. 25)

Objectives

1. Contrast inductive reasoning with deductive reasoning.

2. Outline the principal steps generally used in the scientific method of investigating a problem.

3. Explain how observations differ from conclusions and how a hypothesis differs from a theory.

4. Explain why one or more control groups are used in an experiment.

5. Explain why most biologists regard the modern theory of evolution by natural selection as "scientific" and the idea that life has always existed in the forms we see today as "not scientific."

Self-Quiz Questions

True-False If false, explain why.

___ F (1) Deduction is a form of reasoning that proceeds from specific observations to the development of a generalized concept.

___ (2) Quantitative studies generally use instruments or tools to obtain precise measurements.

___ (3) A control group helps to establish how far a variable deviates from standard values.

___ F (4) When any test group is not equivalent to a natural population, the test group is said to be randomized.

Sequence Arrange the following steps of the scientific method in correct chronological sequence from first to last:

(5) b___   (6) d___   (7) f___   (8) e___   (9) a___   (10) c___

(a) Carry out the tests. Repeat as often as necessary to find out whether results consistently will be as predicted.

(b) Use trained judgment in selecting and summarizing the relevant preliminary observations from what could be nearly infinite observational trivia.

(c) Report objectively on the results of the tests and the conclusions drawn from them.

(d) Review all available preliminary observations. Be sure to note the range of conditions under which they have been made.

(e) Devise ways to test whether the explanation is valid. Think through how different but related conditions might affect the test outcome. Be sure the test you devise will address these so-called variables.

(f) Work out a hypothesis that seems in line with the observations.

Labeling Assume that you have to determine what object is inside a sealed, opaque box. Your only tools to test the contents of the box are a bar magnet and a triple-beam balance. Label each of the following with an O (for observation) or a C (for conclusion).

(11) O___ The object has two flat surfaces.

(12) O___ The object is composed of nonmagnetic metal.

(13) ___ The object is not a quarter, a half-dollar, or a silver dollar.

(14) ___ The object weighs *x* grams.

(15) ___ The object is a penny.

2-II   EMERGENCE OF EVOLUTIONARY THOUGHT

Summary

2,300 years ago, Aristotle categorized organisms in a hierarchy, according to the view that life proceeded gradually from lifeless matter through ever more complex forms of animal life—the Great Chain of Being. In the 1700s, Linnaeus developed the binomial system of nomenclature, which assigned a two-part (genus + species) name to each recognizably distinct kind of organism, established a concise hierarchical system, and reinforced the prevailing idea that each species is unique and unchanging. By the late 1700s and early 1800s, the discoveries of homologous structures, vestigial structures, and index fossils in the earth's geological strata caused biologists to question the idea that all forms of life had persisted unchanged since the earth's creation. Works published by Cuvier and Lamarck proposed that organisms have changed over time (have evolved), but extensive testing failed to demonstrate support for their theories of catastrophism and the inheritance of acquired characteristics.

Key Terms

Aristotle (p. 26)	family (p. 27)	vestigial structures (p. 28)
species (p. 26)	order (p. 27)	Buffon (p. 28)
Linnaeus (p. 26)	class (p. 27)	Cuvier (p. 28)
binomial system of nomenclature (p. 26)	phylum (p. 27)	catastrophism (p. 28)
genus (p. 27)	division (p. 27)	Lamarck (p. 29)
species epithet (p. 27)	kingdom (p. 27)	inheritance of acquired characteristics (p. 29)
	homologous structures (p. 27)	

Objectives

1. State the contributions of Aristotle, Linnaeus, Buffon, Cuvier, and Lamarck to our current views of the origins of the earth's species.

2. Arrange, in order of fewer and fewer organisms included, the following categories of classification: class, family, genus, kingdom, order, phylum, species.

3. Explain why the global explorations of the sixteenth century created problems for biologists, and how the binomial system of nomenclature not only helped to solve some of those problems but also encouraged biologists from countries with different languages to compare information and solve problems.

4. Define and contrast homologous and vestigial structures.

5. Explain how the discovery of homologous and vestigial structures caused biologists to reappraise their views on the unchanging nature of species.

6. Explain how the geologic record of fossils in rock layers differed from the idea that species are fixed and unchanging.

7. Connect the idea of use and disuse of organs with the theory of inheritance of acquired characteristics.

Self-Quiz Questions

Fill-in-the-Blanks

The theory of inheritance of acquired characteristics is primarily linked with the name of (1) _____. (2) _____ believed that there had been several "centers of creation," that the origin of species had been spread out geographically, and that species might have become changed through time.

Arrange to correct hierarchical order with the largest, most inclusive category first and the smallest, most exclusive category last:

(3) ___   (4) ___   (5) ___   (6) ___   (7) ___   (8) ___   (9) ___

A. class   B. family   C. genus   D. kingdom   E. order   F. phylum   G. species

True-False If false, explain why.

___ (10) The most inclusive (largest) taxonomic category is the phylum.

___ (11) Linnaeus was one of the first biologists to believe that species evolve substantially as time passes.

___ (12) There are more different species in a class than in an order.

___ (13) The binomial system developed by Linnaeus was a phylogenetic system based on the evolutionary relationships of organisms.

___ (14) Homologous structures are body parts that have no apparent role in the functioning of an organism.

___ (15) The organisms that were present in the earliest rock layers are essentially the same as those in the most recent (topmost) strata.

___ (16) According to Lamarck, parts of the body that have been stimulated by constant use grow larger than normal, and the greater development of those body parts is somehow transmitted to the offspring.

___ (17) Cuvier developed the theory that an organism's free will acts together with environmental influences to excite "vital fluids" that enlarge specific body parts.

___ (18) Malthus believed that the food supply of a population increases faster than the population increases.

Multiple Choice Choose the one most appropriate answer.

___ (19) Which of these men believed that life evolved *and* that new species are formed?

   (a) Aristotle   (b) Linnaeus   (c) Cuvier   (d) Lamarck

Summary

Spurred by the ideas of Lyell and Malthus and his experiences on the voyage of the *Beagle*, Darwin developed the theory of evolution by natural selection; he was later joined by Wallace in suggesting that species do change over time. Heritable variations occur among members of a species. Each population produces more offspring than can survive to reproduce. Bearers of the traits that improve chances for surviving and reproducing under prevailing environmental conditions tend to produce more descendants than other members of the population; this differential reproduction ensures that the most adaptive ("fit") traits will show up more frequently in the next generation.

Key Terms

Lyell (p. 30)

uniformitarianism (p. 30)

breeds (p. 31)

selection (p. 31)

Malthus (p. 32)

theory of natural selection (p. 32)

population (p. 32)

adaptive (p. 32)

differential reproduction (p. 32)

evolution of new species (p. 32)

Objectives

1. Distinguish between *population* and *species*.

2. List in sequential order the statements that compose the Darwin-Wallace theory of evolution by natural selection.

3. Indicate how the writings of Lyell, Hutton, and Malthus helped shape Darwin's early formulation of his theory.

4. Outline briefly how new species can evolve.

5. Explain how scientists determine which individuals should be grouped into the same species.

6. Describe Wallace's role in the formulation of the Darwin-Wallace theory.

Self-Quiz Questions

   True-False   If false, explain why.

____ (1) If two or more individuals from different populations can interbreed and produce offspring that can survive and reproduce, those individuals should be grouped in the same species.

____ (2) If two organisms are in the same species, they also must be part of the same population.

____ (3) When resources become scarce, competition for similar resources promotes greater specialization among the competitors.

PERSPECTIVE

Summary

A phylogenetic system categorizes organisms by using data from the fossil record, comparative anatomy, genetics, biochemistry, reproductive biology, behavior, ecology, geology, and geography to construct the major lines of evolutionary descent. Whittaker's five-kingdom system classifies organisms according to their structure and means of obtaining nutrients and energy.

Key Terms

Haeckel (p. 33)
phylogenetic system of
  classification (p. 34)
Whittaker (p. 34)
Monera (p. 34)

Protista (p. 34)
Plantae (p. 34)
Fungi (p. 34)
Animalia (p. 34)

autotroph (p. 34)
heterotroph (p. 34)
prokaryotic (p. 34)
eukaryotic (p. 34)

Objectives

1. Explain what is meant by a phylogenetic system of classification and how it differs from the way that organisms were originally classified by Linnaeus.

2. Characterize the kingdoms in Whittaker's five-kingdom system and distinguish each from the other four kingdoms. In order to do this, you may have to consult Table 2.2.

Self-Quiz Questions

   True-False If false, explain why.

____ (1) Phylogenetic classification is a process by which organisms inherit acquired characteristics.

____ (2) Modern classification systems try to categorize organisms phylogenetically whenever the necessary background information is available.

____ (3) All prokaryotes are monerans.

INTEGRATING AND APPLYING KEY CONCEPTS

(1) Suppose you are an ecologist working in an African game preserve that includes elephants. You want to discover the precise migratory habits of the elephants so that they can encounter humans less frequently and thus have better chances to survive. (a) What sources would you go to in order to review all available observations? (b) Which hypothesis might you try to test? (c) What tools might you use in testing your hypotheses? (d) Name three variables that might affect the migratory movements of elephants. (e) How could you isolate each of these variables and study the effects of just one variable at a time so as to deduce its contribution to elephant migratory behavior?

(2) What sorts of topics are generally regarded by scientists as untestable by the kinds of methods that scientists generally use?

(3) Do you think that all humans on earth today should be grouped in the same species?

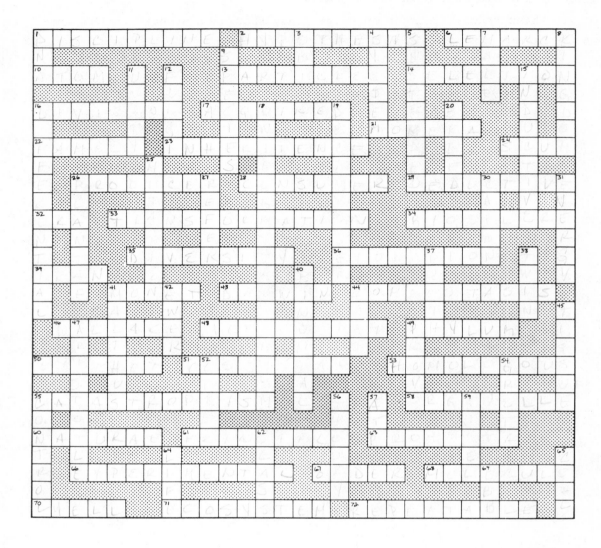

Across

1. a state of order based on submission to rules and authority
2. an educated guess
6. proposed that characteristics acquired during a lifetime could be inherited by offspring
10. smallest unit of a pure substance that has the properties of that substance
13. subatomic ___ ; very small parts
14. choosing or choice
16. the largest classification category of plants
17. not inherited
21. includes blue-green algae, bacteria, and their allies
22. a taxonomic group of related genera
23. the process of genetic transmission of characters or characteristics
24. a filmy layer of extraneous or impure matter that forms on the surface of a liquid or body of water

26. an organism that can synthesize its own food from appropriate inorganic substances
28. an organism that must depend on other organisms for its food
29. reasoning that proceeds from the general to the specific
32. age; geological span of time
33. change from one form to another
34. all of the regions on earth that support self-sustaining and self-regulating ecological systems
35. variety; different kinds
36. change in the frequency that a gene occurs in a population
39. Persia, in older days, is ___ now
41. French impressionist painter, 1840-1926
43. coformulator of the theory of natural selection
44. a state of physiological equilibrium produced by a balance of functions and of chemical composition within an organism
46. coformulator of the theory of evolution by natural selection
48. the physical and chemical processes that maintain life
49. a group of closely related classes of organisms
50. a term that characterizes the nature of a person or thing
51. occurring or persisting as a rudimentary or degenerate structure
53. corresponding in structure and evolutionary origin
55. the theory that geological processes such as mountain building, flooding, earthquakes, etc. occur through time in violent, disruptive surges rather than at a constant rate
58. discrete structure that has a specific function within a cell
60. *Scala* _____, by Aristotle
61. a fundamental doctrine on which new concepts are based or from which they are drawn
63. taxonomic group that includes unicellular or colonial eukaryotes
66. relating to being tested
67. a taxonomic group of closely related families
68. two or more atoms bonded together
70. wrote *Principles of Geology*
71. a biological community interacting with its nonliving environment
72. capable of being done again

Down

1. the molecular blueprint for life
3. Japanese sash
4. organ ___
5. held in abeyance; maintained in an undecided state
7. medieval war club with a spiked or flanged head; an aromatic spice
8. most inclusive (largest) category of classification
9. least inclusive category of classification of organisms
11. oneness; commonality
12. a group of organisms that absorb food they have digested outside their bodies
15. reasoning that proceeds from the specific to the more general
16. not proceeding at the same rate
18. the theory that the erosive action of wind and water and the gradual uplifting of mountain ranges have occurred at fairly constant rates through time
19. an organism that breaks down the remains of dead organisms
20. a variety or subspecies of organism
25. a heritable alteration of the genes or chromosomes of an organism
26. the kingdom that includes multicellular photosynthetic organisms
27. to direct to a source for help or information
30. a castrated rooster
31. the capacity to do work

37. a hard bonelike structure in a jaw or skeleton used to seize, hold, or chew
38. one of two offspring born at the same birth
40. having two names or a two-part name
41. said that populations increase faster than their food supply
42. a large, wide-mouthed pitcher or jug
44. an organism that harbors and provides nourishment for a parasite
45. a group of cells that cooperate to carry out a particular function
47. Greek philosopher (384-322 B.C.) who attempted to categorize all organisms according to a "great chain of being"
49. resort city in Utah
52. a volatile, highly flammable liquid that consists of two hydrocarbon groups linked by an oxygen atom
54. ___ Garbo
55. used as a standard for judging the results of an experiment
56. a coherent set of ideas that form a general frame of reference for further studies in the field of inquiry
57. the energy "currency" of a cell; a temporary energy-storage molecule
59. ___ Adams, American photographer of natural landscapes
62. a taxonomic group of closely related orders
64. the opposite of death
65. the smallest unit of a living organism that is capable of independent function
69. fade away, diminish

3
ATOMS, MOLECULES, AND CELL SUBSTANCES

3–I ORGANIZATION OF MATTER   38–42

The Nature of Atoms   39

Summary

All events in the living world depend on the organization and behavior of atoms and molecules. All energy transfers and transformations within and between living things occur at the level of atoms and molecules. The shapes and behavior of cells are determined by molecules contained within.

    The structure of each atom in a given element makes it different from atoms of other elements. The number and arrangement of its subatomic particles (protons, neutrons, and electrons) dictate the behavior of the atoms of a specific element. Uncharged neutrons in the nucleus, together with the protons, account for the *mass* (weight) *number* of an element. The number of protons alone dictates the *atomic number* of the element and helps, indirectly, to establish how that atom will react with other atoms.

    The number of positively charged protons in an *atom* always is equal to the number of negatively charged electrons; thus, atoms have no net charge. In an *ion* the number of protons is not equal to the number of electrons; ions are electrically charged.

Key Terms

element (p. 38)   electron (p. 39)   mass number (p. 39)
compound (p. 38)   molecule (p. 39)   isotopes (p. 39)
atom (p. 39)   electric charge (p. 39)   ion (p. 39)
proton (p. 39)   nucleus (p. 39)
neutron (p. 39)   atomic number (p. 39)

Objectives

1. Define *energy* and explain what it means for a living organism.

2. Explain how an organism that could obtain energy from different sources (a food generalist) might have an advantage over organisms that obtain energy from only one source (food specialists).

3. Indicate the relationships between atoms and elements by comparing the definition of each term.

4. Compare the definitions of *atom* and *molecule*.

5. Define *electric charge* and state how (a) two identically charged particles react to each other and (b) two oppositely charged particles react to each other.

6. List and describe the three types of subatomic particles in terms of location (inside or outside the nucleus) and charge.

7. Define and distinguish *atomic number* and *mass number*.

8. Show how an atom of carbon is different from atoms of oxygen, hydrogen, and nitrogen.

9. Explain what isotopes of an element are and how differences in the number of neutrons and the mass numbers distinguish each isotope.

10. Define *ion* and explain how an atom can be converted into an ion.

11. Name the four most abundant elements in the human body.

Self-Quiz Questions

True-False If false, explain why.

____ (1) Conversion of light energy into chemical energy, as occurs in plants, is an energy transformation.

____ (2) The number of protons in the nucleus of an atom is equal to the number of electrons arranged in orbitals outside the nucleus.

____ (3) The number of protons in a nucleus helps, indirectly, to determine how that atom will react with other atoms.

____ (4) If two particles bear the same type of electric charge, they will be attracted to each other.

____ (5) Carbon 12 and Carbon 14 are isomers of each other.

____ (6) A sodium atom can be converted into a sodium ion if the sodium atom can lose an electron.

Matching Choose the one most appropriate answer for each.

(7) ____ atom

(8) ____ atomic number

(9) ____ electric charge

(10) ____ electron

(11) ____ element

(12) ____ energy

(13) ____ ion

(14) ____ isotope

(15) ____ mass number

(16) ____ matter

(17) ____ molecule

(18) ____ neutron

(19) ____ proton

(20) ____ pure

A. a charged atom

B. an uncharged subatomic particle

C. the number of protons in the nucleus of one atom of an element

D. the smallest neutral unit of an element that shows the chemical and physical properties of that element

E. that which occupies space and has mass

F. one or more atoms linked together by one or more chemical bonds

G. contributes the capacity to do work, that is, to move a certain amount of material across a specific distance

H. a positively charged subatomic particle

I. composed of only certain atoms or molecules

J. protons + neutrons = _____

K. a negatively charged subatomic particle

L. ninety-two different ones occur in nature

M. a form of an element, the atoms of which have a number of neutrons different from other forms of the same element

N. pushes away similar particles but attracts oppositely charged particles

3–II   How Electrons are Arranged in Atoms

COMMENTARY: Keeping Track of Electrons: Some Options

Electron Excitation

Summary

An *orbital* is a specific region of space outside the nucleus in a particular *energy level* where an electron is most likely to be located. An orbital can contain no more than two electrons. Negatively charged electrons, arranged in orbitals outside the nucleus, tend to get as close as possible to the positively charged protons in the nucleus and stay as far away as possible from each other. As soon as the inner orbitals are each filled with two electrons, orbitals of the next highest energy level are the next to fill. Elements composed of atoms that have two electrons in each occupied orbital tend to be chemically nonreactive substances; atoms that have one or more orbitals containing only one electron tend to react chemically with other such atoms.

Electrons occupying the orbitals in the highest occupied (outermost) energy level of an atom are the ones that interact with other atoms to form either a molecule or some other group of associated atoms. Orbitals have specific shapes, which determine the shapes of the molecules that the orbitals compose. An electron can absorb certain amounts of incoming energy and as a result spend more of its time farther out from the pull of the nucleus. If the outside energy source is removed, the excited electron eventually gives off the extra energy it absorbed and returns to whichever orbital closest to the nucleus can accommodate it.

Key Terms

orbital (p. 40)   reactive (p. 40)   electron dot model (p. 41)
electron energy   probability distribution (p. 40)   absorb (p. 41)
  level (p. 40)   orbital model (p. 41)   release (p. 41)
nonreactive (p. 40)   shell model (p. 41)   increment (p. 42)

Objectives

1. Define *orbital* and indicate how electrons are related to orbitals.

2. Describe the relationships among energy levels, orbitals, and electrons. Explain what happens when an orbital is filled with two electrons.

3. Examine Figure 3.1 and state which of the elements listed tend to be chemically reactive with other substances. Which elements listed do not tend to react chemically?

4. Describe the situation that enables atoms to be reactive.

5. Explain how the shapes of orbitals can influence the specific shapes of molecules that are formed by several atoms becoming linked together.

6. Distinguish the orbital model, the shell model, and the electron dot model of electron distribution from each other.

7. Explain what happens to the behavior of one or more of its electrons (a) when an atom absorbs energy and (b) when the atom releases energy.

Self-Quiz Questions

True-False If false, explain why.

____ (1) In an atom the volume of space that can accommodate two electrons at most is called an energy level.

____ (2) When an atom absorbs energy, electrons that are part of that atom move faster and tend to spend more of their time farther from the nucleus than they did before the energy was absorbed.

____ (3) In order for an atom to react with another atom, all orbitals of that atom must already contain two electrons.

____ (4) Atoms that have four orbitals that each contain only one electron can form more bonds, and thus more molecules with different shapes, than can atoms with only two orbitals that each contain only one electron.

3–III   INTERACTIONS AMONG ATOMS IN THE CELLULAR WORLD

Summary

In a chemical bond one or more orbitals of one atom become linked with one or more orbitals of another atom (or atoms), establishing an energy relationship between the participating atoms, which holds the atoms at certain distances from each other and makes the group of atoms acquire a specific three-dimensional shape.

In cells, molecules are held together by strong covalent bonds that involve a sharing of electron pairs, and by weaker interactions (ionic bonds and hydrogen bonds) between atoms or ions having opposite charge.

Key Terms

chemical bond (p. 42)	electronegative (p. 43)	hydroxyl group (p. 44)
energy relationship (p. 42)	organic (p. 43)	acetyl group (p. 44)
ionic bond (p. 42)	inorganic (p. 43)	carboxyl group (p. 44)
ionic crystal (p. 42)	hybrid (p. 43)	disulfide group (p. 44)
covalent bond (p. 43)	tetrahedron (p. 43)	amino group (p. 44)
nonpolar covalent bond (p. 43)	hydrocarbon (p. 43)	phosphate group (p. 44)
polar covalent bond (p. 43)	functional groups (p. 44)	hydrogen bond (p. 44)

Objectives

1. Define and use in a sentence the term *chemical bond*.

2. Explain how a chemical bond can be formed between two atoms.

3. Identify which bonds are strong and which bonds are weak.

4. Describe fully covalent, ionic, and hydrogen bonds so as to distinguish each from each other.

5. Contrast nonpolar and polar covalent bonds.

6. Explain the reason(s) two single atoms, each with a lone electron in one of its outermost orbitals, would tend to become bonded together. Explain what is gained by forming bonds.

7. Explain why carbon atoms are part of so many more substances than atoms of any other element.

8. Distinguish the terms *organic* and *inorganic*.

9. Explain why some carbon-containing molecules are flat and others are more three dimensional.

10. Describe and draw a tetrahedron. Explain how tetrahedra are related to carbon skeletons.

11. Illustrate, by drawing their carbon skeletons, a linear chain, a branched chain, and a ring structure.

12. (a) Distinguish the hydroxyl group from the carboxyl group (Fig. 3.4).
    (b) Describe the amino and phosphate groups.

13. State one way that living organisms use hydrogen bonding.

Self-Quiz Questions

True-False If false, explain why.

____ (1) A chemical bond will be formed when two atoms share a pair of electrons, and a chemical bond may be formed by atoms losing or gaining one or more electrons.

____ (2) An ionic bond is stronger than a covalent bond.

____ (3) Covalent bonds typically occur within molecules; weaker bonds usually form between two or more different molecules but may also form within the same molecule.

____ (4) The atoms in table salt (sodium chloride) are linked together with covalent bonds.

____ (5) In order to form a covalent bond, two orbitals from different atoms, each containing one electron, must overlap so that a pair of electrons may be shared between the two nuclei.

____ (6) Chemical bonds form between atoms because the two atoms bonded together are more stable than the two atoms are when they are separate.

____ (7) Table sugar dissolves in water because its constituent atoms have been ionized.

____ (8) Carbon atoms are part of so many different substances because, in one of their electron configurations, there are four unpaired electrons in the outermost occupied energy level. Thus each unpaired electron can form a covalent bond with a variety of other atoms, including other carbon atoms.

____ (9) All organic molecules contain carbon.

____ (10) When a ring like this ⬡ is shown, it's understood that a carbon atom occurs at every corner and that no other atoms are attached.

____ (11) In a triple bond three electrons are shared by two atoms.

____ (12) In a branched chain each carbon atom is never bonded to more than one other carbon atom.

____ (13) The orbitals of a carbon atom project outward from its nucleus to the four corners of a tetrahedron; the atom can bond covalently at all four of its "corners."

Short-Answer Essay

(14) Why does a carbon atom generally form four bonds? (Consult Figure 3.2 remembering what is required to form a bond, and page 43.) In order for carbon to form four bonds consistently, what must happen to the distribution of electrons on the second energy level as carbon is about to bond with other atoms?

(15) What is a hydrocarbon? Give an example of a hydrocarbon.

(16) Interpret the meaning of ⬡ .

(17) How are double and triple bonds formed within molecules?

Fill-in-the-Blanks

A(n) (18) _____ group contains an oxygen atom and a hydrogen atom.

A(n) (19) _____ group contains two oxygen atoms, a carbon atom, and a hydrogen

atom. A(n) (20) _____ group contains four oxygen atoms, two hydrogen atoms, and

a phosphorus atom. A(n) (21) _____ group contains one nitrogen atom and two

hydrogen atoms.

Summary

Certain substances dissolve in water because they can become dispersed among and form
weak bonds with the polar water molecules. Such substances are hydrophilic. Substances
that cannot form weak bonds with water molecules are hydrophobic. Hydrophobic molecules (such as oil) tend to cluster together in a watery environment.

Covalent bonds formed between atoms of carbon, nitrogen, oxygen, and hydrogen
typically contain bond energies between 80 and 110 kilocalories per mole. (A mole is
a specific number (6.023×10^{23}) of atoms or molecules of any substance.) Under conditions that generally prevail in cells, ionic bonds are much weaker than covalent bonds.

Water can dissociate into hydroxide ions and hydrogen ions. A substance whose
molecules release hydrogen ions in solution is known as an acid. Any substance that
combines with hydrogen ions in solution is a base. pH values are a shorthand measure
of the degree to which a solution is acidic, basic, or neutral. All biochemical reactions that usually take place in water are sensitive to changes in pH. It is for-
tunate that the pH of pure water is 7, or neutral, for that contributes to the stability of biochemical processes in cells.

Inside cells, buffer molecules help to maintain pH values within rather narrow
tolerance ranges. Cell functioning also depends on essential ions such as K^+, Na^+,
Ca^{++}, Mg^{++}, and Cl^-.

In the molecules of life, each carbon atom is typically linked by four covalent
bonds to other atoms—most often, other atoms of carbon or atoms of hydrogen, oxygen,
nitrogen, sulfur, and phosphorus. Short chains or ring-shaped carbon-containing molecules may in turn be joined together by *condensation* to form the larger molecules
characteristic of living organisms. When the time comes to break large molecules down
into their subunits, *hydrolysis* (which is essentially the reverse of condensation)
occurs.

Key Terms

dissolve (p. 45)	dissociation (p. 46)	alkaline (p. 47)
hydrophilic (p. 45)	reversible (p. 46)	buffers (p. 47)
hydrophobic (p. 45)	bicarbonate (p. 46)	salt (p. 47)
bond energy (p. 45)	ions (p. 46)	calmodulin (p. 47)
kilocalorie (p. 45)	acid, acidic (p. 46)	condensation (p. 47)
mole (p. 45)	base, basic (p. 46)	polymer (p. 47)
hydrogen ion, H^+ (p. 45)	neutral (p. 46)	subunits (p. 47)
ionized (p. 46)	concentration (p. 46)	hydrolysis (p. 48)
hydroxide ion, OH^- (p. 46)	pH value (p. 46)	

Objectives

1. Explain what "to dissolve" means.

2. Distinguish between hydrophobic and hydrophilic substances.

3. Give an example of something that is influenced by hydrophobic interactions.

4. Compare covalent, ionic, and hydrogen bonds in terms of their typical bond energies.

5. Explain how water dissociates into its ions by writing a chemical reaction like that shown on page 45. State whether you think the reaction is reversible.

6. Describe the pH scale, specifying the acidic range, the basic range, and the point of neutrality.

7. Distinguish between an acid and a base. Give three examples of acids and three examples of bases (see Fig. 3.6).

8. Explain why the pH of substances is important to living organisms.

9. Define the term *salt* and state how a salt is related to ions.

10. List four essential salts.

11. (a) Define and contrast condensation and hydrolysis.
    (b) Show how one process resembles the other in reverse.
    (c) State the class of proteins that participates in both processes.
    (d) State the role of water molecules in each case.

Self-Quiz Questions

True-False If false, explain why.

_____ (1) A kilocalorie is equal to 6.023 ~~atoms~~ of a substance. [handwritten: mole]

_____ (2) The shapes of lipid-based molecules are influenced by their hydrophobic interactions with water molecules.

_____ (3) Hydrophilic substances generally dissolve in water.

_____ (4) Ionic bonds and hydrogen bonds contain approximately similar bond energies.

_____ (5) Molecules are assembled into polymers by means of hydrolysis.

_____ (6) Hydrolysis in reverse resembles condensation as it generally occurs in living cells.

_____ (7) Neither condensation nor hydrolysis occurs in living systems without the participation of enzymes.

Fill-in-the-Blanks

Two examples of essential ions are (8) ____Na+____ and (9) _____ .

H_2CO_3 (H-O-C $\overset{O}{\underset{O-H}{\diagup}}$) dissociates to form H^+ plus (10) ___HCO___ , the (11) ___bicarbonate___

ion. Ionic substances that lack hydrogen and hydroxyl ions are called (12) ___salt___ .

Sometimes polar molecules (13) ___dissociate___ ; that is, they separate into two (or more)

ions.

Matching Choose the one most appropriate answer for each.

(14) ___ acid

(15) ___ base

(16) ___ essential ion

(17) ___ hydrogen ion

(18) _A_ hydroxyl ion

(19) ___ neutral

(20) ___ salt

A. OH^-

B. Na^+, Cl^-

C. H^+

D. NaCl

E. combines with hydrogen ions in solution

F. releases hydrogen ions in solution

G. 7

3–V   CARBOHYDRATES

Summary

Carbohydrates form some of the main structural materials (e.g., cellulose, chitin) and foods (sucrose, glucose, starch, glycogen) that serve as reservoirs of energy. *Monosaccharides* are simple sugars that are the basic subunits of the larger carbohydrates, the polysaccharides. *Polysaccharides*, such as starch, are digested by hydrolysis into simple sugars (i.e., glucose). Many simple sugars can be linked together by condensation to make cellulose or starch.

Key Terms

carbohydrate (p. 48)
monosaccharide (p. 48)
glucose (p. 48)
fructose (p. 48)
galactose (p. 49)
ribose (p. 49)
deoxyribose (p. 49)
structural isomers (p. 49)

aldehyde (p. 48)
ketone (p. 48)
disaccharide (p. 49)
sucrose (p. 49)
lactose (p. 49)
maltose (p. 49)
polysaccharide   (p. 49)
glycogen (p. 49)

amylose (p. 49)
chitin (p. 49)
cellulose (p. 49)
starch (p. 49)
microfibrils (p. 50)
macrofibrils (p. 50)

Objectives

1. Explain how carbohydrates differ from hydrocarbons.

2. Define *carbohydrate*. State what the basic molecular subunits of carbohydrates are.

3. Distinguish among mono-, di-, and polysaccharides and name as many of each as you can.

4. Illustrate how a carbohydrate is assembled from its basic subunits.

5. Indicate how mono-, di-, and polysaccharides are used in living organisms. Explain how the structure of each allows organisms to use that form of carbohydrate for a specific purpose.

Self-Quiz Questions

   True-False If false, explain way.

____ (1) Glucose, a crystalline sugar, has the molecular formula $C_6H_{12}O_6$. The mass numbers of hydrogen, carbon, and oxygen are 1, 12, and 16, respectively. The molecular weight of glucose is 29.

____ (2) $(CH_2O)_6$ is just another way of writing $C_6H_{12}O_6$.

____ (3) Peptide bonds link monosaccharides into polysaccharides called starches.

____ (4) In living organisms carbohydrates serve as structural supports and as food reserves.

____ (5) Polysaccharides can be broken down into simpler sugars if the appropriate enzymes are present.

____ (6) Condensation is the process that assembles simple sugars into starches.

____ (7) Carbohydrates contain oxygen; hydrocarbons do not.

   Matching Match *all* applicable letters with the appropriate terms. A blank may contain more than one letter.

(8) ____ amylose

(9) ____ cellulose

(10) ____ chitin

(11) ____ fructose

(12) ____ galactose

(13) ____ glucose

(14) ____ glycogen

(15) ____ lactose

(16) ____ maltose

(17) ____ ribose

(18) ____ sucrose

A. monosaccharide

B. disaccharide

C. polysaccharide

D. used as a structural support

E. used as a food reserve

F. table sugar

G. milk sugar

H. used in brewing and in germinating seed

I. a five-carbon sugar

J. a six-carbon sugar

3–VI LIPIDS

Summary

Lipids are oily or waxy substances such as true fats, waxes, steroids (such as cholesterol and estrogen), terpenes, phospholipids, and glycolipids. True fats are compact forms of stored energy. Waxes serve in waterproofing and protecting surfaces from potentially damaging agents. Phospholipids and glycolipids are important features of cell membranes.

Key Terms

lipid (p. 50)	unsaturated fat (p. 51)	sex hormones (p. 52)
true fat (p. 50)	wax (p. 51)	terpene (p. 52)
glycerol (p. 50)	cutin (p. 51)	chlorophyll (p. 52)
carbon skeleton (p. 50)	suberin (p. 52)	carotenoids (p. 52)
fatty acid (p. 51)	steroid (p. 52)	phospholipid (p. 52)
saturated fat (p. 51)	cholesterol (p. 52)	glycolipid (p. 52)
triglyceride (p. 51)	bile acids (p. 52)	

Objectives

1. Explain how lipids differ from carbohydrates.

2. Define *lipid* and identify the molecular subunits of lipids.

3. Distinguish true fats, waxes, steroids, terpenes, phospholipids, and glycolipids in terms of structure and function.

4. Illustrate how the various classes of lipids function in living organisms. Also state the principal sources of each class of lipids.

Self-Quiz Questions

True-False If false, explain why.

____ (1) Lipids contain oxygen, but carbohydrates lack oxygen.

____ (2) A true fat is broken down by hydrolysis into three fatty acid molecules and one molecule of glycerol.

____ (3) Saturated fats contain the maximum possible number of hydrogen atoms that can be covalently bonded to the carbon skeleton of their fatty acid "tails."

____ (4) A polyunsaturated fat is usually an oil at room temperature and generally contains many double bonds in the fatty acid component.

____ (5) Waxes are long-chain alcohols combined with long-chain fatty acids. They help organisms prevent water loss and predation.

____ (6) Some of the male and female sex hormones belong to the group of lipids known as *terpenes*.

Matching Choose _all_ the appropriate answers for each.

(7) ___ carotenoid

(8) ___ chlorophyll

(9) ___ cholesterol

(10) ___ cutin

(11) ___ glycolipid

(12) ___ phospholipid

(13) ___ saturated fat

(14) ___ unsaturated fat

A. basic fabric for all cell membranes

B. butter and bacon

C. vegetable oil

D. wax

E. terpene

F. steroid

G. covers and waterproofs plant surfaces

H. involved in chemical recognition between cells and substances present in their surroundings

3–VII PROTEINS

Summary

Some proteins, such as enzymes, speed up chemical reactions. Other proteins transport substances or cause cells (or parts of cells) to move. Yet other proteins are structural elements in bone and cartilage, transmit chemical information, or protect vertebrates from disease agents. Each protein may be broken down into its own particular assemblage of some twenty different _amino acids_. The chemical nature, size, shape, and ordering of side groups projecting from the amino acids dictate how a protein will interact chemically with other substances and, hence, the role that the protein will play in a living system. The ultimate structure of a protein can be investigated on three (or four) levels: primary, secondary, tertiary, and quaternary. Disruption of the weak bonds on which the secondary, tertiary, and quaternary structures are based causes _denaturation_: a dramatic and irreversible loss of the three-dimensional structure of the protein.

Key Terms

proteins (p. 52)
enzymes (p. 52)
hemoglobin (p. 52)
hormones (p. 53)
amino acid (p. 53)
amino group (p. 53)
carboxyl group (p. 53)
peptide bond (p. 53)

polypeptide chain (p. 53)
R group (p. 53)
primary structure (p. 53)
secondary structure (p. 54)
tertiary structure (p. 54)
quaternary structure (p, 54)
collagen (p. 54)

denaturation (p. 55)
albumin (p. 55)
cytochromes (p. 55)
glycoproteins (p. 55)
gamma globulin (p. 55)
lipoproteins (p. 56)
nucleoproteins (p. 56)

Objectives

1. Define *protein* and identify the basic building blocks of proteins.

2. Describe the structure that all amino acids have in common and the structures that make each of the twenty amino acids different.

3. Explain how a peptide bond is formed.

4. Indicate the specific factors that determine (a) how a protein will react with other substances and (b) the role that the protein will play in a living organism.

5. List some specific proteins.

6. Illustrate how the primary, secondary, tertiary, and quaternary structures of a protein are formed.

7. Define *denaturation* and explain how it occurs.

8. Explain what enzymes do.

9. Contrast the basic structure and function of glycoproteins, lipoproteins, and nucleoproteins.

Self-Quiz Questions

    <u>*True-False*</u> If false, explain why.

____ (1) Amino acids are linked together by hydrolysis, a process that splits out molecules of water as the amino acid subunits are linked together.

____ (2) Bone and cartilage are constructed, in part, of specific proteins.

____ (3) R groups projecting from the main carbon skeleton determine how a long-chain protein will interact chemically with other substances.

____ (4) The primary structure of a protein is formed principally by hydrogen bonds linking together various amino acids.

____ (5) An amino group contains a nitrogen atom and two hydrogen atoms; a carboxyl group contains two oxygen atoms, a carbon atom, and a hydrogen atom.

____ (6) A lipoprotein contains a sugar component plus a protein component.

    <u>*Matching*</u> Choose the one most appropriate answer for each. Note that no answer may be appropriate.

(7) ____ cytochromes

(8) ____ enzymes

(9) ____ glycoprotein

(10) ____ hemoglobin

A. gamma globulin

B. speed up a metabolic reaction

C. compounds of proteins and nucleic acids

D. iron-containing proteins that transport electrons in most kinds of cells

Summary

Each nucleotide contains at least a five-carbon sugar, a nitrogen-containing base, and a phosphate group. Nucleotide-based molecules serve as chemical messengers between cells (cAMP), as energy carriers (ATP), as transporters of hydrogen ions and electrons in metabolic reactions (NAD^+ and FAD), and as molecules that encode genetic instructions (DNA) or help translate these instructions (RNA) into the proteins upon which all forms of life are based.

Key Terms

nucleotide (p. 56)	phosphate group (p. 56)	ATP (p. 56)
nitrogen-containing base (p. 56)	adenosine phosphates (p. 56)	NAD^+ (p. 56)
pyrimidine (p. 56)	nucleotide coenzymes (p. 56)	FAD (p. 56)
purine (p. 56)	nucleic acids (p. 56)	DNA (p. 57)
	cAMP (p. 56)	RNA (p. 57)

Objectives

1. Describe a nucleotide by discussing its components and the way they are linked together.

2. Define *purine* and *pyrimidine* and distinguish them from each other.

3. List the three principal nucleotide-based molecules, give two examples of each, and tell how each functions in living organisms.

4. Define *nucleic acid* and describe its basic structure.

Self-Quiz Questions

   True-False   If false, explain why.

____ (1) ATP is a temporary energy-storage molecule.

____ (2) A purine contains one carbon skeleton bent around to form a ring, whereas a pyrimidine contains two such rings.

____ (3) Adenosine phosphates act as chemical messengers between cells and as energy carriers.

____ (4) Nucleic acids are long chains of nucleotides strung together, with nitrogenous bases connecting the phosphates and with sugars sticking out to the side.

____ (5) DNA consists of two parallel nucleic acid strands twisted about each other and cross-linked by hydrogen bonds.

____ (6) DNA contains the code for constructing one or more proteins.

____ (7) RNA molecules translate the DNA code into a protein product.

Matching Choose *all* the appropriate answers for each.

(8) _____ adenosine phosphates

(9) _____ nucleotide coenzymes

(10) _____ nucleic acid

A. long, single-stranded or double-stranded

B. ATP

C. transport hydrogen ions and their associated electrons

D. cAMP, a chemical messenger

E. NAD^+ and FAD

F. RNA and DNA

INTEGRATING AND APPLYING KEY CONCEPTS

(1) Humans can obtain energy from many different food sources. Do you think this ability is an advantage or a disadvantage in terms of long-term survival? Why?

(2) If the ways that atoms bond affect molecular shapes, do the ways that molecules behave toward one another influence the shapes of organelles? Do the ways that organelles behave with one another influence the structure and function of cells?

(3) If proteins have the most diverse shapes and the most complex structure of all molecules, why do you suppose that proteins are not the code molecules used to construct new proteins?

4
CELL STRUCTURE AND FUNCTION: AN OVERVIEW

GENERALIZED PICTURE OF THE CELL

Emergence of the Cell Theory
Basic Aspects of Cell Structure
    and Function
Cell Size

PROKARYOTIC CELLS

EUKARYOTIC CELLS

THE NUCLEUS

Nuclear Envelope
Chromosomes
Nucleolus

ORGANELLES OF AN ENDOMEMBRANE
SYSTEM

Endoplasmic Reticulum and
    Ribosomes
Golgi Bodies
Lysosomes
Microbodies

OTHER CYTOPLASMIC ORGANELLES

Mitochondria
Chloroplasts and Other Plastids
Central Vacuoles in Plant Cells

INTERNAL FRAMEWORK OF EUKARYOTIC
CELLS

The Cytoplasmic Lattice
Microfilaments and Microtubules

STRUCTURES AT THE CELL SURFACE

Cell Walls and Other Surface
    Deposits
Microvilli
Cilia and Flagella
Cell-to-Cell Junctions

SUMMARY OF MAJOR CELL STRUCTURES
AND THEIR FUNCTIONS

Summary

Crude systems of lenses developed early in the seventeenth century by Galileo, Hooke, and van Leeuwenhoek enabled observations of cells as small as bacteria and sperm. Since the 1820s, specific technological advances in lens design have enlarged our understanding of cellular structure and behavior. Various kinds of light and electron microscopes have their own particular uses. Low-resolution light microscopes are used to observe the cells of living organisms, whereas high-resolution electron microscopes form images of structures as small as 0.2 nanometers. Scanning electron microscopes have lower resolving power but remarkable depth of field; surface features of cells and organisms can be examined.

Scientists know that all organisms are composed of one or more cells, that the cell is the basic living unit of organization for all organisms, and that all cells arise from preexisting cells. Most cells are quite small but nevertheless highly organized. The surface-to-volume ratio determines the size and shape a cell reaches before it either stops growing or divides. Pathways of chemical communication link the plasma membrane, cytoplasm, and nuclear regions.

Key Terms

Galileo (p. 58)
Hooke (p. 58)
van Leeuwenhoek (p. 59)
Brown (p. 59)
Schwann (p. 59)
Schleiden (p. 59)
cell theory (p. 59)
Virchow (p. 59)
plasma membrane (p. 59)
nucleoid (p. 59)

nucleus (p. 59)
cytoplasm (p. 59)
cytosol (p. 59)
micrometers (p. 60)
nanometers (p. 60)
diffusion (p. 60)
surface-to-volume
  ratio (p. 60)
compound light
  microscope (p. 62)

spherical aberration (p. 62)
chromatic aberration (p. 62)
phase contrast microscope (p. 62)
resolution (p. 62)
transmission electron
  microscope (p. 62)
high-voltage electron
  microscope (p. 63)
scanning electron microscope
  (p. 63)

Objectives

1. Briefly describe the contributions made by each of the following scientists to the modern understanding of cell biology: Galileo, Hooke, van Leeuwenhoek, Brown, Schwann, Schleiden, Virchow.

2. (a) Describe how each of these microscopes works:
   conventional compound light microscope
   phase contrast light microscope
   transmission electron microscope
   scanning electron microscope

   (b) Tell whether or not living cells can be viewed by each type of microscope and state the limitations of each type.

3. List the basic ideas of the cell theory stated by Schleiden, Schwann, and Virchow.

4. Consult Appendix II and determine how you would explain the range of cell sizes to a person who is unfamiliar with the metric system.

5. State what the plasma membrane, the nucleus, and the cytoplasm do in a cell.

Self-Quiz Questions

True-False If false, explain why.

____ (1) The cell is the smallest independent living unit.

____ (2) The average cell is approximately as large as your thumbnail.

____ (3) A plasma membrane permits some substances to cross and restricts other substances from crossing.

Matching Choose the one best answer for each.

(4) ___ Galileo

(5) _A_ Hooke

(6) ___ Schleiden and Schwann

(7) ___ van Leeuwenhoek

(8) ___ Virchow

A. originated the term *cell*

B. all cells must come from preexisting cells

C. the first person to observe a great diversity of microscopic organisms

D. said that all organisms are constructed of cells

E. the first person to record any biological observations under the microscope

Fill-in-the-Blanks

One problem with compound (9) _____ microscopes is spherical aberration: tiny objects appear blurred when brought close to the objective lens. If you wish to observe a living cell, it must be small or thin enough for (10) _____ to pass through. (11) _____ is the property that dictates whether small objects close together can be seen as separate things. A (12) _____ is one one-billionth of a meter. With a (13) _____ electron microscope, a beam of electrons is transmitted through a prepared, thinly sliced section of a cell or organism. With a (14) _____ electron microscope, a narrow beam of electrons is played back and forth across a specimen's surface, which has been coated with a thin metal layer. The (15) _____ is a membrane-bound zone of hereditary control.

4–II PROKARYOTIC CELLS

EUKARYOTIC CELLS

64

64-67

Summary

Every prokaryotic cell has a plasma membrane, cytoplasm, a nucleoid, and ribosomes; most also produce a cell wall and have a few internal membranes. Prokaryotic DNA in the nucleoid does not coil into chromosomes.

In eukaryotic cells, a true nucleus replaces the nucleoid and additional membrane-bounded organelles are distributed throughout the cytoplasm. Unlike animals, many protistans, true fungi, and land plant cells have a cell wall surrounding the plasma membrane.

Key Terms

prokaryotic (p. 64)	Golgi bodies (p. 64)	cytoplasmic lattice (p. 64)
cyanobacteria (p. 64)	lysosomes (p. 64)	dictyosome (p. 65)
ribosomes (p. 64)	microbodies (p. 64)	microtubules (p. 65)
cell wall (p. 64)	mitochondria (p. 64)	microfilaments (p. 65)

eukaryotic (p. 64)   plastids (p. 64)   centriole (p. 65)
organelles (p. 64)   chloroplast (p. 64)   microvilli (p. 65)
endoplasmic   central vacuole (p. 64)
  reticulum (p. 64)

Objectives

1. Look at Figures 4.7 and 4.8 and describe how plant cells differ from animal cells. Then look at Figure 4.6a and describe how a generalized moneran cell differs from plant and animal cells.

2. Explain how the term *prokaryote* is related to the moneran kingdom and how the term *eukaryote* is related to the other four kingdoms.

3. List the types of organisms that produce cell walls. State what cell walls are made of and explain what cell walls do for cells that have them.

Self-Quiz Questions

Matching Select the single best answer. A letter may be used more than once.

(1) ____ endoplasmic   A. photosynthesis occurs here
        reticulum
                              B. digestion and disposal
(2) ____ Golgi complex
                              C. energy extraction
(3) ____ lysosomes
                              D. material synthesis, modification and distribution
(4) ____ microbodies
                              E. hereditary instructions for synthesis and cell
(5) ____ mitochondria   operation

(6) ____ nucleus   F. material conversions and disposal

Fill-in-the-Blanks

(7) _____ have a membrane-bound nucleus and other membrane-bound organelles;

(8) _____ lack such structures. (9) _____ cells do not produce walls,

although some secrete products to the surface layer of tissues in which they are

formed. (10) The _____ is the region of hereditary control in prokaryotic cells.

4–III THE NUCLEUS

Summary

All eukaryotic cells contain a nucleus that is the storehouse of information that promotes survival and reproduction. A *nuclear envelope* encloses the *nucleoplasm* and is regularly traversed by pores.

Chromatin is a mass of DNA and associated proteins that are actively regulating the metabolic behavior of the cell; during cell division, chromatin becomes tightly coiled into wormlike structures called *chromosomes* for efficient distribution into the daughter cells. Prokaryotic DNA has few associated proteins and cannot condense into chromosomes.

The *nucleolus* is a dense region of nucleoplasm where the subunits that will later be constructed into ribosomes are made.

Key Terms

nucleus (p. 68)
nuclear envelope (p. 68)
nucleoplasm (p. 68)
latticelike
  framework (p. 68)

pores (p. 68)
annulus (p. 68)
chromatin (p. 68)
chromosomes (p. 69)

nucleoprotein (p. 69)
nucleolus (p. 69)
precursor (p. 69)
ribosomal subunits (p. 69)

Objectives

1. Describe the basic organization of the nuclear envelope, the nucleolus, and the chromatin/chromosome system.

2. State the processes that occur in each of the three nuclear components mentioned in Objective 1.

3. Explain why it is advantageous that chromatin condenses into chromosomes at the onset of cell division.

4. State the principal difference between prokaryotic and eukaryotic DNA.

Self-Quiz Questions

True-False If false, explain why.

____ (1) The nucleolus is part of the nucleoplasm.

____ (2) If the chromatin network did not condense into chromosomes during cell division, many more tangles and breakage of the DNA material would occur.

____ (3) The nucleolus produces the materials from which the Golgi complex is later constructed in the cytoplasm.

Fill-in-the-Blanks

The (4) _____ is the region where ribosomal subunits are synthesized. Masses of

DNA and its associated proteins that extend throughout the nucleoplasm when it's not

undergoing division are called (5) _____. During cell division this material

condenses into (6) _____.

4–IV ORGANELLES OF AN ENDOMEMBRANE SYSTEM

Summary

Various organelles in eukaryotic cells affect intracellular substances in different ways. *Ribosomes* are the sites of protein synthesis. If ribosomes are on the *rough endoplasmic reticulum*, the proteins they synthesize will be exported from the cell. Smooth endoplasmic reticulum lacks ribosomes but contains enzymes that help to assemble fats. The smooth endoplasmic reticulum also participates in isolating and transporting materials and in breaking down storage materials and potentially toxic materials. *Golgi bodies* compose a membrane system (generally located near the nucleus) that receives materials from the endoplasmic reticulum, packages them, and transports them. Most polysaccharides are synthesized in the Golgi bodies, and the finishing touches are put on glycoproteins here. *Lysosomes* are sacs of digestive enzymes in animal cells; they sometimes merge with membrane-enclosed materials and digest them. *Microbodies* convert excess amounts of some substances to different substances that are required but not available.

Key Terms

endomembrane system (p. 70)
ribosomes (p. 69)
endoplasmic
  reticulum (p. 70)
cisternal space (p. 70)
rough ER (p. 70)
smooth ER (p. 70)

sac (p. 70)
secretory (p. 70)
sarcoplasmic
  reticulum (p. 71)
Golgi bodies (p. 71)
dictyosome (p. 71)
vesicle (p. 71)

lysosome (p. 72)
endocytotic vesicles (p. 72)
autophagic vacuoles (p. 72)
microbodies (p. 73)
peroxisomes (p. 73)
glyoxysomes (p. 73)

Objectives

1. Distinguish between rough and smooth endoplasmic reticulum in structure and the substances each produces.

2. Draw a diagram of a Golgi body and explain how Golgi bodies accomplish their functions.

3. Trace the pathway traveled by a protein that has just been exported from a cell. Start at the point where its constituent amino acids are in the cytoplasm.

4. Explain how microbodies (peroxisomes) can be useful to the cells of your tissues.

5. Describe how you would be able to distinguish microbodies from ribosomes in an electron micrograph.

6. Explain how (a) animal cells and (b) plant cells dismantle and dispose of waste materials.

True-False   If false, explain why.

____ (1) If you ate only fats and proteins and excluded all carbohydrates from your diet, microbodies in your cells would attempt to convert excess proteins and fats into carbohydrates in an attempt to provide what you lacked.

____ (2) Rough endoplasmic reticulum is a membrane system that has many ribosomes associated with it.

____ (3) A protein due to be exported from the cell is made by ribosomes located in the rough endoplasmic reticulum and exported in sacs pinched off from the endoplasmic reticulum.

____ (4) Hydrolytic enzymes that are usually stored in chromoplasts are able to break down virtually every large molecule used in cell architecture.

Matching   Choose the one best answer for each.

(5) ____ Golgi body

(6) ____ lysosome

(7) ____ microbody

(8) ____ sarcoplasmic reticulum

(9) ____ rough endoplasmic reticulum

(10) ____ smooth endoplasmic reticulum

A. stores and releases calcium ions in muscle cells

B. synthesizes proteins to be exported from that cell

C. converts excess amounts of some substances to different substances that are needed but not available

D. assembles fats and fat derivatives; breaks down glycogen

E. assembles, packages, transports, and exports substances

F. a small bag of hydrolytic enzymes; found in animal cells

4–V   OTHER CYTOPLASMIC ORGANELLES

Summary

All eukaryotic cells contain *mitochondria*, which have membrane systems similar to those of chloroplasts except that the inner membrane forms projecting shelves (cristae) rather than stacks of coins. Mitochondria convert energy stored in carbon compounds to forms that the cell can use, principally ATP.

   In plants and some of the protistans, the reactions of photosynthesis occur in organelles called *chloroplasts*. Each chloroplast has an outer membrane and a complex, much-folded inner membrane thrown into configurations that resemble stacks of coins (*grana*). Apparently the grana are the sites where some of the energy of sunlight is absorbed by chlorophyll and passed on to be stored in the bonds of ATP and $NADPH_2$. The energy released by ATP and $NADPH_2$ can subsequently be used in the *stroma* to build carbon compounds, such as sugar molecules, from carbon dioxide and water.

In photosynthetic prokaryotes, light-trapping reactions and ATP formation occur on plasma membrane that has been folded back into the cytoplasm, and the remainder of the energy-transformation reactions occur in the cytoplasm.

Storage organelles concentrate materials not intended for export in places that are out of the way of metabolic activity. *Plastids* are plant cell storerooms that may contain starch grains, plant pigments, oil, and other high-energy food reserves. The storage organelles in animal and protistan cells are generally known as *vacuoles*, as are the large fluid-filled sacs that promote environmental contact in many plant cells.

Key Terms

aerobic (p. 74)	plastids (p. 74)	chlorophyll (p. 75)
mitochondrion (p. 74)	chloroplast (p. 74)	stroma (p. 75)
-dria	chromoplasts (p. 74)	granum, grana (p. 75)
crista, cristae (p. 74)	amyloplasts (p. 74)	starch grains (p. 75)
matrix (p. 74)	central vacuole (p. 75)	

Objectives

1. Describe the basic structure of the mitochondrion. Indicate the processes that occur in mitochondria, the approximate size of mitochondria, and the role of the cristae.

2. State which types of cells contain (a) chloroplasts and (b) mitochondria.

3. Describe the basic structure of the chloroplast, indicate how the grana differ from the stroma and what happens in each, and tell where the chlorophyll is located.

4. Explain how cells store materials.

5. Describe how cells can increase their size and interior surface area through the use of vacuoles.

Self-Quiz Questions

True-False If false, explain why.

____ (1) Muscle cells and other cells that demand high-energy output generally have many more chloroplasts than less active cells.

____ (2) Mitochondria and chloroplasts are organelles that have both inner and outer membranes.

____ (3) Enzymes and molecules involved in ATP formation are located on the outer membranes of mitochondria.

____ (4) Chloroplasts always appear green due to the chlorophyll they contain.

____ (5) Chlorophyll is found in all chloroplasts.

____ (6) Root hairs lengthen through vacuole enlargement; this increase in surface area promotes the absorption of essential ions and nutrients that may be scarce in the environment.

<u>*Matching*</u>   Choose the one best answer for each.

(7) ___ chromoplast

(8) ___ amyloplast

(9) ___ vacuole

(10) ___ vesicle

(11) ___ mitochondrion

A. colorless; accumulates starch grains

B. a tiny bag that transports substances through the cytoplasm

C. stores plant pigments

D. a large storage compartment that generally stores fluids

E. narrow, hollow cylinder

F. tubulins

G. provide the ATP necessary for rapid movements

<u>*Fill-in-the-Blanks*</u>

In photosynthetic prokaryotes such as bacteria and blue-green algae, light-trapping reactions and ATP formation occur on the (12) _____ _____ _____.

In photosynthetic eukaryotes these reactions occur in an organelle called the (13) _____. Two energy-rich molecules produced in the grana are (14) _____ and (15) _____. An energy-rich molecule produced by a mitochondrion is (16) _____.

4–VI   INTERNAL FRAMEWORK OF EUKARYOTIC CELLS

76-78

The Cytoplasmic Lattice

76-77

Microfilaments and Microtubules

77-78

Summary

Cellular contents stream about in living eukaryotic cells, and proper cell functioning requires organization based on a three-dimensional lattice that pervades the cytoplasm.

Microfilaments are extremely long, thin structural elements composed of contractile proteins. The activities of microfilaments can cause cells to pinch in two, crawl across a culture dish, change shape, or can make organelles move around in the cytoplasm. *Microtubules* are tiny hollow cylinders that are thicker and more rigid than microfilaments. Microtubules help to dictate the shape of many cells and cell extensions, but they are also involved in the movement of chromosomes during nuclear division and in the beating of *cilia* and *flagella* that move cells through their environment.

Key Terms

cytoplasmic lattice (p. 76)
actin (p. 77)
myosin (p. 77)
contraction (p. 77)
microfilament (p. 77)

cytoplasmic
  streaming (p. 78)
microtubule (p. 78)
cilia (p. 78)

flagella (p. 78)
microtubule organizing
  centers, MTOCs (p. 78)
tubulins (p. 78)

1. Distinguish microfilaments from microtubules in structure and function. Mention actin, myosin, tubulins, and cytoplasmic streaming in your explanations.

2. Describe the process of cytoplasmic streaming and explain how it can benefit plant cells and amoebalike cells.

3. Explain how microfilaments and microtubules influence cell shape, motion, and growth.

4. Describe the cytoplasmic lattice and explain why you think that it cannot be seen very well using ordinary transmission electron microscopy.

Self-Quiz Questions

 Fill-in-the-Blanks

The principal subcellular structures that are involved in establishing and maintaining

the shapes of cells and extensions from cells are (1) _____. (2) _____

are long, thin structural elements composed of contractile proteins. The (3) _____

_____ is a three-dimensional network that pervades the cytoplasm. (4)

_____ are assembled from protein subunits called (5) _____.

4–VII   STRUCTURES AT THE CELL SURFACE

Summary

Cell walls provide support, resist mechanical pressure, and confer tensile strength when cells absorb water and expand. Porous cell walls occur in all kingdoms except Animalia and are quite diverse structurally. *Microvilli* are fingerlike projections of the plasma membrane that aid the cell in absorbing or secreting substances. *Bacterial flagella* are threadlike extensions of the cell surface that behave like propellers. *Eukaryotic flagella* and *cilia* are assembled from microtubules that are anchored in basal bodies; they help small organisms to move through liquid environments. Cilia and flagella also help to attract food particles to nonmotile animals such as some mollusks and all sponges; they also move fluids and suspended particles along the surfaces of epithelium lining the digestive, respiratory, and reproductive tracts of many types of animals. Cell-to-cell junctions enable cells to interact with their cellular neighbors.

cell walls (p. 78)   flagellin (p. 80)   dynein (p. 81)
primary cell wall (p. 79)   flagella (p. 80)   tight junctions (p. 81)
secondary cell wall (p. 79)   cilia (p. 80)   adhering junctions (p. 82)
lignin (p. 79)   9 + 2 array (p. 80)   gap junctions (p. 82)
microvillus (p. 79)   centrioles (p. 80)   desmosome (p. 82)
bacterial flagellum (p. 80)   basal body (p. 81)   plasmodesmata (p. 82)

Objectives

1. State the benefits and limitations imposed on a cell by the presence of cell coats, capsules, sheaths, and walls.

2. List the major groups of organisms that have walls surrounding their cells.

3. Describe the structure of cell walls.

4. Contrast the roles of microvilli with those of cilia and flagella. Describe and distinguish among the structures of each.

5. Explain how prokaryotic flagella differ from eukaryotic flagella.

6. Describe the relationship between centrioles and basal bodies.

7. List the major groups of organisms that have eukaryotic cilia and flagella, and state some of the functions provided by those organelles.

8. Compare several aspects of structure and function in the various types of cell-to-cell junctions.

Self-Quiz Questions

Matching Link each letter with its appropriate blank(s). Each blank should have only one letter.

____ (1) adhering junction

____ (2) basal body

____ (3) bacterial flagellum

____ (4) cell wall

____ (5) centriole

____ (6) cilium

____ (7) eukaryotic flagellum

____ (8) gap junction

____ (9) microvillus

____ (10) tight junction

A. Short, barrel-shaped organelle that organizes the interior of a cilium or flagellum

B. a slender fingerlike extension of the plasma membrane that functions in absorption

C. a whiplike extension that functions like a propeller; flagellin present

D. lignin and cellulose; cutin

E. assembled from microtubules in a 9 + 2 array within a sheath that is continuous with the plasma membrane

F. plasmodesmata; interconnect all living cells in multicelled plants

G. rows of membrane proteins match up and form sealing strands between adjacent epithelial cells

H. desmosome; abundant in epithelial cells of animals

I. exists in a pair that is pushed apart by spindle formation

Summary

As time passed, cells became more complex. There were selective advantages for cells that increased their internal surface area with infoldings and outfoldings of membrane on which metabolic reactions could occur. Such developments have enabled cells to acquire energy and materials in highly controlled and specialized ways.

Key Terms

No new ones.

Objectives

1. List the basic minimal parts a cell required in order to carry on metabolism and live.

2. Explain the advantages that accrue for cells that develop many internal compartments.

3. Consult and memorize Table 4.1. Decide which kingdom has the least complex cells, and which has the most complex cells. Indicate which other kingdom shares the most features with plants, and which other kingdom most strongly resembles the animals.

4. Explain what in the life-styles of plant and protistan cells gives them the most diverse organelles of all cells.

Self-Quiz Questions

    Fill-in-the-Blanks

(See next page.) Identify each structure in its accompanying blank.

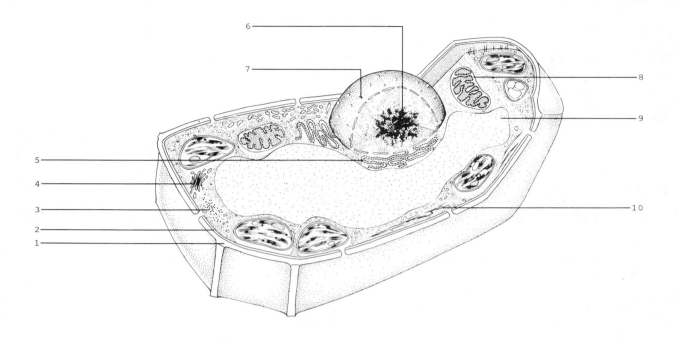

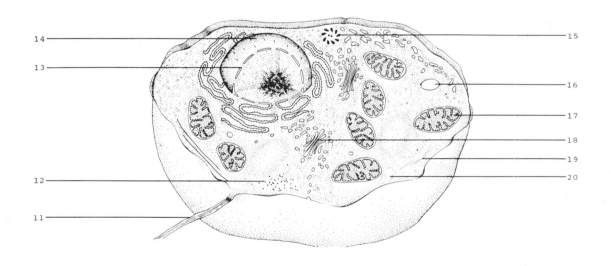

Figures 4.8 and 4.9

If the cell structure is present in all or most members of that group, put a check (√) in its box; if the cell structure is present in some of that group, put a cross (+) in its box. If it is not present, leave the box blank.

	Monera	Protista	Fungi	Plantae	Animalia
Cell wall	(1)	(2)	(3)	(4)	(5)
Plasma membrane	(6)	(7)	(8)	(9)	(10)
Photosynthetic pigments	(11)	(12)	(13)	(14)	(15)
Membrane-bound chloroplasts	(16)	(17)	(18)	(19)	(20)
Mitochondria	(21)	(22)	(23)	(24)	(25)
Ribosomes	(26)	(27)	(28)	(29)	(30)
Endoplasmic reticulum	(31)	(32)	(33)	(34)	(35)
Microtubules, microfilaments	(36)	(37)	(38)	(39)	(40)
Complex cilia or flagella	(41)	(42)	(43)	(44)	(45)
DNA molecules	(46)	(47)	(48)	(49)	(50)
Chromatin condensed into chromosomes	(51)	(52)	(53)	(54)	(55)

True-False   If false, explain why.

____ (56) Some cells can live if they contain only a plasma membrane, DNA molecules, and cytoplasm that lacks other cell structures.

____ (57) All living organisms are made of one or more cells.

____ (58) New cells can arise only from cells that already exist.

____ (59) There are no essential differences in the ways that prokaryotes and eukaryotes acquire and process energy and materials.

INTEGRATING AND APPLYING KEY CONCEPTS

(1) Which parts of a cell constitute the minimum necessary for keeping the simplest of living cells alive?

(2) How did the existence of a nucleus, compartments, and extensive internal membranes confer selective advantages on cells that developed these features?

5
WATER, MEMBRANES, AND CELL FUNCTIONING

ON CELLULAR ENVIRONMENTS

WATER AND CELL FUNCTIONING

   The Importance of Hydrogen
     Bonds
   Solvent Properties of Water

CYTOPLASMIC ORGANIZATION OF
WATER

CELL MEMBRANES: FLUID STRUCTURES
IN A LARGELY FLUID WORLD

MOVEMENT OF WATER AND SOLUTES
ACROSS MEMBRANES

   Overview of Membrane Transport
     Systems
   Diffusion
   Osmosis
   Active Transport
   Endocytosis and Exocytosis

MEMBRANE SURFACE RECEPTORS

PERSPECTIVE

Summary

Each type of cell has a range of structural and functional responses to some range of conditions typical of the environment in which its predecessors evolved. Even though some organisms live in habitats that periodically dry up, none of the *activities* associated with the term *living* can proceed without water. Water has many chemical and physical properties that help to protect cells both from fluctuations in the environment outside cell boundaries and from internal changes. Water has temperature-stabilizing and cohesive properties because of the hydrogen bonding that occurs between water molecules. Water is an extraordinary solvent, and most substances that dissolve in water remain unchanged by their association with water molecules.

Key Terms

tolerance (p. 85)
temperature-stabilizing
  properties (p. 86)
temperature (p. 86)
specific heat (p. 86)

heat of vaporization (p. 86)
calorie (p. 86)
evaporation (p. 86)
heat of fusion (p. 86)
cohesion (p. 86)

surface tension (p. 87)
sphere of hydration (p. 88)
dissolve (p. 88)
solvent (p. 88)
solute (p. 88)
inert (p. 88)

Objectives

1. Explain how the water that composes the bulk of most organisms helps those organisms to tolerate diverse variable environmental conditions.

2. Explain why none of the activities associated with the term *living* can occur without water.

3. Distinguish between specific heat and heat of vaporization.

4. Give two specific examples of organisms that depend directly on the cohesive properties of water and describe the nature of that dependence. Speculate about what those organisms would do if water molecules were *not* cohesive.

5. Explain why the fact that water is an outstanding solvent is of such importance to living organisms. State how you think life would be limited if water could dissolve only a few different substances.

6. Explain the role that hydrogen bond formation plays in some of the physical and chemical properties of water.

Self-Quiz Questions

Fill-in-the-Blanks

(1) _____ is the amount of heat energy a single gram of a substance can absorb before its temperature increases by 1°C. As water absorbs heat and the molecules move more quickly, (2) _____ _____ between neighboring water molecules are more readily broken and formed anew. The amount of heat energy that must be absorbed before a single gram of liquid will be converted to gaseous form is the (3) _____ _____ _____ of that substance. When molecules at the surface of a liquid begin moving fast enough to escape into the atmosphere, the process is called (4) _____. The surface of the liquid left behind drops in temperature because much of its energy has been (5) _____ by the escaping molecules and converted into their energy of (6) _____. In pure water each molecule is attracted to and hydrogen bonded with other water molecules; such an attraction between like molecules is called (7) _____. Water is an outstanding (8) _____ because it dissolves many different types of (9) _____.

Summary

Water molecules have negatively charged ends and positively charged ends. The polar nature of water encourages spheres of hydration to be formed around ions, preventing ions from reacting with each other and forcing them to remain dispersed in the cytoplasm rather than concentrating in some part of the cell. Thus water molecules help keep vital ions available for reactions throughout the cell. Large molecules may have charged regions that tend to attract either the negative or the positive ends of water molecules (that is, to be hydrophilic) and other regions that bear no net charge (in other words, are hydrophobic). The organization of membranes and of sol and gel states depends on the availability of water molecules in the cell environment. Thus regional systems of the cell are formed in response to the structure of water in the cytoplasm.

Key Terms

colloidal particles (p. 88)   gel state (p. 89)   fil (p. 89)
colloids (p. 88)   phospholipids (p. 89)   cytoplasmic water (p. 89)
sol state (p. 89)   lipid bilayer (p. 89)

Objectives

1. Describe how spheres of hydration are formed around simple ions (due to the polar nature of water), dispersing ions in the cytoplasm and keeping the ions from interacting.

2. Compare the formation of spheres of hydration around small ions with the organization of water on the surfaces of proteins that have ionic regions (are hydrophilic).

3. Describe how a phospholipid bilayer comes to form in a watery solution.

4. Explain how a gel region in a cell could change into a sol state region and vice versa. Speculate about how an amoeba moves from one watery region to another.

Self-Quiz Questions

Fill-in-the-Blanks

Nonpolar regions of large molecules have no net charge and show little tendency to

form hydrogen bonds with water; they are said to be (1) _____ (water dreading).

(2) _____ have fatty acid "tails" (which repel water) and "heads" with (3)

_____ and alcohol groups (which dissolve in water). When ions and water mole-

cules surround large molecules so as to form an electrically charged "cushion," the

substance remains in solution and is said to be in the (4) _____ state.

True-False If false, explain why.

___ (5) In a plasma membrane the hydrophilic tails point inward, tail to tail, and form a region that excludes water.

___ (6) When proteins of overall like charge don't repel each other, but adhere in a continuous, spongy network, the system is said to be in a gel state.

___ (7) Cytoplasm is a formless substance with little structure.

___ (8) Water molecules tend to cluster around each positively charged ion with their positive ends pointing toward the ion, thus forming a sphere of hydration.

___ (9) The tails of phospholipids are hydrophobic.

5–III CELL MEMBRANES: FLUID STRUCTURES IN A LARGELY FLUID WORLD   89-91

MOVEMENT OF WATER AND SOLUTES ACROSS MEMBRANES   91-95

Overview of Membrane Transport Systems   91-92

Summary

Cell membranes are phospholipid bilayers studded with diverse protein particles. Each bilayer forms a sort of fluid sea in which the proteins are suspended like icebergs. Some of the phospholipids involved have short or kinky tails that impart fluidity to the lipid matrix. Because proteins are embedded within or bonded weakly to either surface of this fluid foundation, the model that describes membranes is referred to as the *fluid mosaic model*. Some of the proteins help to transport substances across the membrane; others attach to the cytomusculature within the cell or form sites where substances attach to the membrane surface. The plasma membrane is a regulatory structure; it is differentially permeable and helps to maintain appropriate amounts of ions and molecules inside the cell even when such are scarce or too abundant. In this way, homeostasis is ensured at the cellular level.

Maintaining internal concentrations of water and solutes depends on "passive" transport mechanisms such as diffusion, osmosis, and bulk flow, which do not alter in any way the direction in which a substance is moving on its own. No direct energy outlay by the cell is required for these processes.

Maintaining internal concentrations of water and solutes also depends upon "active" transport mechanisms such as active transport, exocytosis, and endocytosis, all of which work to move a substance in a direction contrary to its spontaneous direction of movement. They cannot operate without direct energy outlays by the cell.

Key Terms

freeze-fracturing (p. 89)	concentration gradient (p. 92)	passive transport systems
freeze-etching (p. 89)	pressure gradient (p. 92)	(p. 92)
fluid mosaic model (p. 89)	temperature gradient (p. 92)	active transport systems
differentially permeable (p. 91)	electric gradient (p. 92)	(p. 92)

1. Explain why the "sandwich" model of the plasma membrane was rejected in favor of the fluid mosaic model of membrane structure.

2. Describe what constitutes the "mosaic."

3. Explain how the plasma membrane is connected to the "musculature" of the cell.

4. Distinguish between *permeable* and *differentially permeable*.

5. Distinguish between active and passive forms of transport and give two examples of each. Describe how cells expend energy.

6. Explain what is meant by *concentration gradient* and explain how such gradients are formed.

Self-Quiz Questions

Fill-in-the-Blanks

(1) _____ - _____ and (2) _____ - _____ are two relatively new methods of preparing membranes to study with the electron microscope. A (3) _____ _____ serves as a sort of fluid sea matrix in which diverse (4) _____ are suspended like icebergs. Together, the two components form the (5) "_____."

The plasma membrane is (6) _____ _____; some molecules travel rapidly across the membrane, others cross it more slowly, and some are kept from crossing it at all. (7) _____ is one of the few molecules that can move freely into and out of the cell. (8) _____ _____ _____ cannot operate without direct energy outlays by the cell.

5–IV Diffusion   92–93
Osmosis   93–94

Summary

Diffusion is a random movement of like molecules along a concentration gradient from their region of greater concentration to a region of lesser concentration. It accounts for the greatest volume of substances being moved into and out of cells, and it is also an important transport process within cells. *Facilitated diffusion* involves proteins embedded in the plasma membrane assisting in the passage of small molecules across the membrane. In *bulk flow*, different ions and molecules present in a fluid move together in the same direction—often in response to a pressure gradient. *Osmosis* is the movement of water across differentially permeable plasma membranes in response to solute concentration gradients and/or pressure gradients.

Key Terms

passive (p. 92)   simple diffusion (p. 92)   osmosis (p. 93)
energy outlay (p. 92)   rate of diffusion (p. 92)   hypotonic (p. 94)
active (p. 92)   facilitated diffusion (p. 93)   hypertonic (p. 94)
active transport (p. 92)   channel (p. 93)   isotonic (p. 94)
exocytosis (p. 92)   bulk flow (p. 93)
endocytosis (p. 92)   pressure gradient (p. 93)

Objectives

1. First define diffusion, then distinguish between free diffusion and facilitated diffusion. Explain what causes diffusion of any sort.

2. List three factors that influence the rate of diffusion and state what sorts of substances diffuse readily across plasma membranes.

3. Distinguish bulk flow from osmosis.

4. Understand the effects of osmosis on living cells.

Self-Quiz Questions

    Fill-in-the-Blanks

(1) _____, (2) _____, (3) _____ _____, and a few other

simple molecules and some (4) _____ diffuse readily across plasma membranes.

Diffusion is driven by the (5) _____ _____ _____ inherent in all

individual molecules as they move from a region of (6) _____ concentration to

a region of (7) _____ concentration. In (8) _____ _____, differ-

ent ions and molecules present in a fluid move together in the same direction, often

in response to a pressure gradient.

    True-False If false, explain why.

____ (9) Diffusion accounts for the greatest volume of substances being moved into and out of cells.

____ (10) Facilitated diffusion is a type of diffusion that requires the cell to expend ATP molecules.

____ (11) Osmosis occurs in response to a concentration gradient that involves unequal concentrations of water molecules.

Endocytosis and Exocytosis

Summary

When ions and molecules required by a cell are scarce, the cell must expend energy in
order to stockpile these nutrients by getting them to move against their concentration
gradient. *Active transport* is such a process that depends upon proteins serving as
carriers or as fixed channels to move ions or molecules across the plasma membrane. As
much as 70 percent of the energy readily available in a cell may be devoted to active
transport. Cells that must move relatively large amounts of solids or fluids across
the plasma membrane use *bulk transport*, in which the substances to be transported are
enclosed in membrane-bound compartments and moved through the membrane to the opposite
side. *Endocytosis* and *exocytosis* are forms of bulk transport that move substances into
and out of the cell, respectively.

Key Terms

active transport (p. 94)	sodium-linked	endocytosis (p. 95)
sodium-potassium	cotransport (p. 95)	vesicle (p. 95)
pump (p. 94)	hydrogen-linked	phagocytosis (p. 95)
calcium pump (p. 94)	cotransport (p. 95)	pinocytosis (p. 95)
		exocytosis (p. 95)

Objectives

1. Explain how a cell gets the ions and molecules it requires if they aren't abundant
    in the environment.

2. Explain how a cell excretes ions and molecules it doesn't need when there are
    many more of them outside the cell than inside.

3. Distinguish active transport from bulk transport, endocytosis from exocytosis,
    and phagocytosis from pinocytosis.

4. State which cells depend on the sodium-potassium pump and which depend on the
    calcium pump.

Self-Quiz Questions

   True-False If false, explain why.

____ (1) Active transport depends on proteins that serve either as carriers or as
         fixed channels across the plasma membrane.

____ (2) Bulk flow requires an expenditure of ATP to move large amounts of materials
         across membranes and may occur in response to a pressure gradient.

____ (3) Nerve cells depend on the sodium-potassium pump mechanism in order to re-
         ceive messages from other cells.

____ (4) The secretion of mucus is achieved by exocytosis.

Summary

Membrane surface receptors enable cells to recognize chemically and attach to extracellular substances; these receptors are generally *glycoproteins* or *glycolipids*, which
regulate specific changes in cellular metabolism, supply nutrients to cells, remove
harmful substances or function in cell-to-cell identity. In many eukaryotic cells surface receptors are attached to and interact with the cytoskeleton.

Key Terms

membrane surface receptors (p. 95)   glycolipids (p. 96)
activate (p. 96)   aggregation (p. 96)
glycoproteins (p. 96)

Objectives

1. List three ways that membrane surface receptors contribute to cellular function.

2. Explain what would result if membrane surface receptors suddenly were to vanish
    from all of your cells.

3. List six functions of the plasma membrane (see Perspective).

Self-Quiz Questions

  Fill-in-the-Blanks

Membrane surface receptors are often (1) _____ groups attached to regions of

membrane (2) _____ and membrane lipids. When activated, some receptors bring

about changes in cell (3) _____ or behavior. In multicelled organisms, surface

receptors also function in identifying cells of like type during (4) _____

_____. Among many eukaryotic cells, surface receptors transmit signals that

seem to alter the distribution of (5) _____ and microtubules.

INTEGRATING AND APPLYING KEY CONCEPTS

(1) Name three aspects of the biological world that would be changed if water were not
    a polar molecule.

(2) If there were no such thing as active transport, how would the lives of organisms
    be affected?

6
ENERGY TRANSFORMATIONS IN THE CELL

6–I ON THE AVAILABILITY OF USEFUL ENERGY

Two Energy Laws

Summary

All events in our universe are governed by two laws of energy. The first law of thermodynamics states that there is some total amount of energy in the universe, and that
amount never changes. Energy may be converted from one form to another, but it can be
neither created nor destroyed. Energy exists in two general forms: *potential energy* (a
potentially usable form that is not, for the moment, being used) and *kinetic energy*
(the energy of motion). Kinetic energy can be transferred from one region (or system)
to another, and it can also be changed from one form to another.
    The second law of thermodynamics says that, left to itself, any system and its sur-
roundings spontaneously undergo conversions to less-organized form. Each time that happens, some energy gets randomly dispersed in a form that is not readily available to do
work. Thus, although the total amount of energy in the universe stays the same, the
amount available in useful forms is dwindling. *Entropy* (that is, energy so dispersed
that it's no longer available to do work) is constantly on the increase. The entropy of
any local region can be lowered as long as that region is being resupplied with usable
energy being lost from some other place. Through energy transfusions from the sun, the
universal trend toward increased entropy can be postponed here on earth.

Key Terms

first law of
  thermodynamics (p. 98)
potential energy (p. 99)

kinetic energy (p. 99)
second law of
  thermodynamics (p. 100)

spontaneous (p. 100)
randomly dispersed (p. 100)
entropy (p. 100)

1. State the first and second laws of thermodynamics.

2. Explain energy conversion by giving an example.

3. Explain what *spontaneous* means in relation to energy.

4. Trace the path of energy flow from the sun to your muscle movement.

5. Explain how, if the universe is becoming progressively disordered, a human embryo can grow into an infant and an infant can grow into an adult.

6. Distinguish potential energy from kinetic energy and give an example of each.

Self-Quiz Questions

> *True-False* If false, explain why.

____ (1) The first law of thermodynamics states that entropy is constantly increasing in the universe.

____ (2) Your body steadily gives off heat equal to that from a 100-watt light bulb.

____ (3) When you eat a potato, some of the stored chemical energy of the food is converted to mechanical energy that moves your muscles.

____ (4) A reservoir containing water is an example of a system containing gravitational potential energy.

6–II   Metabolic Reactions: Energy Changes in Cells

Equilibrium and the Cell

Summary

A *metabolic reaction* is a form of internal energy change in the cell. According to the first and second laws of thermodynamics, a chemical reaction tends to proceed spontaneously (that is, on its own, with no energy boost) toward a state of minimum energy and maximum disorder. How efficiently energy is used in any system depends on the precise route by which one energy form is transferred or converted to another form. If the route consists of a big explosive reaction that occurs in one step, much of the potential energy is converted to entropy rather than energy that can be used to do work. Cells generally convert potential chemical energy stored in nutrient molecules in a *series* of steps, each of which releases tiny amounts of kinetic energy. The energy from such *exergonic reactions* is used to drive energy-requiring (*endergonic*) reactions in the cell. By relying on such metabolic pathways, a cell temporarily conserves some energy that might otherwise be lost as entropy during its continual chemical conversions. Products from one or more reactions can serve as intermediates for subsequent reactions in the pathway.

Chemical reactions are generally reversible under certain conditions. An increase in concentration, temperature, or pressure may cause some "*product*" molecules to revert to *reactant* molecules. A chemical reaction that is occurring in a closed system (that is, where neither the reactant nor the product molecules are free to go elsewhere) will reach a state of equilibrium in which the reaction is proceeding as quickly in reverse as it is forward, and there is no further change in the net concentrations of reactants or products. At *equilibrium*, product concentrations may be greater or lower than reactant concentrations, depending on how much energy is fed into or released from the reaction. In cells, the availability of raw materials (reactants) shifts, and requirements for different products may vary from minute to minute. In a changing environment, cellular homeostasis is maintained only through constant adjustments in the metabolic pathways that sustain life.

Key Terms

dissipation (p. 100)
metabolic
   reaction (p. 100)
reactants (p. 100)
products (p. 100)
spontaneously (p. 100)

minimum energy (p. 100)
maximum disorder (p. 100)
exergonic reactions (p. 101)
endergonic
   reactions (p. 101)
synthesis (p. 101)

metabolic pathways (p. 101)
intermediate product (p. 101)
equilibrium (p. 102)
net change (p. 102)
cellular homeostasis (p. 102)

Objectives

1. Explain what might happen to cells if they tried to use the most direct route in releasing the potential chemical energy stored in their nutrient molecules.

2. Explain what causes a chemical reaction to occur.

3. Distinguish between exergonic and endergonic reactions by defining and giving an example of each.

4. Draw a metabolic pathway and show how exergonic reactions can encourage an endergonic reaction to occur.

5. Show how chemical potential energy could be converted into mechanical kinetic energy (a) in a reaction that would occur in one explosive step, and (b) in a series of reactions that release small amounts of usable energy.

6. State what the requirements are for a system to achieve equilibrium. Then decide whether you think a living system can ever be in a true state of equilibrium.

7. List some ways that cells can manage their metabolic pathways when the availability of raw materials shifts.

Self-Quiz Questions

True-False If false, explain why.

____ (1) Energy-requiring reactions generally provide the energy for exergonic reactions to occur.

____ (2) The most direct route used in converting one form of energy to another form is generally the most efficient; more usable energy is likely to be derived from that route than from any other.

____ (3) At equilibrium, the amount of reactants is equal to the amount of products.

____ (4) A chemical reaction cannot achieve equilibrium if the reactants and products are free to go elsewhere.

6–III ENZYME FUNCTION 102–105

Activation Energy and Enzymes 102–103

Enzyme Structure and Functioning 103–104

Summary

An enzyme is a protein catalyst that speeds up the rate of a chemical reaction by low-
ering the activation energy that must be reached before the reaction will occur. In
living systems metabolic reactions must occur within a certain range of low temperatures so that substances with shapes that depend on hydrogen bonding are not altered
beyond redemption. An enzyme holds the reacting molecules in an orientation so favorable that collisions between the molecules are more likely to break certain bonds and
cause the atoms to become rearranged into the product molecules. In addition, an enzyme bonding to a reactant molecule sometimes can strain some of the bonds so they are
forced to break apart. Enzymes accelerate the rate at which a reaction approaches
equilibrium, but they do not alter the proportions of reactants and products that will
be present once "equilibrium" is reached.

Key Terms

activation energy (p. 102)	polypeptide chain (p. 103)	induced-fit model (p. 103)
collision rate (p. 102)	catalytic site (p. 103)	transition state (p. 103)
average kinetic energy (p. 102)	cleavage (p. 103)	optimum fit (p. 104)
spontaneous (p. 102)	substrates (p. 103)	unstable (p. 104)
enzymes (p. 102)	active site (p. 103)	induce (p. 104)
catalysts (p. 102)	complementary (p. 103)	orient (p. 104)
reaction rate (p. 102)	reversible (p. 103)	hydrolytic enzymes (p. 104)
carbonic anhydrase (p. 103)		

Objectives

1. Describe what must happen in order for reactant molecules to become product
   molecules.

2. Explain what is meant by a *spontaneous reaction*.

3. Describe the precise role that enzymes play in speeding up chemical reactions and
   explain why enzymes make particularly effective catalysts.

4. List three ways that enzymes encourage reactions to occur more quickly.

5. State which processes in your body use chemical reactions that form either carbonic acid or bicarbonate ions. Then state which processes in your body use reactions that convert bicarbonate ions or carbonic acid into CO_2 and water.

Self-Quiz Questions

    True-False If false, explain why.

____ (1) In order for two reactant molecules to become product molecules, the reactant
       molecules must first collide with a certain minimum energy.

____ (2) The striking of a match to start its wood burning is an example of a spon-
       taneous reaction.

_____ (3) Enzyme shape may change during catalysis.

_____ (4) The appropriate enzyme lowers the amount of energy that must be supplied to enable the reactants to be converted to products, and it also lowers the amount of energy required by the reverse reaction.

_____ (5) The active site is a groove on the reactant molecule.

_____ (6) When blood reaches your lungs, carbonic acid is converted to CO_2 and H_2O.

Summary

Enzymes work only on certain sets of substrates and only under certain environmental conditions. Most enzymes are effective at or near pH 7. Most enzymes cannot tolerate high temperatures because their structures change. Without their characteristic structures, particularly of the active site, enzymes lose their catalytic ability. Without enzyme activity, metabolism grinds to a halt and cells die. Because enzymes govern the flow of substances through the cell, controls over enzyme activity are vital for all forms of life; these controls help regulate which sets of substances will be formed in a cell, and in what amounts. Controls over enzyme activity include mechanisms that inhibit the activity of enzymes already formed, mechanisms that activate precursor forms of enzymes, and mechanisms that control synthesis of enzymes. *Allosteric enzymes* are control agents involved in *feedback inhibition,* which is an important mechanism for controlling the activity of metabolic pathways.

Key Terms

optimum pH (p. 104)
optimum temperature (p. 105)
denaturation (p. 105)
allosteric enzyme (p. 105)

substrate molecule (p. 105)
end-product molecule (p. 105)
feedback inhibition (p. 105)
inactive precursor forms (p. 105)

Objectives

1. Name two factors that influence rates of enzyme activity.

2. Describe what happens when an enzyme is denatured.

3. Explain why parents become very concerned when small children develop very high body temperatures as a result of severe infections.

4. Suppose you switched to a diet that contained only fats, proteins, and nucleic acids and omitted carbohydrates. Suppose also that your body can make the several different kinds of carbohydrates it needs from fats and proteins. Explain how allosteric enzymes might be of use to you in making the required carbohydrates.

Fill-in-the-Blanks

(1) _____ and (2) _____ are two factors that influence the rates of enzyme activity. When denaturation occurs, (3) _____ bonds holding the protein in its coiled secondary structure break, (4) _____ interactions shift, and the protein chains unwind.

During severe viral infections, extremely high fevers can cause (5) _____ of proteins, which leads to cell death. When carbohydrates are in your diet, there is no need to construct them; thus the (6) _____ enzyme should be rendered inactive by the (7) _____ binding to the inhibitor site, changing the shape of the (8) _____ site so that no substrate molecules can bind to it to be processed into carbohydrate. When the carbohydrate concentration falls, what is bound to the enzyme lifts off, the shape of the (9) _____ site is again restored, and substrate can again bond to the enzyme and be converted into carbohydrate.

6–V   TYPES OF ENERGY TRANSFERS IN CELLS   106–109

Formation and Use of ATP   106–108

Summary

Energy carriers are compounds that help to transfer the potential energy stored in molecules such as glucose to places where the energy is needed. Hydrolysis of the "high-energy" phosphate bonds of ATP releases larger amounts of useful energy than hydrolysis of other kinds of covalent bonds. The greater the energy change during a reaction, the more energy becomes available for transfer to other substances in the cell. Enzymes couple the hydrolysis of ATP and other energy-yielding reactions to the energy-requiring reactions involved in performing work in the cell. Thus reactions that don't readily proceed on their own can be driven indirectly by energy-releasing reactions such as the splitting of ATP. ATP molecules are used over and over again in such reactions. ATP is sometimes called the "universal energy currency" of cells because it transfers usable energy to reactions concerned with energy metabolism, biosynthesis, active transport, and cellular movement.

Key Terms

energy carriers (p. 106)
electron carriers (p. 106)
phosphate bonds (p. 106)
nucleotide (p. 106)

pyrophosphate bonds (p. 106)
"high-energy" bonds (p. 106)
hydrolysis (p. 106)
adenosine diphosphate, ADP (p. 106)

adenosine triphosphate, ATP (p. 106)   adenosine monophosphate AMP (p. 107)
adenine (p. 106)   phosphorylation (p. 108)
ribose (p. 106)   precursors (p. 107)
triphosphate (p. 106)   coupled reactions (p. 107)

Objectives

1. Explain how the hydrolysis of ATP coupled to a reaction in which two simple molecules are synthesized into one larger molecule resembles the connection of an electric motor to a sewing machine.

2. Describe the process of phosphorylation and show how it is important to cells.

3. Name three energy-requiring activities of cells.

4. Describe the structure of ATP and name its five components.

5. Explain why the energy associated with pyrophosphate (high-energy) bonds is transferred easily to other molecules in the cell.

6. Show how ATP can be broken down into its products; then show how ADP can be further degraded.

Self-Quiz Questions

   Fill-in-the-Blanks

(1) _____, (2) _____ _____, and (3) _____ are three

energy-requiring activities of cells. ATP is constructed of (4) _____,

(5) _____. and three (6) _____ groups. When ATP is hydrolyzed, a

molecule of (7) _____ in the presence of an appropriate (8) _____

is used to split ATP into (9) _____, a (10) _____ group, and, most im-

portant, usable (11) _____, which is easily transferred to other molecules in

the cell.

6–VI Electron-Transfer Reactions   108–109

Summary

Energy is embodied not only in phosphate bonds but also in some of the electrons associated with carrier molecules such as NADH, NADPH, $FADH_2$, and the cytochromes. An *oxidation* reaction strips from an atom or molecule one or more electrons, which are simultaneously gained by other atoms or molecules in a *reduction* reaction. Sometimes hydrogen ions or atoms are transferred as well as the electron. Free-moving electron carriers such as NAD^+ and $NADP^+$ accept electrons and hydrogen at one reaction site in the cell and transfer them to different reaction sites concerned with ATP production or biosynthesis. Because the loss and gain happen simultaneously, these two events are regarded as being coupled in an oxidation-reduction reaction. Sometimes electron carrier molecules such as the cytochromes are arranged in a series known as an electron transport chain, and oxidation-reduction reactions can occur one right after another in organized sequence. These chains of membrane-bound electron carriers are prominent features in the metabolic pathways of photosynthesis and cellular respiration where ATP and other energy storage molecules are synthesized.

Key Terms

oxidized (p. 108)
reduced (p. 108)
oxidation-reduction
   reaction (p. 108)
NAD^+ (p. 108)

NADH (p. 108)
$NADP^+$ (p. 108)
NADPH (p. 108)
flavin adenine
   dinucleotide (p. 108)

flavin mononucleotide, FMN (p. 109)
cytochromes (p. 109)
electron transport systems (p. 109)
analogy (p. 109)
electric gradients (p. 109)

Objectives

1. Distinguish an oxidation reaction from a reduction reaction, then show how the two must be coupled into an oxidation-reduction reaction.

2. Explain how phosphorylation reactions and oxidation-reduction reactions are related to the transfer of energy.

3. Describe an electron transport chain. Tell which types of molecules participate in such chains and how they participate.

4. State what electron transport chains accomplish for the cells that contain them.

5. Describe how NAD^+, $NADP^+$, FAD, and the cytochromes take part in energy transfers. Write an equation for a generalized reaction that shows these energy-poor molecules becoming energy-rich molecules. Do this by translating each of the preceding molecules into a capital letter. Then show what the letter combines with to yield a product molecule.

Self-Quiz Questions

   True-False If false, explain why.

____ (1) A substance that has been oxidized has gained one or more electrons.

____ (2) Phosphorylation is a chemical reaction in which a phosphate group is attached to another molecule, thus increasing the potential chemical energy of that molecule.

____ (3) Reduction is a chemical reaction in which one or more electrons or hydrogen atoms is attached to a molecule, thus increasing its potential chemical energy.

   Fill-in-the-Blanks

An (4) _____ _____ chain is a series of (5) _____ carrier molecules that carries out an organized sequence of (6) _____ - _____ reactions. The first molecule in line accepts an excited (7) _____ from a donor molecule outside the chain. It gets transferred from the first molecule to a series of (8) _____ molecules, releasing usable (9) _____ at each step. At certain transfer points, the amount of energy given off is sufficient to do useful work, such as attaching a (10) _____ group to ADP. The last molecule in the chain gives up the electron to an (11) _____ acceptor molecule. Removing the electron in this way keeps the transport chain clear for operation. (12) _____, (13) _____, and (14) _____ can accept two electrons and then later release those electrons as part of a step-by-step reaction series. There are also some iron-containing protein compounds that are referred to collectively as (15) _____ and are located largely in chloroplasts and mitochondria.

INTEGRATING AND APPLYING KEY CONCEPTS

A piece of dry ice left sitting on a table at room temperature vaporizes. As the dry ice vaporizes into CO_2 gas, does its entropy increase or decrease?

7
ENERGY-ACQUIRING PATHWAYS

7–I FROM SUNLIGHT TO CELLULAR WORK: PREVIEW OF THE MAIN PATHWAYS

Summary

The three major pathways of photosynthesis, glycolysis, and cellular respiration are linked by energy flowing through them. In *photosynthesis*, sunlight energy is trapped by pigment molecules (such as chlorophyll) and is used to form ATP and NADPH, which are intermediate energy carriers. These carriers transfer some of their energy to reactions in which carbon dioxide and water from the environment are converted into food molecules. Glycolysis and aerobic respiration are two interconnected ways of releasing for use energy stored in food molecules.

Key Terms

autotrophic organism (p. 111)
photosynthetic
   autotroph (p. 111)
chemosynthetic
   autotroph (p. 111)

inorganic compound (p. 111)
heterotrophic
   organism (p. 111)
glycolysis (p. 111)

respiration (p. 111)
photosynthesis (p. 111)
cycling (p. 111)

Objectives

1. Study Fig. 7.1 until you can remember how the reactants and products of each of the three major energy-trapping and energy-releasing pathways are interrelated. Then produce the diagram from memory on another piece of paper.

2. Name the principal raw materials needed to begin each of the three processes and the products formed at the end of each process.

Self-Quiz Questions

Fill-in-the-Blanks

(1) _____ _____ obtain energy from sunlight. (2) _____ _____ oxidize inorganic compounds that contain sulfur and ammonium to obtain energy. Photosynthetic autotrophs include all plants, some protistans, and some (3) _____. Chemosynthetic autotrophs are limited to a few kinds of (4) _____. (5) _____ organisms feed on autotrophs, each other, or organic wastes. Energy stored in organic compounds such as glucose may be released by the two interconnected pathways (6) _____ and (7) _____, which release carbon dioxide and water.

7–II PHOTOSYNTHESIS

Summary

Most autotrophs carry out photosynthesis of one sort or another. The simplest and most ancient photosynthetic process is *cyclic photophosphorylation*, in which pairs of electrons from the photolysis of water molecules are attracted to a single photosystem (a collection of pigments), which, some of the time, is busy absorbing specific wavelengths of sunlight. Some of this energy is transferred to the electrons, which become excited, move faster, and are passed to an electron acceptor molecule and then through a transport chain. As electrons are transferred down through the chain, some of the energy released is used to form ATP from ADP and inorganic phosphate. Eventually, the electrons lose their excess energy and return to the photosystem. Every two electrons that enter the cyclic photophosphorylation pathway can cause one ATP molecule to be formed. The plant can then use this ATP to do various forms of work.

Most modern autotrophs carry out *noncyclic photophosphorylation*, which includes two photosystems and two transport chains. When the two photosystems function together, electrons do not flow in a cycle. Instead, every two new electrons from water molecules are used to make two ATP molecules via the first transport chain. Then they are reexcited in the second photosystem, and, as they pass down the second transport chain, the two electrons end up in a molecule of NADPH and are exported from the light reactions and used to drive one of the dark reactions. Because they are exported, the electrons do not follow a cyclic path. Some of the ATP is also used to drive one of the dark reactions.

Key Terms

photosynthesis (p. 111)
absorbed (p. 111)
chlorophyll (p. 111)
pigment molecules (p. 111)
chloroplasts (p. 111)
light-dependent
   reactions (p. 111)
NADPH (p. 111)
light-independent
   reactions (p. 111)
stroma (p. 112)

granum (p. 112)
thylakoids (p. 113)
photosystems (p. 113)
photons (p. 113)
chlorophyll a (p. 114)
chlorophyll b (p. 114)
carotenoids (p. 114)
Photosystem I,
   P700 (p. 115)
Photosystem II,
   P680 (p. 115)

nanometers (p. 115)
transmit (p. 115)
cyclic photo-
   phosphorylation (p. 115)
noncyclic photo-
   phosphorylation (p. 116)
photolysis (p. 117)
concentration gradient
   (p. 117)
electric gradient (p. 117)
chemiosmotic theory
   (p. 117)

Objectives

1. Distinguish autotrophs from heterotrophs and give two examples of each group.

2. Consider the internal structure of a chloroplast and contrast the grana with the stroma by describing their differences in structure and function.

3. Describe the role that chlorophyll and the accessory pigments play in the light reactions. After consulting Fig. 7.3, state which colors of the visible spectrum are absorbed by (a) chlorophyll, (b) phycocyanin, and (c) carotenoids.

4. Contrast cyclic and noncyclic photophosphorylation in terms of the substances produced, the number of photosystems, and the number of transport chains.

5. Explain what the water split during photolysis contributes to both cyclic and noncyclic photophosphorylation.

6. Name the two energy carrier molecules produced during noncyclic photophosphorylation and indicate how they will be used later.

Self-Quiz Questions

Fill-in-the-Blanks

(1) _____ organisms can synthesize their own organic food molecules from inorganic raw materials; (2) _____ organisms cannot. In photosynthesis, (3) _____ _____ and (4) _____ are the raw materials that are converted into (5) _____ and (6) _____-_____ _____ _____; energy from (7) _____ is trapped by (8) _____ _____ and is used to form the intermediate energy carriers (9) _____ and (10) _____.

(11) _____ photophosphorylation is the simplest and most ancient photosynthetic process; one photosystem and one electron transport chain are involved, and (12) _____ is the sole energy carrier molecule that is produced. Most modern autotrophs carry out (13) _____ _____, which includes two photosystems and two electron transport chains. (14) _____ is the other energy carrier molecule that is formed from (15) _____ ions and (16) _____ (obtained from water) that become linked to the nucleotide energy carrier called (17) _____. Because they are exported from the light reactions, the (18) _____ do not follow a cyclic path.

7–III   THE LIGHT-INDEPENDENT REACTIONS

Summary

During the light-independent reactions, a photosynthetic cell can use energy carriers such as ATP and NADPH to produce organic food molecules. In the first stage of these reactions, carbon dioxide from the air is combined with (that is "fixed" to) ribulose biphosphate and incorporated into stable intermediate compounds (PGA). In the second stage, ATP and NADPH supply chemical energy to convert PGA to PGAL, some of which is then used to form glucose and some of which is used to make more ribulose biphosphate.

Tropical grasses that are adapted to high light levels, high temperatures and limited water have complex leaf structures and a carbon fixation system that precedes the Calvin-Benson cycle. Such plants are called C_4 plants, in which one leaf cell type specializes in preliminary CO_2 uptake and rapid, active transport of the fixed carbon dioxide to another leaf cell type where the Calvin-Benson cycle operates. C_4 plants can continue to photosynthesize even on hot, dry days, when sunlight is intense and when water must be conserved.

Key Terms

light-independent
   reactions (p. 118)
ribulose biphosphate,
   RuBP (p. 118)
intermediate (p. 118)

phosphoglyceric acid,
   PGA (p. 118)
carbon dioxide
   fixation (p. 118)
phosphoglyceraldehyde,
   PGAL (p. 118)

Calvin-Benson cycle (p. 118)
sucrose (p. 119)
tuber (p. 119)
C_4 plant (p. 120)
photorespiration (p. 120)

1. Explain why the light-independent reactions are called by that name.

2. Describe the process of carbon dioxide fixation by stating which reactants are necessary to get the process going and what are the stable products of this process alone.

3. Describe the Calvin-Benson reaction series in terms of its reactants and products.

4. Explain what happens to each of the products of photosynthesis.

Self-Quiz Questions

> Fill-in-the-Blanks

The light-independent reactions can proceed without sunlight as long as (1) _____

and (2) _____ are available. The reactions begin when an enzyme links (3)

_____ _____ to (4) _____ _____, a five-carbon compound.

The resulting six-carbon compound is highly unstable and breaks apart at once into two

molecules of a three-carbon compound, (5) _____. This entire reaction sequence

is called (6) _____ _____ _____. For every (7) _____ car-

bon dioxide molecules fixed, twelve molecules of the three-carbon compound are pro-

duced. The twelve (8) _____ molecules now enter a reaction series which uses

twelve (9) _____ and twelve (10) _____ molecules from the (11) _____

_____ _____ to convert PGA into (12) _____. Two of these mole-

cules are rearranged and linked together to form one (13) _____ molecule, which

is regarded as the principal end-product of the dark reactions. Six more ATP molecules

are used to regenerate (14) _____ _____, the five-carbon molecule to

which CO_2 was initially linked.

7–IV CHEMOSYNTHESIS 121

Summary

Some bacteria can harness energy released during the oxidation of such inorganic sub-stances as ammonium ions (NH_4^+) and sulfur compounds. Such chemosynthesizers are found in the soils of terrestrial ecosystems and in the sediments of aquatic ecosys-tems where their activities enhance the cycling of nitrogen and sulfur through natural communities.

Objectives

1. Distinguish chemosynthesis from photosynthesis and state which groups of organisms carry on each process.
2. Consult Chapter 38 and predict what would happen if there were no organisms that could assist in the cycling of nitrogen and sulfur.

Self-Quiz Questions

Fill-in-the-Blanks

Chemosynthetic bacteria are (1) _____, for they can live completely on an (2) _____ diet. Unlike photosynthesizers, they harness energy released during the (3) _____ of such inorganic substances as (4) _____ ions and (5) _____ compounds.

Some (6) _____ bacteria use ammonia molecules as their energy source, stripping them of hydrogen ions and their associated electrons. The cycling of (7) _____ through natural communities depends in large part on their activities, for (8) _____ is found in the proteins of organisms.

INTEGRATING AND APPLYING KEY CONCEPTS

Suppose that humans acquired all of the enzymes needed to carry out photosynthesis. Speculate about the changes in anatomy, physiology and behavior that would have to accompany humans having those enzymes in order for them to actually carry out photosynthetic reactions.

8
ENERGY-RELEASING PATHWAYS

OVERVIEW OF THE MAIN
ENERGY-RELEASING PATHWAYS

    Glycolysis: First Stage of
      Carbohydrate Metabolism
    Aerobic Respiration
    Anaerobic Respiration
    Fermentation Pathways

A CLOSER LOOK AT AEROBIC
RESPIRATION

    First Stage: Glycolysis
    Second Stage: The Krebs Cycle

Third Stage: Electron Transport
    Phosphorylation
Glucose Energy Yield:
    Summing Up

FUELS OR BUILDING BLOCKS? CONTROLS
OVER ENERGY METABOLISM

PERSPECTIVE

Summary

Energy-rich carbohydrates and other food molecules are stockpiled in autotrophic organisms. Eventually, autotrophic cells and the cells of the heterotrophs that dine on them will break down these molecules and use the stored energy to drive life processes. Although many different types of molecules are stored and eventually used, glucose breakdown and the associated energy transfers leading to ATP formation are central to the functioning of almost all prokaryotic and eukaryotic cells. Glucose is partially dismantled by the glycolytic pathway; during this process, some of its stored energy ends up in two ATP molecules. At the end of glycolysis, glucose has been converted to two molecules of pyruvic acid (pyruvate); also, substrates give up electrons and hydrogen ions to two NAD^+, which yields two molecules of the electron carrier NADH. If oxygen--O_2--is not present in sufficient amounts, pyruvate enters fermentative pathways in which pyruvate is converted into fermentation products such as lactate or ethanol and carbon dioxide. Fermentative processes ensure that the essential carrier molecule NAD^+, is regenerated and glycolysis can continue. Some prokaryotes use another kind of pathway, *anaerobic respiration*, in which an inorganic compound other than oxygen is the final acceptor of electrons from NADH. If sufficient oxygen is present, pyruvate enters the *aerobic pathways* (the Krebs cycle + electron transport phosphorylation) and 36 additional ATP molecules may be generated.

The entire aerobic pathway can generate 19 times as many ATP molecules as does glycolysis plus any of the fermentative processes, so cells and organisms that carry
out aerobic glucose metabolism have the means to accomplish much more work than
anaerobic organisms.

Key Terms

organism (p. 122)
photosynthesis (p. 122)
chemosynthesis (p. 122)
carbohydrate metabolism (p. 122)
oxidation-reduction reaction (p. 122)
electron acceptor molecules (p. 123)
NAD^+ (p. 123)
NADH (p. 123)
pyruvic acid, pyruvate (p. 124)
glycolysis (p. 124)

aerobic respiration (p. 124)
electron transport system (p. 124)
anaerobic respiration (p. 126)
fermentation (p. 126)
denitrifying bacteria (p. 126)
methanogenic bacteria (p. 126)
alcoholic fermentation (p. 126)
ethanol (that is, ethyl alcohol) (p. 126)
lactate fermentation (p. 126)
lactate (p. 126)

Objectives

1. Contrast the anaerobic and aerobic pathways of glucose metabolism in terms of
    (a) the organisms that use the pathways, (b) the number of ATP molecules gener-
    ated by each, (c) the names of the components of each type of pathway, and (d)
    the products of each type of pathway.

2. Explain why glucose breakdown and ATP formation are so important to the func-
    tioning of almost all prokaryotic and eukaryotic cells.

3. List some places where there is very little oxygen present and anaerobic organisms
    might be found.

4. Describe what happens to pyruvate in anaerobic organisms. Then explain the neces-
    sity for pyruvate to be changed to any of a variety of fermentative products.

5. State which factors determine whether the pyruvate (pyruvic acid) produced at the
    end of glycolysis will enter into the alcoholic fermentation, the lactate fer-
    mentation, or the acetyl-coenzyme A formation pathway.

6. Calculate how many net ATP molecules are produced during the glycolysis + fer-
    mentation path for each glucose molecule that began the trip.

7. Calculate how many net NADH molecules are produced during glycolysis + fermenta-
    tion for every glucose molecule that enters the pathway.

Self-Quiz Questions

Fill-in-the-Blanks

(1) _____ organisms can synthesize and stockpile energy-rich carbohydrates and

other food molecules from inorganic raw materials. (2) _____ is partially dis-

mantled by the glycolytic pathway, during which process some of its stored energy ends

up in two (3) _____ molecules. Some of the energy of glucose is released during

the breakdown reactions and is used in forming the energy carriers (4) _____ and (5) _____. These reactions take place in the cytoplasm. At the end of glycolysis the starting molecule has been converted to two molecules of (6) _____. If (7) _____ is not present in sufficient amounts, the end product of glycolysis enters (8) _____ pathways, in which it is converted into such products as (9) _____ or (10) _____ and (11) _____ _____. If sufficient oxygen is present, the end product of glycolysis enters (12) _____ pathways (the (13) _____ cycle + (14) _____ _____ _____), during which process (15) _____ additional (16) _____ molecules are generated. In the (17) _____ _____, the food molecule fragments are further broken down into (18) _____ _____ bits. During the reactions, hydrogen atoms (with their (19) _____) are stripped from the fragments and are transferred to the energy carriers (20) _____ and (21) _____. The electrons are then sent down a (22) _____ _____, and this energy they give off is used in forming (23) _____. Electrons leaving the chain combine with hydrogen ions and (24) _____ to form water. These reactions occur only in (25)_____.

8–II   A CLOSER LOOK AT AEROBIC RESPIRATION

Summary

During glycolysis, which occurs in the cytoplasm, one molecule of glucose is broken down to two molecules of pyruvic acid. In the process, four ATP molecules are made for every two ATPs used, and two NADH molecules bear away from glycolysis hydrogen ions and electrons recently acquired from an intermediate of glucose breakdown, PGAL. Glycolysis would quickly halt if the process ran out of NAD^+, which serves as the hydrogen and electron acceptor. When sufficient oxygen is present in eukaryotic cells, *pyruvate* from glycolysis is completely oxidized to carbon dioxide during two phases called acetyl-coenzyme A formation and the Krebs cycle (Fig. 8.5). In these reactions, which occur in the matrix within the inner membrane of the mitochondrion, NAD^+ and FAD serve as temporary acceptors for hydrogen ions and electrons released during the oxidations and are reduced to NADH and $FADH_2$. NADH and $FADH_2$, in turn, feed their hydrogen ions and electrons into the process known as electron transport phosphorylation, which occurs on the surface of the inner mitochondrial membrane, where NADH and $FADH_2$ are stripped of hydrogen ions and the high-energy electrons associated with those ions. The hydrogen ions are pumped into a reservoir that includes all of the space between the outer and inner mitochondrial membranes; thus a proton gradient is established, the energy from which drives the phosphorylation of ADP and ATP. Energy released in the passage of the electrons down an electron transport chain is used to form ATP molecules. [Most NADH molecules that enter the oxidative-phosphorylation sequence yield three ATP molecules; every $FADH_2$ molecule yields two. In heart and liver cells, all incoming NADH enters the electron transport chain at the "top" and every NADH generates three ATP molecules; in other eukaryote cells, NADH molecules made *during glycolysis* must be transported into the mitochondrion in order to enter the electron transport chain at the same point as $FADH_2$, and so each NADH generates only two ATP molecules (like $FADH_2$)].

Key Terms

Krebs cycle (p. 126)
electron transport
  phosphorylation (p. 126)
phosphoenolpyruvate, PEP (p. 127)
phosphoglyceraldehyde, PGAL (p. 128)
dihydroxyacetone phosphate, DHAP (p. 128)
structural isomers (p. 128)
pyruvic acid, pyruvate (p. 128)
acetyl coenzyme A, acetyl co-A (p. 128)

chemiosmotic theory of ATP
  formation (p. 129)
membrane-bound transport system (p. 129)
matrix (p. 129)
channel protein (p. 129)
oxaloacetate (p. 130)
citrate (p. 130)
kilocalorie (p. 132)
energy-releasing efficiency (p. 132)

Objectives

1. Explain the purpose served by ATP reacting first with glucose, then with fructose-6-phosphate in the early part of glycolysis.

2. Explain why your text says that *four* ATP molecules are made for every two used during glycolysis. Consult Fig. 8.3.

3. Consult Fig. 8.5. State the events that happen during acetyl-coenzyme A formation and explain how the process of acetyl-CoA formation relates glycolysis to the Krebs cycle.

4. State the factors that cause pyruvic acid to enter the acetyl-CoA formation pathway.

5. State what happens to the CO_2 produced during acetyl-CoA formation and the Krebs cycle.

6. Consult Fig. 8.6 and predict what will happen to the NADH produced during acetyl-CoA formation and the Krebs cycle.

7. Calculate the number of ATP molecules made during acetyl-CoA formation and during the Krebs cycle for *each glucose* molecule that entered glycolysis.

8. Summarize the basic ideas of the chemiosmotic theory and tell what the theory helps to explain.

9. Briefly describe the process of oxidative phosphorylation by stating what reactants are needed and what the products are. State the exchange value of NADH and $FADH_2$.

Self-Quiz Questions

(1) What is achieved by ATP reacting first with glucose, then with fructose-6-phosphate, in the early part of glycolysis?

(2) What happens to dihydroxyacetone phosphate midway through glycolysis?

(3) What happens to the ADP produced during the first part of glycolysis?

(4) What happens to the ATP produced during the last part of glycolysis?

(5) What are the potential uses of NADH produced during glycolysis?

(6) What role do NAD^+ and FAD serve in acetyl-coenzyme A formation and the Krebs cycle?

(7) Write the reduced forms of NAD^+ and FAD.

(8) Into which process are the reduced forms of NAD^+ and FAD fed?

(9) How many ATP molecules are made during acetyl-coenzyme A formation for every glucose molecule that enters glycolysis? During the Krebs cycle? During glycolysis (net)?

(10) How many NADH molecules are made during acetyl-coenzyme A formation for every glucose molecule that enters glycolysis? During the Krebs cycle? During glycolysis?

(11) How many $FADH_2$ molecules are made during acetyl-coenzyme A formation for every glucose molecule that enters glycolysis? During the Krebs cycls?

(12) In heart and liver cells, how many ATP molecules can each NADH molecule generate during electron transport phosphorylation? How many ATP molecules can each $FADH_2$ molecule generate during electron transport phosphorylation?

(13) How many ATP molecules (net) are generated directly during glycolysis and the Krebs cycle?

(14) How many ATP molecules are generated in heart and liver cells as a result of NADH acting as an energy-carrier molecule?

(15) How many ATP molecules are generated as a result of $FADH_2$ acting as an energy-carrier molecule?

(16) What happens to aerobic respiration when molecular oxygen becomes scarce?

(17) How many more ATP molecules can be made as a result of glycolysis plus aerobic respiration in ordinary eukaryote cells than can be made as a result of glycolysis plus fermentation?

8–III   FUELS OR BUILDING BLOCKS? CONTROLS OVER ENERGY METABOLISM   132-133

Summary

Many energy sources other than glucose can be fed into the glycolytic, CoA formation, or Krebs cycle pathways. Carbohydrates can be converted to glucose or some other compound and thus enter glycolysis. Fats are first digested into glycerol and fatty acids, which enter glycolysis and CoA formation. Proteins are digested into a variety of amino acids; some enter glycolysis, but many enter the Krebs cycle at various points by being converted into Krebs cycle intermediates. Metabolizing a gram of fat generally yields about twice as much ATP as metabolizing either a gram of carbohydrate or a gram of protein, so fats are efficient long-term energy storage molecules. Cells generally use simple sugars, amino acids, nucleotides, fatty acids, and glycerol to synthesize as many of the more complex molecules as they need; the remainder of that taken in is sent down respiratory pathways to make ATP and waste products or else is converted to lipids or starches for longer term energy storage. Enzymes catalyzing key reactions in glycolysis and the Krebs cycle govern whether molecules are degraded as an ATP-generating source, or are converted to intermediate forms that can be used in biosynthesis.

Key Terms

biosynthesis (p. 132)   phosphofructokinase (p. 132)   inhibitor (p. 132)
catalyze (p. 132)   cleavage (p. 132)

Objectives

1. List some sources of energy other than glucose that can be fed into the respiratory pathways.

2. Explain what cells do with simple sugars, amino acids, fatty acids, and glycerol that exceed what the cells need for synthesizing their own assortments of more complex molecules.

3. Predict what your body would do to synthesize its needed carbohydrates and fats if you switched to a diet of 100 percent protein.

Self-Quiz Questions

True-False   If false, explain why.

____ (1) Glucose is the only carbon-containing molecule that can be fed into the glycolytic pathway.

____ (2) Fats are efficient long-term energy storage molecules.

____ (3) Simple sugars, fatty acids, and glycerol that remain after a cell's biosynthetic needs have been met are generally sent to the cell's respiratory pathways for energy extraction.

____ (4) Carbon dioxide and water, the products of aerobic respiration, generally get into the blood and are carried to gills or lungs, kidneys, and skin, where they are expelled to the outside of an animal's body.

____ (5) Humans can synthesize twelve essential amino acids as long as appropriate carbohydrates are included in the diet.

8–IV PERSPECTIVE

133-135

Life apparently originated when the environment was rich in molecules containing carbon, hydrogen, oxygen, and nitrogen. Presumably the first forms of life were heterotrophic prokaryotes that utilized glycolysis to extract energy from the early carbon compounds. As increasing demands were made on these carbon resources, alternative forms of harnessing energy, such as photosynthesis and chemosynthesis, evolved among the prokaryotes. Glycolysis followed by fermentation is an inefficient method of energy extraction that produces lactic acid or alcohol as accumulating by-products. Photosynthesis produces oxygen as a waste product.

Advances in the complexity of cells caused cells to become compartmentalized, and a variety of complex processes could occur in these well-organized regions. Cellular respiration was one such complex process that used the waste products of glycolysis and photosynthesis to extract much more energy and produce carbon dioxide and water, the raw materials of photosynthesis. Thus the cycling of carbon, hydrogen, and oxygen came full circle and life became more self-sustaining and balanced with the "carbon cycle." Other similar cycles have come to exist for other essential elements, such as nitrogen and phosphorus.

Once molecules are assembled into the cells of organisms, their organization must be sustained by outside energy derived from food, water, and air. All living forms are part of an interconnected web of energy use and materials cycling that permeates all levels of biological organization. Should energy fail to reach any part of these levels, life will become increasingly disordered. Energy flows only from forms rich in potential energy to forms having fewer and fewer usable stores of it. Life is no more, and no less, than a wonderfully complex system of prolonging order. It can do so because hereditary instructions are passed from generation to generation, so that, even though individual organisms die, life and order continue.

Key Terms

carbon cycle (p. 133)   phospholipids (p. 133)   enzymes (p. 134)
aqueous (p. 133)   bilayer (p. 133)

Objectives

1. Outline the supposed evolutionary sequence of energy-extraction processes.

2. Scrutinize Fig. 8.8 closely; then reproduce the carbon cycle from memory.

Self-Quiz Questions

True-False   If false, explain why.

____ (1) Energy is recycled along with materials.

____ (2) The first forms of life on Earth were most probably photosynthetic eukaryotes.

____ (3) Photosynthesis produces molecular oxygen as a by-product.

____ (4) Energy flows only from forms rich in potential energy to forms with fewer usable stores of energy.

INTEGRATING AND APPLYING KEY CONCEPTS

How is the "oxygen debt" that occurs in runners and sprinters related to aerobic and anaerobic forms of respiration in humans?

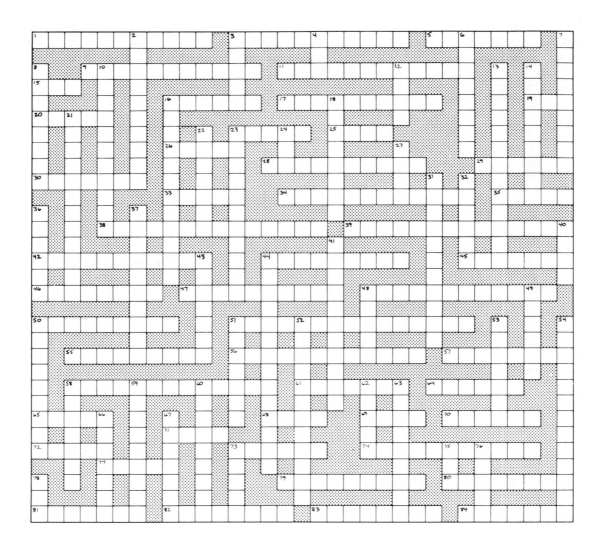

Across

1. organelle in which photosynthetic reactions occur
3. substances composed of fat and carbohydrate that are important constituents of cell membranes
5. lipid assembled from five-carbon subunits
9. the principal light-trapping pigment
11. organelles that release ninety percent of the energy stored in glucose
15. a liquid dispersion of ions and water molecules around cell components, particularly proteins

16. a polysaccharide that forms the exoskeleton of insects and their relatives
17. the enzyme-controlled process that breaks down large molecules into their subunits by reaction with water molecules
19. one of a series of teeth on a gear
20. surname of English royal family from Henry VII (1485) through Elizabeth I (1603)
23. both ___ and flagella have the 9 + 2 arrangement of microtubules
25. ___ flow; different kinds of molecules present in a fluid move together in the same direction in response to a gradient
26. a length of time established by the moon's revolution around the earth
28. an enzyme that works as a control agent by distorting the active site
29. fluid ___ model of membrane structure
30. glucose + fructose = ____
33. gradient
34. an anaerobic pathway that generally converts pyruvic acid to either lactic acid or ethyl alcohol and carbon dioxide
35. a kingdom is subdivided into ___
38. harnessing energy released during the oxidation of such inorganic substances as ammonium ions and sulfur compounds
39. the study of heat and other forms of energy
42. net protein ___, a means of comparing proteins from different sources
44. able to let some substances pass through
45. a pure substance composed of only one kind of atom
46. negatively charged particle located outside the nucleus
47. consists of a nitrogenous base + a phosphate group + a five-carbon sugar
48. sugar or starch is a ___
50. a gain of electrons
51. attaching a phosphate group to a substance
55. chromosomes disperse into the ___ network
56. a substance composed of simple sugars and amino acids; important in cell recognition
57. a six-carbon sugar found in fruits and elsewhere
58. an extremely fine threadlike structural element involved in cell movement
61. a ___ bond covalently links two amino acids
64. a system that prevents large changes in pH
65. Syringa vulgaris, widely cultivated for its clusters of fragrant purplish or white flowers
68. a town in Buckinghamshire, England, on the Thames opposite Windsor; the site of a public school for boys
69. Greek goddess of victory
70. this is one of two substances that compose microfilaments
71. Fe; an important constituent of hemoglobin
72. aboriginal people of New Zealand, of Polynesian-Melanesian descent
73. not basic; caustic
74. accessory light-trapping pigment molecules
77. French composer, 1866-1925
79. the passive movement of particles of a substance from a region of high concentration to a region of low concentration
80. a membranous bag of hydrolytic enzymes
81. a complex substance composed of amino acids
82. sticking together
83. a ___ bond is formed when two atoms share a pair of electrons
84. passive transport of water across membranes in response to concentration gradients and/or pressure gradients

Down

2. all monerans are organisms of this type
3. ___ complex; a membrane system that assembles, packages, and transports substances
4. permit
6. having the capacity to combine chemically with some other atoms
7. releasing energy
8. atoms that have the same atomic number but different atomic weights
10. substances that do not mix easily with water
12. contains organ of smell and the beginning of the respiratory tract
13. P-O-P linkages; ___ bonds
14. the ___ structure of a protein is established by repeated hydrogen bonding
16. the ___ theory explains how H^+ ions are involved in oxidative phosphorylation
18. an organelle involved in protein synthesis
21. a substance that contains two simple sugars
22. a measure of the randomness, disorder, or chaos in a system
23. a steroid component of most cell membranes
24. sick
27. Bhagavad ___
31. vesicles of extracellular fluid are brought into a cell by bulk transport
32. a protein catalyst
36. that which is dissolved
37. all of the physical and chemical processes involved in the maintenance of life
40. place
41. an organism that cannot synthesize its own food
43. electrically neutral particle in the nucleus of an atom
44. the principal energy-trapping pathway used by autotrophs
48. an underground vault or chamber
49. osmotically induced internal pressure
50. endoplasmic ___, a network of membranes
51. a substance that reflects certain colors of light
52. assembled from glycerol, fatty acids, a phosphate group and usually a nitrogen-containing alcohol
53. the control center of a cell
54. forms of the same kind of molecule (e.g., glucose) that differ in the way certain functional groups are covalently bonded to a carbon atom
58. a double sugar composed of two glucose subunits
59. the loss of electrons
60. an organism that can synthesize its own food
62. of, or relating to, atoms that have gained or lost one or more electrons
63. the cells of such an organism have membrane-bound nuclei and other organelles
66. projecting shelves formed by the inner membrane of the mitochondrion
67. of, or related to, motion
73. the smallest units of an element that still have the physical and chemical properties of that element
75. a fish of the genus Anguilla
76. a fertile or green area with water in a desert or wasteland
78. a derivative of a nucleotide used in energy storage

9
CELL REPRODUCTION

Summary

Whether single-celled or multicellular, each living organism follows a life cycle characteristic of its species: it undergoes sequential changes in form and function from the moment it emerges as a distinct entity until its death. The point on which all life cycles pivot is reproduction: making a copy of oneself. These copies repeat the changes in form and function that must unfold during the prescribed life cycle for a species. Instructions for producing each new cell reside in DNA, in which the sequence of nucleotide bases represents instructions for assembling certain RNA molecules, which in turn contain instructions for building proteins. Enzymes are proteins that give the cell a means to control the assembly of organic molecules from simple building blocks present in its surroundings. Thus, directly or indirectly, DNA governs which organic materials will occur in a cell and, in many cases, how those materials will become arranged as the cell grows and develops.

The continuity of life depends on the replication of a cell's set of DNA molecules before that cell reproduces. When the cell does divide, one DNA set ends up in each daughter cell, along with a portion of the parental cell cytoplasm that contains enzymes involved in reading the instructions contained in DNA and RNA.

In prokaryotic cell reproduction an original DNA molecule and its replica are both attached to the plasma membrane; *membrane growth* between their attachment sites separates them and allows distribution into daughter cells. In eukaryotes hereditary instructions are divided up among a set of two or more different DNA molecules, which vary in length and shape. Also, there can be one, two, or more of these sets in a cell. With *mitosis* the number of DNA sets in the nucleus does not change. With *meiosis* the number of DNA sets in each daughter nucleus is exactly half of what it was in the parent nucleus.

Key Terms

reproduction (p. 138)
life cycle (p. 138)
"cell cycle" (p. 138)
nucleotides (p. 139)
gene (p. 139)
DNA replication (p. 139)
RNA molecules (p. 139)

translation (p. 140)
prokaryotic fission (p. 140)
binary fission (p. 140)
membrane growth (p. 140)
chromosomes (p. 141)
diploid (p. 141)
polyploid (p. 141)

mitosis (p. 141)
meiosis (p. 141)
haploid (p. 141)
cytokinesis (p. 141)
gamete (p. 141)
meiospores (p. 141)

Objectives

1. Define the term *life cycle* and use it in an example.

2. Explain how DNA is related to reproduction in cells.

3. Explain how DNA is related to protein synthesis.

4. Describe how cells use the instructions in DNA to synthesize molecules such as carbohydrates and lipids.

5. Define replication and tell why it must occur before a cell reproduces.

6. Explain why it is important that daughter cells receive a significant portion of the parental cytoplasm during reproduction.

7. List in order the events that occur during binary fission.

Self-Quiz Questions

Fill-in-the-Blanks

Each living organism undergoes sequential changes in form and function characteristic

of its species called a (1) _____ _____. Making a copy of oneself is

(2) _____, and the process begins at the molecular level. Instructions for pro-

ducing each new cell reside in (3) _____. (4) _____, which are specific

DNA regions that tell how to build different kinds of (5) _____ _____

(hence different kinds of proteins), are activated from the whole collection. The ones

activated function in particular cells at particular times. (6) _____ instructs only how to directly build proteins, some of which are the cellular catalysts called

(7) _____. These proteins may assist reactions that lead to other nonprotein

molecules needed by the cell—for instance, (8) _____ and (9) _____. The

continuity of life depends on the (10) _____ of a cell's set of DNA molecules

before that cell reproduces so that the cell acquires two complete sets.

Matching Choose the one most appropriate answer for each.

(11) ___ binary fission

(12) ___ conjugation

(13) ___ diversification

(14) ___ replication

A. a process that yields several different types from one original type

B. One set of DNA becomes two sets.

C. Segments of DNA are transferred from one cell to another across a bridge composed of cellular material.

D. an asexual process in which partitioning leads to two separate cells having basically similar parts

True-False If false, explain why.

___ (15) Prokaryotes reproduce by binary fission.

___ (16) Asexual processes such as binary fission do not promote diversification in prokaryotes.

9–II COMPONENTS OF EUKARYOTIC CELL DIVISION

Summary

Parceling out DNA without errors to a new generation of cells is a remarkable feat, even for prokaryotes that have a single circular DNA molecule. Unwinding, unfolding, untwisting, assembling a duplicate strand, rewinding, refolding, and separating must occur in long, easily tangled DNA molecules. More than a billion years ago, some fundamental changes in structure arose, which improved the cell's ability to manage increased amounts of DNA. These changes occurred in the ancestors of modern eukaryotes. The physical isolation of DNA in the nucleus, chromosome packaging, and a microtubular system for moving many chromosomes at a time form the basis of reproduction in single-celled eukaryotes, as well as physical growth in multicelled eukaryotes. Each chromosome contains a DNA molecule and different proteins grouped together as a single, intact fiber—a nucleoprotein. During cell construction and maintenance activities, the nucleoprotein is unfolded and uncoiled so that replication of DNA and synthesis of proteins can occur in accord with DNA's instructions; in this extended state, the nucleoprotein is referred to as *chromatin*. Before and during nuclear division, the nucleoprotein is folded and coiled into the condensed form called the *chromosome*. The number of chromosomes differs from one species to the next, but within one multicellular organism, each cell (except the sex cells) contains the same number of chromosomes as the others. Mitosis is only a small slice of a larger series of events known as the *cell cycle*. Cells spend most of their time in interphase, when two phases of protein synthesis and one phase of DNA replication occur. Interphase may last for only a few minutes, though in some cell types, it can last for more than a century!

Key Terms

chromosomes (p. 142)
microtubules (p. 142)
nucleoprotein (p. 142)
condensation (p. 142)
sister chromatids (p. 142)

centromere (p. 142)
somatic (p. 142)
nematode (p. 143)
microtubular spindle (p. 143)
polarity (p. 143)
cell cycle (p. 144)

interphase (p. 144)
differentiation (p. 144)
G_1 period (p. 144)
S stage (p. 144)
G_2 stage (p. 144)

Objectives

1. List two of the changes that occurred in cellular structure that allowed increased amounts of hereditary material to be managed more efficiently.

2. Scrutinize Fig. 9.9; then, from memory, reproduce the entire figure of the cell cycle.

3. List the events of interphase.

Self-Quiz Questions

True-False If false, explain why.

____ (1) Within a single organism, each cell except the sex cells contains the same number of chromosomes as the others.

____ (2) Immediately prior to cell division, the hereditary material becomes extended and less condensed.

____ (3) Each prokaryote generally has a single, circular chromosome.

____ (4) Replication is the process that builds cellular proteins.

Fill-in-the-Blanks

Two changes in cellular construction that promoted more efficient management of

hereditary material were the packaging of (5) _____, and the development of

a system of (6) _____ for moving many (7) _____ at a time.

Summary

DNA replication precedes mitosis and is followed at some point by condensing of the duplicated material, the sister chromatids, into chromosomes. The chromosomes are guided into separate daughter cells in two equivalent parcels. The microtubules forming the mitotic spindle are responsible for separating sister chromatids from each other. Once the two parcels of separated chromatids arrive at opposite poles, newly forming nuclear membranes surround each parcel, and two nuclei eventually form. The chromosome content of each daughter nucleus is identical with that of the original nucleus. As soon as nuclear division (mitosis) is complete, cytokinesis (apportioning cytoplasm to each of the daughter nuclei) generally occurs. Cytokinesis is the basis for reproduction in single-celled eukaryotes and the basis for physical growth in multicellular eukaryotes.

Key Terms

mitosis (p. 144)	metaphase (p. 146)	cleavage furrow (p. 148)
prophase (p. 144)	chromatid (p. 147)	microfilaments (p. 148)
spindle microtubules (p. 144)	spindle equator (p. 147)	contractile proteins (p. 148)
astral spindle (p. 144)	kinetochore (p. 147)	actin (p. 148)
centriole (p. 145)	anaphase (p. 147)	myosin (p. 148)
anastral spindle (p. 145)	telophase (p. 148)	cell plate formation (p. 148)
nucleolus (p. 146)	cytokinesis (p. 148)	vesicles (p. 148)

Objectives

1. Describe the process of mitosis and what it accomplishes.

2. Describe the events that must occur in order for DNA to duplicate itself without error.

3. Describe the appearance of the hereditary material (a) immediately before and during nuclear division, and (b) when the cell is carrying out construction and maintenance activities.

4. List the events of each phase of mitosis.

5. Describe microtubule behavior during metaphase and anaphase as chromatids are lined up and then pulled apart.

6. Contrast the way cytokinesis occurs in animals with the way it occurs in plants.

Self-Quiz Questions

Matching Choose the one most appropriate answer for each.

(1) ___ anaphase

(2) ___ centromere

(3) ___ chromatid

(4) ___ cytokinesis

(5) ___ G_1 phase

(6) ___ G_2 phase

(7) ___ metaphase

(8) ___ prophase

(9) ___ S phase

(10) ___ telophase

A. first phase of protein synthesis

B. cytoplasm allotted to and divided between two nuclei

C. chromosomes condense; mitotic spindle starts to form

D. DNA synthesis occurs now

E. final phase of mitosis; nuclear membranes re-form

F. connects two sister chromatids

G. second phase of protein building

H. elongate half of a chromosome

I. sister chromatids separate, move to opposite spindle poles now

J. chromosomes lined up along equator of cell

True-False If false, explain why.

___ (11) In plant cells a cleavage furrow forms around the cell's midsection.

___ (12) Mitosis *must* be followed by cytokinesis so that the cytoplasm is apportioned equally between two daughter nuclei.

___ (13) During early metaphase two sets of microtubules become attached to each chromosome at the centromere; one set extends from one pole to a chromatid, and one set extends from the opposite pole to the other chromatid.

___ (14) The centrioles divide during anaphase.

___ (15) During anaphase the microtubules that connect the centromere to its pole shorten.

9–IV MEIOSIS: HALVING THE NUMBER OF CHROMOSOME SETS

Summary

In organisms that reproduce sexually, the diploid parent organisms must, at some point, produce haploid reproductive cells (gametes), which find each other and fuse (an event known as *fertilization*), thus forming a new diploid cell, the *zygote*. *Mitosis* occurs in diploid as well as haploid organisms and maintains whatever number of chromosomes the parent cell had (either 2N or 1N), but *meiosis* reduces the parental diploid number (2N) to the haploid state (1N). If there were no such reduction, fertilization would result in a zygote with twice the amount of hereditary information required to build a new individual.

Complete meiosis is composed of two sequences of events. The first sequence halves the number of chromosomes, most of which have undergone some changes in their assortments of genes through processes known as crossing-over and genetic recombination (to be discussed later). The second sequence then separates the chromatids in a process that resembles mitosis mechanically, although four haploid daughter cells form rather than the two diploid cells that result from mitosis.

Key Terms

morphological (p. 149)
homologues (p. 149)
homology (p. 150)
sexual reproduction (p. 150)
gametes (p. 150)
fertilization (p. 150)
interkinesis (p. 151)
prophase I (p. 152)
condensation (p. 152)
homologue pairing (p. 152)

synapsis (p. 152)
tetrad (p. 152)
crossing over (p. 152)
nonsister chromatids (p. 152)
gene transcription (p. 152)
chiasmata (p. 152)
recondensation (p. 152)
alleles (p. 153)
genetic
  recombination (p. 153)

metaphase I (p. 153)
anaphase I (p. 153)
telophase (p. 153)
separation of
  homologues (p. 153)
separation of sister
  chromatids (p. 153)
decondense (p. 153)

Objectives

1. Contrast the life cycle of a haploid prokaryote with the life cycle of a diploid animal or plant.

2. List the differences between prophase I of meiosis and prophase of mitosis. State how the events of meiotic prophase I increase diversity within a species.

3. List the events that occur in the eight stages of meiosis.

4. Scrutinize Fig. 9.13 and note the stage in which crossing over and genetic recombination occur. State how gene transcription during prophase I prepares a new individual for events immediately following fertilization.

5. Contrast the overall achievements of mitosis with those of meiosis.

Self-Quiz Questions

Matching Assume for the following linkages that the cell starts with two *pairs* of already duplicated chromosomes, and that each chromosome consists of two chromatids. Choose the one most appropriate answer for each term.

(1) ____ Anaphase I

(2) ____ Anaphase II

(3) ____ Metaphase I

(4) ____ Metaphase II

(5) ____ Prophase I

(6) ____ Prophase II

(7) ____ Telophase I

(8) ____ Telophase II

A. Meiosis I is completed by this stage.

B. One member of each of the homologous pairs is separated from its mate; both are guided to opposite poles.

C. Two pairs of chromosomes are lined up at the spindle apparatus equator.

D. Chromosomes become clearly visible; nuclear region has two pairs of already-duplicated chromosomes.

E. Each centromere splits; sister chromatids are separated and moved to opposite poles.

F. Each daughter cell ends up with one chromosome of each type and may function as a gamete.

(continued)

G. Spindle apparatus re-forms; each cell has two chromosomes, each of which consists of two chromatids.

H. Two chromosomes are lined up at spindle apparatus equator.

True-False If false, explain why.

____ (9) The first diploid cell formed after fertilization is called a gamete.

____ (10) Most eukaryotic cells are diploid.

____ (11) Binary fission is a form of sexual reproduction.

____ (12) Haploid organisms have only a single copy of each of the genes needed for cell construction, growth, and maintenance.

9–V WHERE CELL DIVISIONS OCCUR IN PLANT AND ANIMAL LIFE CYCLES   156–158

Summary

Eukaryotes rely on both sexual and asexual reproductive processes at various stages of the life cycle. The sexual process requires both gamete formation and fertilization, but the timing of these separate events varies considerably during the life cycles of different species. Most animals form haploid gametes; soon thereafter, one male and one female gamete generally fuse during fertilization. Gametogenesis occurs only in special sex cells in male and female reproductive organs. Spermatogenesis produces four tiny, equally sized haploid spermatids; oogenesis produces one larger functional egg and three tiny nonfunctional polar bodies.

Plants differ from animals in certain features of their life cycles. The body of the more highly evolved plants is diploid, with multicellular organs in which meiosis occurs. But in plants, meiosis does not result directly in the formation of gametes. Instead, haploid spores are produced, each of which, if it encounters suitable germinating conditions, grows into a multicellular body composed exclusively of haploid cells. On one or more of these multicellular gametophytes, multicellular sex organs form, in which eggs or sperm are produced by *mitosis*. The gametes are haploid, like the gametophyte. When two unlike gametes fuse, a diploid zygote is formed that will grow into the diploid spore-producing sporophyte generation. Plant life cycles, then, contain alternating prolonged haploid and diploid stages, known as alternation of generations.

All of the various types of life cycles have become responsive to such environmental conditions as seasonal changes in the availability of food, water, and living sites. And all life cycles must overcome the constraints of the surface-to-volume ratio, must control and integrate the activities of differentiated body parts, and must handle enough hereditary material to ensure reliable growth, development, and reproduction of large, differentiated bodies.

Key Terms

prokaryotic fission (p. 156)
alternation of
  generations (p. 156)
gametogenesis (p. 156)
spermatogenesis (p. 157)
spermatogonium (p. 157)
primary spermatocyte (p. 157)
secondary spermatocyte (p. 157)

spermatid (p. 157)
sperm (p. 157)
oogenesis (p. 157)
oogonium (p. 157)
primary oocyte (p. 157)
secondary oocyte (p. 157)
egg (p. 157)
ovum (p. 157)

polar bodies (p. 157)
zygote (p. 157)
differentiate (p. 157)
meiospores (p. 157)
sporophyte (p. 157)
gametophyte (p. 157)
pollen grains (p. 158)

Objectives

1. Describe and distinguish the two forms of gametogenesis. Explain why unequally sized daughter cells are produced during oogenesis.

2. Contrast the generalized life cycle for plants with that for animals. Make sure that terms such as *sporophyte* and *gametophyte* are used and defined in your explanation.

3. Explain what is meant by *alternation of generations*.

4. Describe how *you* are a continuum in time and space of patterns of growth, development, and reproduction.

Self-Quiz Questions

  Sequence

Arrange in correct order of development the following:

(1) ____ primary spermatocyte, (2) ____ sperm, (3) ____ spermatid, (4) ____ spermatogonium, (5) ____ secondary spermatocyte

Put a 1 by the stage that appears first, and a 5 by the stage that finishes the process of spermatogenesis.

  True-False If false, explain why.

____ (6) The stage that ends meiosis II in male animals is the spermatid stage.

____ (7) The spore in a plant life cycle grows into the sporophyte generation.

____ (8) The gametophyte generation of more highly evolved plants produces haploid gametes as a result of mitosis.

____ (9) All cells in you are diploid except for any gametes you may produce.

____ (10) In plants the haploid sporophyte generation alternates with the diploid gametophyte generation.

INTEGRATING AND APPLYING KEY CONCEPTS

What do you think is the environment to which humans are best adapted? Did you base your answer on biological features or cultural features?

10
OBSERVABLE PATTERNS OF INHERITANCE

Summary

In 1865, Gregor Mendel presented the results of his now-famous experiments establishing the physical and mathematical rules that govern inheritance. Through his combined talents in plant breeding and mathematics, he perceived patterns in the expression of traits from one generation to the next. Mendel examined seven different traits (flower color, for instance), each trait with two different strains (white-flowered and purple-flowered), in sexually reproducing garden pea plants. Garden pea flowers can not only fertilize themselves, but the gametes from one individual can fertilize the gametes of another individual (in a process known as cross-fertilization). Mendel obtained seeds from strains in which all self-fertilized offspring displayed the same form of a trait, generation after generation; these are known as true-breeding strains. When two true-breeding organisms displaying different forms of a trait are crossed, the offspring are called *hybrids* because the two genetic units inherited, one from each parent, are unlike each other. Because Mendel first concentrated on studying a *single* trait, his initial experiments were called *mono*hybrid crosses. The results of monohybrid cross experiments demonstrated that, in sexually reproducing organisms, two units of heredity control each trait. During gamete formation, the two units of each pair are separated (segregated) from each other and end up in different gametes. This statement is the Mendelian principle of segregation.

trait (p. 160)
the "blending" hypothesis (p. 160)
Mendel (p. 160)
cross-fertilization (p. 161)
true-breeding strains (p. 162)
hybrids (p. 162)
monohybrid crosses (p. 162)
dihybrid cross (p. 162)

parent generation, P (p. 162)
first-generation offspring, F_1 (p. 162)
second-generation offspring, F_2 (p. 162)
recessive (p. 162)
dominant (p. 162)
the Mendelian principle of
  segregation (p. 163)

Objectives

1. Explain why the blending hypothesis is not regarded as an acceptable explanation of genetic behavior today.

2. Distinguish self-fertilizing from cross-fertilizing organisms. Explain the relation of true-breeding strains to self-fertilization.

3. Describe Mendel's monohybrid cross experiments with different true-breeding strains of garden peas (Fig. 10.2). Give the numbers and ratios he obtained in his F_1 and F_2 generations when he crossed a round-seeded parent with a wrinkled-seeded parent.

4. Distinguish the terms *dominant* and *recessive* and state what happens to the recessive gene in a hybrid.

5. Distinguish the P, F_1, and F_2 generations from each other.

6. Explain why, as a result of monohybrid crosses, Mendel always obtained an approximate 3:1 phenotypic ratio of dominant to recessive traits in the F_2 generation.

7. Define *probability* and explain what it has to do with genetics.

8. State the Mendelian principle of segregation and explain what would happen if segregation did not occur.

Self-Quiz Questions

   Fill-in-the-Blanks

Mendel perceived patterns in the expression of (1) _____ from one generation to

the next. He examined seven different traits, each with two different (2) _____

in sexually reproducing garden (3) _____ _____ plants, which can

carry out both self-fertilization and (4) _____-_____. Self-fertilized

plants that all display the same form of a trait generation after generation are

(5) _____-_____ strains. When two of these strains that display differ-

ent forms of a trait are crossed, the offspring are called (6) _____ because

the two genetic units inherited, one from each parent, are unlike each other. If two

of the hybrid offspring are mated, one-fourth of their offspring are (7) _____;

that is, they again show the trait which seemed to have disappeared in their parents' generation. Through such studies, Mendel demonstrated that the widely believed (8) _____ theory was untrue. Mendel's initial experiments were called (9) _____ crosses because he concentrated on examining a single trait. With such experiments, he was able to demonstrate that, in sexually reproducing organisms, two units of heredity control each trait. During gamete formation, the two units of each pair are separated from each other and end up in different gametes; the statement of this fact is the Mendelian principle of (10) _____.

Summary

Biochemical experiments and improvements in microscopy have advanced our understanding of the segregation process since Mendel's work in the mid-1800s, and the terms used to describe genetic events have changed accordingly. *Genes* are distinct units of heredity that dictate the traits of a new individual. There is one gene for each trait in a gamete, which is haploid. Following the fusion of two gametes (egg and sperm), the new individual has two genes for each trait (that is, is diploid).
    Alternative forms of genes for a given trait are known as *alleles*. In a diploid individual, two alleles for a given trait may be identical (and the individual is said to be *homozygous* for that trait) or two alleles may be different (and the individual is *heterozygous*). In heterozygous individuals, expression of one allele may mask expression of the other; the masking allele is dominant and the masked allele is recessive. Even so, both alleles retain their physical identity, and all the pairs of alleles constitute the individual's *genotype* (genotype makeup) for a particular assemblage of traits. The term *phenotype* refers to the collection of structural and behavioral features that are expressed, not masked. To determine whether an individual that shows the dominant form of a trait is heterozygous or homozygous, a *testcross* is performed—the individual is mated with another individual with the homozygous *recessive* genotype. If the unknown individual is homozygous, all offspring will show the dominant trait; if heterozygous, half the offspring will show the recessive trait. The *Punnett-square method* is the easiest way to predict the probable ratio of traits in offspring of monohybrid crosses.

Key Terms

segregation (p. 163)
testcross (p. 163)
backcrossed (p. 163)
genes (p. 163)
locus (p. 163)

alleles (p. 163)
gene pair (p. 163)
homozygote (p. 163)
heterozygote (p. 163)
dominant trait (p. 163)

recessive trait (p. 163)
genotype (p. 163)
phenotype (p. 163)
probability (p. 163)
Punnett-square method (p. 164)

Objectives

1. Define and distinguish *genes* and *alleles*.

2. Contrast the number of genes in a gamete with the number of genes in a zygote.

3. Define and distinguish *homozygous* and *heterozygous*.

4. Define and distinguish *dominant* and *recessive*, giving an example of each.

5. Explain why an individual's phenotype doesn't always express its genotype.

6. State the rules for symbolizing the genotype of a particular pair of alleles in which one allele is dominant and the other is recessive. Then use the stated rules to write the genotype (the pair of letter symbols) for each of the following traits:

    (a) black fur (dominant) vs. white fur (recessive) in rabbits
    (b) freckled (dominant) vs. unfreckled (recessive) in humans
    (c) black (dominant) vs. white (recessive) feathers in Andalusian chickens; the heterozygote appears bluish-gray.

7. In one of Mendel's studies the gene for green pod color was found to be dominant over its allele for yellow pods. Cross a homozygous green-podded plant with a yellow-podded plant. Show by a diagram with the letter symbols both genotype and phenotype for the P, F_1, and F_2 generations. Use the Punnett-square method.

8. If Mendel had crossed one of the F_1 hybrids of the cross in Objective 7 with a yellow-podded plant, what would have resulted? State the general name for such a cross, and state the question answered by such a cross. Show in a diagram the genotype and phenotype of one generation.

Self-Quiz Questions

True-False If false, explain why.

____ (1) There are never more than two alleles for a particular trait in any given population of organisms.

____ (2) The phenotypic ratio of a testcross that involves monohybrids is 1:1 if the dominant-appearing parent is heterozygous.

Solve the Following Problem:

(3) Albinos cannot form the pigments that normally produce skin, hair, and eye color, so albinos have white hair and pink eyes and skin (because the blood shows through). In order to be an albino, one must be homozygous recessive for the pair of genes that codes for the key enzyme in pigment production. Suppose that a woman of normal pigmentation with an albino mother marries an albino man. State the possible kinds of pigmentation possible for this couple's children, and tell the ratio of each kind of children the couple is likely to have. Show the genotype(s) and tell the phenotype(s).

Multiple Choice

____ (4) In genetics, the symbol for the dominant allele is usually the first letter of the trait capitalized. The symbol for the recessive allele is _____.

    (a) the first letter of the recessive trait, capitalized
    (b) the first letter of the recessive trait, in lower case
    (c) the first letter of the dominant trait, in lower case
    (d) none of the above

Summary

Through experiments with dihybrid crosses, Mendel began to perceive how sexual reproduction might foster genetic diversity. The Mendelian principle of independent assortment states that, when gametes are formed, distribution of alleles for a given trait
that have been segregated into one gamete or another is independent of, and doesn't
interfere with, distribution of alleles for other traits that have been segregated.

Key Terms

dihybrid cross (p. 166)   the Mendelian principle of independent
phenotypic ratio (p. 166)   assortment (p. 166)

Objectives

1.   Define *dihybrid cross* and distiguish it from *monohybrid cross.*

2.   Explain why Mendel obtained a 9:3:3:1 phenotypic ratio of characteristics in the
     F_2 generation as a result of a dihybrid cross.

3.   State the principle of independent assortment as formulated by Mendel.

4.   Solve the following problems, all of which deal with two different traits on dif-
     ferent chromosomes, by using either the Punnett-square method or by multiplying
     the separate probabilities.

     (a) The ability to bark and erect their ears are dominant traits in some dogs;
          keeping silent while trailing prey and droopy ears are recessive traits. If a
          barking dog with erect ears (heterozygous for both traits) is mated with a
          droop-eared dog that keeps silent while trailing, what kinds of puppies can
          result?

     (b) A spinach plant with straight leaves and white flowers was crossed with an-
          other spinach plant with curly leaves and yellow flowers. *All* of the F_1 gen-
          eration have curly leaves and white flowers. If any two individuals of the
          F_1 are crossed, what phenotypic ratio will result? Show all gentotypes.

Self-Quiz Questions

Multiple Choice

____ (1) In heterozygous individuals, _____.

     (a) the expression of one allele may mask the expression of the other for any
          given trait
     (b) both alleles for a given trait retain their identity throughout the indi-
          vidual's life cycle
     (c) if one allele masks the expression of the other, the former allele is
          dominant and the latter is recessive
     (d) all of the above are correct

____ (2) When independent assortment of alleles does occur, _____.

     (a) the alleles coding for the differing traits of an individual are parceled
          out independently of one another into separate gametes

(b) the process of meiosis retains the diploid number of chromosomes

(c) the alleles coding for a single trait are parceled out into the same gamete

(d) both a and c are correct

____ (3) A true-breeding red-flowered, tall plant is crossed with a true-breeding dwarfed, white-flowered plant. When two of the resulting F_1 plants are crossed, the phenotypic ratio in the F_2 offspring occur in approximately a _____ ratio. Red flowers and tall plants are dominant.

(a) 1:1:1:1   (b) 1:2:1   (c) 1:3:1   (d) 9:3:3:1

10–IV   VARIATIONS ON MENDEL'S THEMES

Summary

Since Mendel's time, studies have shown that interactions between genes can lead to many different phenotypic possibilities in addition to the clear-cut fully dominant or fully recessive ("all-or-nothing") examples investigated by Mendel. Cases of *incomplete dominance* and *codominance* tell us that there can be a range of dominance when only two alleles are under consideration. Moreover, for one trait, dominance can also be expressed in a variety of forms if the trait is governed by a *multiple allele system*; fruit fly eye color and the blood types that result from the ABO blood group are examples of multiple allele traits. For the sake of simplicity, most examples of heritable traits mentioned in texts are controlled by a single gene locus, but there are many traits (human skin color, for instance) that are governed by several active alleles at different loci; such traits are examples of *polygenic inheritance*.

Epistasis is a situation in which one gene *pair* masks the effect of one or more other pairs; calico cats, in which the "spotting" allele masks either black or yellow, produce white patches in whatever region of the cat's coat the spotting allele is behaving epistatically to the alleles for coat color. Single genes also may have major or minor influences on the expression of more than one trait, a situation known as *pleiotropism*. Sickle-cell anemia and mutations in genes that code for proteins necessary in a variety of structures and/or processes are examples of pleiotropy.

Key Terms

incomplete dominance (p. 168)
mosaic color pattern (p. 168)
codominance (p. 168)
locus (p. 168)
multiple allele
  system (p. 168)
ABO blood group (p. 168)
antigens (p. 169)

antibodies (p. 169)
jaundice (p. 169)
penetrance (p. 170)
variable
  expressivity (p. 170)
Huntington's
  chorea (p. 170)
agglutination (p. 170)

modifier gene (p. 171)
epistatic gene (p. 171)
tyrosinase (p. 171)
melanin (p. 172)
albinism (p. 172)
pleiotropy (p. 172)
sickle-cell anemia (p. 172)
hemoglobin A, HbA (p. 172)

1. Explain why inheritance of many traits is no longer viewed as being under the control of purely dominant or purely recessive agents.

2. List and explain several examples of inheritance in which expression of a given trait occurs as a spectrum of intermediate aspects of the trait rather than just the two extremes (dominant or recessive).

3. State whether or not each of the examples in Objective 2 are controlled by one pair of genes or by more than one pair.

4. Define *epistasis* and cite an example.

5. Explain the meaning of *pleiotropy* and discuss how it is related to sickle-cell anemia. Name some of the collection of phenotypic characteristics that are part of the sickle-cell syndrome.

Self-Quiz Questions

True-False If false, explain why.

____ (1) If a true-breeding red-flowering snapdragon is crossed with a white-flowering snapdragon, all of the F_1 generation will all be red-flowered.

____ (2) If two of the F_1 red-flowered snapdragons are crossed, three-fourths of their offspring (the F_2) will be red-flowered and one-fourth will be white.

____ (3) Hemophilia is an example of a disorder caused by incomplete dominance.

____ (4) The expression of flower color in snapdragons is an example of codominance.

____ (5) A child of blood type AB has a mother with type A blood. His father could not have type O blood.

____ (6) Human skin color is an example of incomplete dominance that demonstrates the validity of the "blending" hypothesis.

Summary

Throughout any individual's life cycle, genes must interact with the environment in the expression of phenotype. Genes provide the chemical messages for growth and development, but neither growth nor development can proceed without environmental contributions to the living form. Gradations of dominance exist between two alleles for a given gene locus, so the expression of one or both may be fully dominant, or one may be incompletely dominant over the other. A hierarchy of dominance can exist among all the alternative forms that may exist for any given type of gene. How a gene behaves in space or time may be influenced by other genes, with all of their positive and negative actions producing some effect on phenotype. The activity of a single gene may have major or minor effects on more than one trait (pleiotropy). The activity of a gene in space and time may be influenced by other genes, with a total of their positive and negative actions producing some effect on phenotype. The environment profoundly influences the expression of all genes and their contributions to phenotype.

Key Terms

melanin (p. 172)
tyrosine (p. 172)
tyrosinase (p. 172)

polydactyly (p. 172)
aging (p. 174)
discontinuous
    variation (p. 175)

quanitative (p. 175)
continuous variation (p. 175)
quantitative inheritance (p. 175)
polygenes (p. 175)

Objectives

1. Consider aging as it occurs in humans and indicate what humans can do to speed or slow the aging process.

2. Cite several examples of environmental alteration of an organism's basic phenotype.

3. Contrast discontinuous and continuous variation. Discuss the relationship between quantitative inheritance and polygenes.

4. Explain why frequency distribution diagrams are used to portray continuous variation in populations.

Self-Quiz Questions

> *True-False* If false, explain why.

____ (1) Becoming thirty pounds overweight is one way that the external environment can influence the phenotypic expression of human genotype.

____ (2) The fact that it is possible to change the extent to which genotypic potential is expressed in the phenotype demonstrates that acquired phenotypic characteristics can be inherited.

INTEGRATING AND APPLYING KEY CONCEPTS

Solve the following genetics problems. Show all set-ups, genotypes, and phenotypes.

(1) A husband sues his wife for divorce, arguing that she had been unfaithful. His wife gave birth to a girl with a fissure in the iris of the eye, a sex-linked recessive trait. Both parents have normal eye structure. Can the genetic facts be used to argue for the husband's suit? Explain why.

(2) Holstein cattle have spotted coats due to a recessive gene; a solid-colored coat is caused by a dominant gene. If two spotted Holsteins are mated, what types of offspring would be likely? Explain your statement with a diagram.

(3) Give the genotypes of the parents in this problem, labeling appropriately. Cocker spaniels are black due to a dominant gene and red due to its recessive allele. Solid color depends on a dominant gene and the development of white spots on its recessive allele. A solid red female was mated with a black and white male, and five puppies resulted: one black, one red, two black and white, and one red and white.

11

EMERGENCE OF THE CHROMOSOMAL THEORY OF INHERITANCE

Summary

During the late 1800s, several cytologists described the events that compose mitosis (Flemming) and meiosis (Weismann, Flemming). Mitotic and meiotic phenomena were linked with Mendel's concepts of segregation and independent assortment, and it was suggested that Flemming's "threads" were likely the carriers of Mendel's "hereditary units."

The chromosomal theory of inheritance states that each chromosome contains a linear sequence of genes that govern the development of phenotype. Diploid organisms inherit one set of chromosomes from each parent; these double sets are mingled as homologous pairs during sexual fusion as a new diploid nucleus is formed. Later, during meiosis, when gametes are formed, the diploid number of chromosomes is halved. Each member of a homologous pair is segregated into different gametes, and the chromosomes of each pair assort independently of the chromosomes of other parts. New genotypic combinations are produced by crossing over, independent assortment, and chromosomal mutations; these chromosomal events give rise to new phenotypes upon which selective agents can act.

Key Terms

cytology (p. 178)
Walther Flemming (p. 178)
mitosis (p. 178)
August Weismann (p. 178)
hybridization (p. 178)
chromosomal theory
  of inheritance (p. 179)

allelic partner (p. 179)
meiosis (p. 179)
independent
  assortment (p. 179)
homologous
  chromosomes (p. 179)

crossing over (p. 179)
deletion (p. 179)
duplication (p. 179)
inversion (p. 179)
translocation (p. 179)
chromosomal
  mutations (p. 179)

Objectives

1. Identify the cytologists that described mitosis and meiosis. State what was so
    useful about Flemming's contribution.

2. List the major tenets of the chromosomal theory of inheritance.

Self-Quiz Questions

Fill-in-the-Blanks

(1) _____ described the behavior of threadlike bodies (chromosomes) during

mitosis in salamander cells. (2) _____ first proposed that chromosomal number

must be reduced by half during gamete formation, but (3) _____ identified the

process; it became known as (4) _____.

Matching Choose the single most appropriate letter.

____ (5) crossing over

____ (6) deletion

____ (7) duplication

____ (8) independent
         assortment

____ (9) inversion

____ (10) segregation

____ (11) translocation

A. chromosomal mutation; segment of chromosome lost

B. homologous chromosomes are channeled into different
    gametes

C. chromosomal mutation; chromosomal segment transferred
    to a nonhomologous chromosomes

D. chromosomal mutation; repeat of DNA sequence in same
    chromosome or in nonhomologous chromosome

E. the exchange of parts of chromatids between nonsister
    chromatids during Prophase I of meiosis

F. chromosomal mutation; chromosomal segment excised and
    reattached backward

G. chromosomes from one pair have no effect on the chan-
    neling into gametes of chromosomes from another pair

Chapter 11   Emergence of the Chromosomal Theory of Inheritance   99

Summary

Thomas Hunt Morgan's research group kept under culture fruit flies with a variety of normal and mutant traits. Morgan and three of his students—Sturtevant, Bridges, and Muller—established the chromosomal basis of inheritance.

They discovered that the X chromosome (the one chromosome in male grasshoppers that doesn't have a partner) carries many genes besides the ones involved in sex determination; any gene on an X chromosome is known as a *sex-linked gene*. In organisms that have only one sex chromosome, all genes on the X chromosome are expressed regardless of whether they are dominant or recessive because there are no homologous genes on a companion chromosome that could mask the recessive instructions. In later studies, they discovered *sex-related genes* that influenced the expression of maleness and femaleness on autosomes (chromosomes other than X or Y chromosomes) as well. Developmental events ensure that these genes are expressed largely through the stimulation or inhibition caused by different sex hormones. Morgan's research group also discovered that some groups of genes do not assort independently as Mendel said because they are physically linked as part of the same chromosome. These and later studies with other organisms have made it clear that the number of *linkage groups* corresponds to the number of chromosomes characteristic of the species, which further confirms Sutton's hypothesis that chromosomes are the vehicles that transport the genes through generations.

Key Terms

Thomas Hunt Morgan (p. 179)	Y-chromosome (p. 179)	Calvin Bridges (p. 182)
Drosophila	multiple allele	testis (p. 182)
melanogaster (p. 179)	systems (p. 180)	X-linked gene loci (p. 182)
sex chromosomes (p. 179)	wild-type allele (p. 180)	linkage (p. 182)
autosomes (p. 179)	spontaneous mutation (p. 180)	linkage group (p. 182)
X chromosome (p. 179)	sex-linked gene (p. 181)	

Objectives

1. Describe the basic types of experiments that Morgan's research group carried out and state two of their major discoveries that dramatically advanced our understanding of genetics.

2. Define *sex chromosomes* and *autosomes*; then distinguish the types of alleles found on each, if possible.

3. Distinguish sex-linked and sex-related genes.

4. Contrast the X and Y chromosomes in terms of length, then explain what, if anything, matches up with or masks the part of the chromosome that is longer than the other chromosome. Finally, explain the consequences of this lone chromosomal segment.

5. Define *linkage* and explain how the concept became related to the chromosomal theory of inheritance.

Self-Quiz Questions

> *Multiple Choice*

____ (1) Most animal and some plant species have sex chromosomes that _____.

> (a) are called autosomes
> (b) are always one X and one Y
> (c) differ in number of in kind between males and females
> (d) All of the above are correct.

____ (2) Normally, all eggs from a human female contain _____.

> (a) one X chromosome   (b) one Y chromosome
> (c) one X and one Y chromosome   (d) only autosomes

____ (3) Which of the following did Morgan and his students not do?

> (a) They isolated and kept under culture fruit flies with the sex-linked recessive white eye trait.
> (b) They developed the technique of amniocentesis.
> (c) They developed the technique of linkage mapping.
> (d) They discovered and described crossing over.

> *True-False* If false, explain why.

____ (4) Linkage of alleles at two or more loci suggests that these alleles are located on the same chromosome.

____ (5) Individual genes always assort independently of one another during meiosis I.

____ (6) In most species males transmit the Y chromosome to their sons but not to their daughters; males receive their X chromosome only from their mothers.

11–III   RECOMBINATION AND LINKAGE MAPPING OF CHROMOSOMES   183

VISUALIZATION OF CHROMOSOMES   184

Summary

Janssens described the crossing over events of the first division of meiosis in which two nonsister chromatids of homologous pairs of chromosomes exchange corresponding segments at breakage points. Crossing over results in genetic recombination: the formation of new combinations of alleles in a chromosome, hence in gametes, and finally in zygotes. Janssens's description of chiasmata (crossovers) formation gave Sturtevant the idea that the locations of genes on their chromosomes could be mapped because the farther apart on the chromosome two linked genes may be, the more likely that chiasma formation and crossing over can disrupt the original combination of linked alleles. The probability of crossing over and recombination occurring at a point somewhere between two genes located on the same chromatid is directly proportional to the distance that separates them. Genes are carried in linear array on chromosomes.

Rapid advances in microscopy enabled cytogeneticists to visualize the structure of chromosomes. Observations of *polytene chromosomes* connected specific chromosome regions with specific phenotypic traits. *G-banding* and the preparation of *karyotypes* permitted ordinary chromosomes to be more clearly identified.

F. Janssens (p. 183)   linkage mapping (p. 183)   centromere (p. 184)
crossing over (p. 183)   standard map unit (p. 183)   G-banding (p. 184)
nonsister chromatids (p. 183) vestigial wings (p. 184)   Giemsa stain (p. 184)
A. Sturtevant (p. 183)   cytogenetics (p. 184)   karyotype (p. 184)
linear sequence (p. 183)   polytene
                                  chromosomes (p. 184)

Objectives

1. Define *crossing over*, state when it occurs, and explain its significance for the chromosomal theory of inheritance.

2. Explain why it is important to note that crossover involves nonsister chromatids rather than sister chromatids. State why "crossover" between sister chromatids would not likely cause evolution in populations.

3. Explain how Janssens's description of chiasmata formation during meiotic prophase gave Sturtevant an idea that has since contributed greatly to our understanding of genetic behavior.

4. State the relationship between the probability of crossing over and subsequent recombination, and the distances between two genes located on the same chromatid.

5. Describe how G-banding and karyotyping are done and tell what the two techniques contributed to cytogenetics.

Self-Quiz Questions

Multiple Choice

____ (1) In the first stage of meiosis, two sister chromatids of one chromosome are drawn into a locus-by-locus alignment with the two sister chromatids of its homologue. This alignment and attachment is called _____.

   (a) crossing over   (b) linkage
   (c) synapsis   (d) polygenic inheritance

____ (2) _____ is a consequence of crossing over.

   (a) A dihybrid cross   (b) Linkage
   (b) Dependent assortment   (d) Genetic recombination

____ (3) A chiasma is _____.

   (a) a type of blindness due to a genetic defect
   (b) a region of apparent contact between nonsister chromatids
   (c) crossing over between sister chromatids
   (d) a linkage group

____ (4) The closer together two genes are on one chromosome, _____.

   (a) the fewer times there will be crossing over and genetic recombination occurring between them
   (b) the more times there will be crossing over and genetic recombination occurring between them
   (c) the farther apart they will appear in a linkage group
   (d) the greater the chance that point mutations will happen to either of them

_____ (5) Genetic recombination occurs as a result of crossing over between nonsister chromatids of homologous chromosomes during _____.

  (a) metaphse of mitosis   (b) prophase I of meiosis
  (c) sexual fusion   (d) prophase II of meiosis

_____ (6) Plotting the positions of genes on chromosomes according to their segregation patterns during meiosis is called _____.

  (a) linkage mapping   (b) karyotyping
  (c) G-banding   (d) transposing

_____ (7) A standard map unit is _____.

  (a) ten chiasma per one hundred bands
  (b) one crossover per hundred gametes
  (c) one mm between the closest G-bands
  (d) the distance between the two closest chiasmata of the C-group

Summary

Crossing-over normally occurs during gamete formation. On very rare occasions, chromosomal rearrangements known as *chromosomal mutations*—deletions, duplications, translocations, and inversions—also occur. They invariably affect fairly large chromosomal segments of the total chromosome number characteristic of the species. In addition to these rare internal arrangements, from time to time whole chromosomes or chromosome sets go astray.

    Aneuploidy (in which a particular type of chromosome is either absent entirely or present three or more times in the diploid chromosome set) arises when homologous chromosomes fail to segregate properly during prophase I of meiosis. *Polyploidy* (in which there are three or more whole sets of chromosomes either in certain tissues or in all somatic cells of an individual) commonly occurs in flowering plants and other eukaryotes; it can be induced by exposing plants to colchicine.

    Domestic hybrids of wheat have been developed by manipulating hybridization, polyploidy, and exposure to colchicine.

    Chromosomes can fuse into fewer chromosomal structures. One chromosome can also split into two (*fission*). Human chromosome 2 may have arisen by chromosome fusion that occurred in ancestral primate populations; all apes have 24 chromosomes; humans have 23. All of these events (crossing over, independent assortment of whole chromosomes, and chromosomal mutations) lead to new genotypic combinations, which, in turn, give rise to new phenotypes upon which selective agents can act. In short, these events are many of the wellsprings of diversity and evolutionary change.

deletion (p. 185)   "jumping genes" (p. 187) tetraploid (p. 187)
synapsis (p. 185)   aneuploidy (p. 187)   propagate (p. 187)
homologue (p. 185)   nondisjunction (p. 187) colchicine (p. 188)
duplication (p. 185)   trisomy (p. 187)   chromosome fusion (p. 188)
phenotypic divergence (p. 186) monosomy (p. 187)   chromosome fission (p.188)
inversion (p. 186)   mosaic tissue (p. 187)   bridging cross (p. 189)
translocation (p. 186)   polyploidy (p. 187)   hybrid (p. 189)
transposition (p. 186)   triploid (p. 187)

Objectives

1. Define *chromosomal mutation* and list four types of such mutations. Explain how such mutations occur.

2. Define *polyploidy*. Tell what agents cause this situation and how it differs from the four types of chromosomal mutations.

3. Explain how the wheat from which bread is made is thought to have evolved.

4. Tell what role all of these chromosomal rearrangements have to do with variation in populations and, thus, with natural selection.

Self-Quiz Questions

Fill-in-the-Blanks

A (1) _____ is the loss of a piece of a chromosome that happens during gamete formation. Segments of chromosomes occasionally break off and move to a nonhomologous chromosome; such an event is known as a(n) (2) _____. A(n) (3) _____ occurs when a deleted piece of chromosome gets turned around and rejoins the chromosome at the same place, but backward. Hybridization in wheat causes multiple sets of chromosomes in gametes, a situation known as (4) _____. (5) _____ is a chemical that induces polyploidy in many plants. (6) _____ causes (7) _____, in which three chromosomes of the same kind are present in the chromosome set. Homologous chromosomes can exchange parts as a result of (8) _____-_____. A cell with four sets of one pair of chromosomes is (9) _____. Two sexually incompatible species can be crossed by first introducing the chromosomes of one into an intermediate species that can successfully interbreed with both; such a cross is a (10)

_____ cross. Some evolutionists suggest that humans may have arisen through

(11) _____ _____ occurring in an ancestor from which both chimpanzees

and humans evolved.

INTEGRATING AND APPLYING KEY CONCEPTS

The parents of a young boy bring him to their doctor. They explain that the boy does not seem to be going through the same developmental stages as his older brother. The doctor orders a common cytogenetics test to be done that reveals that the young boy's cells contain two X chromosomes and one Y chromosome. Describe the test that the doctor ordered and explain how and when such a genetic result--XXY--most logically could have occurred.

12

THE RISE OF MOLECULAR GENETICS

THE SEARCH FOR THE HEREDITARY
MOLECULE

  The Puzzle of Bacterial
    Transformation
  Bacteriophage Studies

THE RIDDLE OF THE DOUBLE HELIX

  Patterns of Base Pairing
  Unwinding the Double Helix: The
    Secret of Self-Replication

DNA REPLICATION

  Meanwhile, Back at the
    Forks . . .

Enzymes of DNA Replication
  and Repair
Summary of DNA Replication

DNA RECOMBINATION

  Molecular Basis of Recombination
  Recombinant DNA Research

CHANGE IN THE HEREDITARY MATERIAL

  Mutation at the Molecular Level

PERSPECTIVE

Summary

During its lifespan, each cell deploys thousands of different enzymes in translating
its hereditary instructions (its genotype) into specific physical traits (its phenotype). Each cell comes into the world with a limited set of specific instructions
that promote survival in a particular kind of habitat. Hereditary instructions must
be of a sufficiently stable chemical nature to ensure constancy in structure and
function, and yet must allow room for subtle change. In the search for the hereditary
molecule, two principal candidates were considered as likely: *proteins*, because they
are diverse and complex; and *nucleic acids*, because, although they have only a simple
repeating-unit structure, they are abundant in the nuclear regions of cells where the
heredity instructions were suspected of being located. After a variety of experiments,
in almost all organisms, DNA (deoxyribonucleic acid) was demonstrated to be the molecule that contains the genetic code.

Key Terms

phenotype (p. 192)
Miescher (p. 192)
deoxyribonucleic
  acid, DNA (p. 192)
transforming principle (p. 193)
Griffith (p. 193)
S form (p. 193)

R form (p. 193)
transformation (p. 193)
<u>assay</u> (p. 193)
Avery (p. 193)
Chargaff (p. 194)
<u>virus</u> (p. 194)
<u>lysis</u> (p. 194)

Hershey and Chase (p. 194)
radioactive
  isotope (p. 194)
<u>bacteriophages</u> (p. 194)
lytic pathway (p. 195)
^{32}P (p. 196)
^{35}S (p. 196)

Objectives

1. State which two classes of molecules were suspected of housing the genetic code before 1952.

2. Identify the types of research carried out by Miescher, Griffith, Avery and colleagues, Chargaff, and Hershey and Chase and the specific advances in our understanding of genetic behavior made by each.

3. Explain how bacteriophages were used to study genetics.

4. State which piece of research demonstrated that DNA, not protein, governed inheritance.

Self-Quiz Questions

True-False If false, explain why.

____ (1) Miescher first isolated and described nucleic acids obtained from pus and fish sperm during the 1920s.

____ (2) ^{35}S can be used to label DNA, and ^{32}P can be used to label almost any protein.

____ (3) DNA is the only hereditary molecule found in viruses.

____ (4) Griffith demonstrated that DNA was the substance that had been permanently transformed when the "rough" strain of bacteria was transformed to the disease-causing "smooth" strain of bacteria.

____ (5) Bacteriophages are bacteria that eat viruses.

____ (6) DNA contains sulfur but not phosphorus.

____ (7) Hershey and Chase were the first to demonstrate that DNA was unquestionably the molecule that contained the genetic code.

12–II   THE RIDDLE OF THE DOUBLE HELIX   194–200

Patterns of Base Pairing   198

Summary

Only four different kinds of nucleotides are found in a DNA molecule, and each has the same phosphate group and the same five-carbon sugar (deoxyribose). Each nucleotide has one of four different nitrogen-containing bases. Cytosine and thymine are single-ring pyrimidines; adenine and guanine are double-ring purines. In a DNA molecule, nucleotides are strung together into a long chain with each phosphate group of one nucleotide linking with the sugar of the next nucleotide; when nucleotides are thus assembled, the purines and pyrimidines project to one side. Chargaff and Franklin, by using a variety of physical and chemical experimental techniques, provided Watson and Crick with enough clues that they were able to build a scale model, which revealed that DNA had to be double-stranded with two nucleotide strands wound helically about each other like a spiral staircase. Watson and Crick stated that the two strands of nucleotides were held together by hydrogen bonds linking purines (A and G) from one strand with pyrimidines (C and T) on the other strand and vice versa; this statement has become known as the *principle of base pairing*. Any purine-pyrimidine pair could follow any other in the DNA chain so that an extremely large number of possible code sequences could result from different arrangements of these two pairs of nitrogenous bases.

Key Terms

nucleotide (p. 194)
nitrogen-containing
  base (p. 194)
deoxyribose (p. 195)
pyrimidines (p. 195)
cytosine (p. 195)
thymine (p. 195)

purines (p. 195)
adenine (p. 197)
guanine (p. 197)
Chargaff (p. 197)
Franklin (p. 197)
x-ray diffraction (p. 197)
nanometer (p. 197)

helix, helical (p. 198)
Watson and Crick (p. 198)
purine-pyrimidine
  pairs (p. 198)
principle of base
  pairing (p. 198)

Objectives

1. Draw the basic shape of a deoxyribose molecule and show how a phosphate group is joined to it when forming a nucleotide.

2. Define, diagram, and distinguish *purine* and *pyrimidine*.

3. Describe how cytosine differs from thymine and adenine from guanine. Then show how each would be joined to the sugar-phosphate combination drawn in the previous objective.

4. List both of the clues that Chargaff's research demonstrated.

5. Describe the technique of x-ray diffraction and state the kinds of information about DNA structure that Rosalind Franklin discovered by using x-ray diffraction.

6. List the particular contributions of Watson and Crick to the modern understanding of DNA structure and behavior.

7. Explain what is meant by the pairing of nitrogen-containing bases (base pairing) and state the feature that causes adenine to pair with thymine, and guanine to pair with cytosine.

8. Explain how the Watson-Crick model obeyed Chargaff's rules and fulfilled Franklin's structural predictions.

9. Describe how the Watson-Crick model reflects the constancy in DNA observed from species to species yet allows for variations from species to species.

Self-Quiz Questions

True-False If false, explain why.

____ (1) Each nucleotide consists of a six-carbon sugar, a phosphate group, and a nitrogen-containing base.

_____ (2) DNA is composed of only four different types of nucleotides, each of which contains either adenine, thymine, cytosine, or guanine.

_____ (3) In every species, the amount of adenine present always equals the amount of thymine, and the amount of cytosine always equals the amount of guanine (A = T, and C = G).

_____ (4) In a nucleotide, the phosphate group is attached to the nitrogen-containing base, which is attached to the five-carbon sugar.

_____ (5) Watson and Crick built their paper model of DNA in the early 1950s.

_____ (6) Rosalind Franklin discovered that DNA was double-stranded.

_____ (7) Guanine pairs with cytosine by forming three hydrogen bonds.

Summary

Once DNA's double-stranded nature had been deduced, Watson and Crick immediately saw how such a molecule could be duplicated prior to mitosis or meiosis. If the hydrogen bond(s) that connect a purine with a pyrimidine in a base pair are broken by the appropriate enzyme and separated, each base can attract and form hydrogen bonds with its complementary nucleotide already synthesized and present in the cell. Each parent DNA strand remains intact, and a new companion strand is assembled on each one. The only sequence that could be attached to one parent strand (say, ATTCGC) would have to be an exact complement of the base sequence (TAAGCG) that occurs on the other parent strand. Each single parent strand ends up forming a new complementary strand, with which it remains linked by hydrogen bonds. In each of the two resulting DNA molecules, one "old" DNA strand has been conserved as a partner for the newly formed strand; this _semiconservative replication_ was demonstrated by the experiments of Meselson and Stahl.

Replication begins at an _origin_ and proceeds simultaneously in two directions from the origin. Cairns demonstrated that DNA synthesis occurs at _replication_ forks. _Discontinuous DNA synthesis_ occurs on both parent strands; new nucleotides are added in wherever an -OH group projects from the 3' carbon atom of a nucleotide already on track.

A variety of enzymes participate in DNA replication; some serve to detach the purine-pyrimidine pairs, and others act to link "new" nucleotides with the parent strands. Still other enzymes form a sort of quality-control inspection team and act to correct errors that may have occurred during replication.

Key Terms

free nucleotide (p. 200)   template (p. 200)   density-gradient
parent strand (p. 200)   complementary,   centrifugation (p. 201)
fully hydrated (p. 200)   complement (p. 200)   cesium chloride (p. 201)
B-form of DNA (p. 200)   Z-form of DNA (p. 200)   centrifugation (p. 201)
double helix (p. 200)   replication (p. 200)   hybrid molecule (p. 201)

semiconservative
  replication (p. 200)
initiation site (p. 200)
origin (p. 200)
bidirectional (p. 200)

John Cairns (p. 200)
replication fork (p. 200)
Reiji Okazaki (p. 201)
Meselson and Stahl (p. 201)

discontinuous DNA
  synthesis (p. 202)
DNA polymerases (p. 202)
DNA ligase (p.203)

Objectives

1. Assume that the two parent strands of DNA have been separated, and that the base sequence on one parent strand is ATTCGC. State the sequence of new bases that will complement the parent strand.

2. Describe how double-stranded DNA replicates itself by complementary base pairing.

3. Outline the fundamental points of Meselson and Stahl's experiments with radioactively labeled bacterial DNA and density-gradient centrifugation. Explain what each of the three zones obtained in the cesium-chloride gradient represented.

4. List four jobs that enzymes perform during the process of replication.

Self-Quiz Questions

  True-False If false, explain why.

____ (1) Adenine hydrogen bonding to guanine is an example of complementary base pairing.

____ (2) The replication of DNA is considered semiconservative because the same four nucleotides are used again and again during replication.

____ (3) Each parent strand remains intact during replication, and a new companion strand is assembled on each one.

____ (4) Meselson and Stahl demonstrated by using density-gradient centrifugation and isotopes that they could create three different types of DNA, which corresponded with three generations of bacterial DNA.

____ (5) Some of the enzymes associated with DNA assembly repair errors that occur during the replication process.

  Multiple Choice

____ (6) A substance acts as a mold for the formation of a substance complementary in form; the mold substance is known as a _____.

    (a) hybrid molecule   (b) plasmid   (c) template   (d) clone

12–IV DNA RECOMBINATION

Summary

Prokaryotes ordinarily reproduce by prokaryotic fission, an asexual process in which the partitioning of a "parental" cell leads to two separate cells having basically similar parts. *Conjugation, transformation,* and *transduction* are additional processes that occur occasionally in prokaryotes. Each of these three processes results in *genetic recombination*—the production of individuals having new assortments of genes that were acquired, one way or another, from more than one parent cell—and they are regarded by many as primitive methods of exchanging genetic information that helped to diversify prokaryotes. Genetic recombination depends in large part on the *same* enzymes that govern DNA replication and repair.

Recombinant DNA is a hybrid molecule that results from breaking apart the DNA from one organism and combining it with DNA from another organism. *Plasmids* are small circles of DNA that are found in some bacteria in addition to the main circular DNA molecule. *Restriction enzymes* can be used to cut open plasmids at a particular site, and *DNA ligase*, an enzyme, can be used to attach some foreign DNA to the opened plasmid; then the plasmid is closed again in its circle. Under certain conditions, the hybrid DNA plasmid can be introduced into new *E. coli* bacterial cells, and it will then be reproduced every time the bacterium undergoes cell division. The bacterium's metabolic machinery can churn out many multiple copies, or *clones*, of the genes carried on the hybrid plasmid ring; with repeated doublings of a bacterial population, millions of these genes can be manufactured overnight. Recombinant DNA techniques hold great promise for transferring genes from one organism to another and for manufacturing proteins such as insulin in a pure form for organisms that cannot make the required substances themselves. Entirely new species with entirely new assemblages of characteristics are now being synthesized in specially equipped laboratories following very strict safety rules.

Key Terms

genetic recombination (p. 203)
recombinant DNA (p. 203)
endonucleases (p. 204)
hydrolyze (p. 204)
Robin Holliday (p. 204)
DNA polymerase (p. 204)
DNA ligase (p. 204)
conjugation (p. 204)

episomes (p. 204)
cytoplasmic bridges (p. 204)
transduction (p. 204)
transformation (p. 204)
pilus (p. 205)
hybridization (p. 205)
vectors (p. 205)
plasmids (p. 205)

restriction
  endonucleases (p. 206)
"sticky ends" (p. 206)
clone (p. 206)
insulin (p. 207)
genetic engineering
  (p. 207)

Objectives

1. Distinguish conjugation, transformation, and transduction from each other.

2. Describe the process by which recombinant DNA is produced. Use terms such as *plasmid* and *restriction enzymes* in ways that make their meaning clear.

3. Define *clone* and explain how clones of recombinant DNA can be useful to people suffering from diabetes.

4. Describe how recombinant DNA techniques theoretically could transfer to crop plants the ability to harness nitrogen (N_2) from the air in order to make proteins, thereby reducing modern agriculture's demand for expensive nitrogen-containing fertilizers.

5. Name one problem that some scientists fear will result from recombinant DNA research. Then state what research geneticists are doing to guard against such a possibility.

True-False If false, explain why.

____ (1) Ligase is an enzyme that links together segments of replicated DNA.

____ (2) Plasmids are the organelles on the surfaces of which amino acids are assembled into polypeptides.

____ (3) Episomes enable conjugation to occur between bacteria of two different mating strains.

____ (4) DNA ligase opens up a plasmid double helix so that a fragment of foreign DNA can base-pair with the plasmid DNA.

Summary

DNA would have to be copied precisely and distributed intact to offspring for genetic information to be transmitted precisely from generation to generation. Occasional errors occur during replication or during the separation of chromosomes during mitosis. In addition, high-energy radiation and mutagenic chemicals that enter cells can damage strands. Repair enzymes are limited in just how well they can fix damage to DNA, so *deletions* (in which one or more base pairs are lost), *insertions* (one or more extra base pairs incorporated into the original sequence), and base substitutions occasionally occur. The overall precision of DNA replication and repair is the source of life's fundamental unity; the rare "mistakes" account for much of its diversity. Although most mutations are harmful, some may confer on their bearers a selective advantage to survive in changing environments and to leave more offspring even as others perish. The mutations that are perpetuated and the increasingly abundant phenotypes that they foster are central to the evolutionary process.

Asexually reproducing haploid organisms are finely tuned for operating under a given range of environmental conditions, but if a mutation occurs in one of their important genes, or if there are radical changes in the environment, their outlook for survival is grim.

Most eukaryotic cells are diploid, with *two* of each type of chromosome characteristic of the species—one of each type of chromosome from each parent. Each parent contributes a complete set of chromosomes to the offspring, which then has two complete sets. For diploid organisms, even if a gene in one chromosome mutates, the equivalent gene on the other chromosome remains functional and may help the organism to survive. With these double sets of directions, a diploid organism can store a variety of instructions, which may be called into play at a later date should environmental conditions change.

<u>crossing over</u> (p. 207) <u>base-pair</u> mutant allele (p. 208)
<u>recombination</u> (p. 207) <u>substitutions</u> (p. 207) repair enzymes (p. 208)
<u>cytological</u> (p. 207) <u>insertion</u> (p. 207) sequences of base
<u>chromosomal mutations</u> (p. 207) <u>deletion</u> (p. 207) pairs (p. 209)
<u>gene mutation</u> (p. 207) <u>frameshift mutations</u> (p. 207)

Objectives

1. List some of the agents that can cause mutations.

2. Explain how repair enzymes act to minimize replication errors.

3. Describe the types of defects that occur in DNA from time to time and indicate whether or not repair enzymes can correct them.

4. Explain the relationship between gene mutations and organismal diversity.

5. Explain why frameshift mutations that occur near the beginning of a DNA sequence of bases almost always result in the production of an unusable protein.

Self-Quiz Questions

Fill-in-the-Blanks

(1) _____-_____ _____ and (2) _____ _____ that enter cells can damage strands of DNA. (3) _____ occur when one or more extra base pairs are incorporated into the original sequence; (4) _____ occur when one or more are left out of the original sequence. (5) _____ mutations generally produce unusable proteins.

INTEGRATING AND APPLYING KEY CONCEPTS

Mention three uses of recombinant DNA that could be immensely beneficial to humans or other life on earth. Then mention a possible scenario for recombinant DNA that could be disastrous to humans.

13

FROM DNA TO PROTEIN: HOW GENES FUNCTION

Summary

Today, we know that genes are portions of DNA that do not code directly for the production of proteins; that is, proteins are not assembled on the DNA. Rather, certain kinds of RNA are intermediaries constructed from ribose-containing nucleotides that are assembled in a complementary fashion on DNA, yet leave the DNA molecules intact in the nuclear region even as the RNA bears genetic messages to the sites of protein synthesis: the ribosomes. In the early 1900s, Garrod studied several examples of inherited metabolic disorders and linked them to the absence or deficiency of an essential enzyme. He suggested that each enzyme was specified by a single unit of inheritance and that, if the unit itself was defective, the enzyme also would be defective. About thirty-five years later, Beadle and Tatum bombarded the spores of red bread mold with x-rays and isolated mutant strains that could grow only on culture media to which specific substances had been added. Analysis of each mutant strain revealed its own specific defective enzyme, so Beadle and Tatum assumed that each gene controls the synthesis of one enzyme. Their assumption became known as the one gene, one enzyme concept.

    Studies using *electrophoresis* to measure the rate and direction of movement of normal and abnormal types of hemoglobin provided a better understanding of the link between genes and proteins. Sickle-cell hemoglobin differs from normal hemoglobin because *one* amino acid (valine) is substituted for another (glutamate). Only one of the 150 amino acids in the beta-polypeptide chains has gone astray in HbS, but that is enough to cause a severe form of anemia. On the basis of such experiments, Beadle and Tatum's hypothesis has been amended as the *one gene, one polypeptide hypothesis*; a protein consists of one or more chemically different polypeptide chains, and a single gene codes for each kind.

Key Terms

ribonucleic acid, RNA (p. 211)
uracil, U (p. 211)
messenger RNA, mRNA (p. 211)
transfer RNA, tRNA (p. 211)
ribosomal RNA, rRNA (p. 211)
transcription (p. 212)
translation (p. 212)
Garrod (p. 212)
phenylalanine (p. 212)
phenylketonuria, PKU (p. 212)
alkaptonuria (p. 212)
inborn errors of metabolism (p. 212)
Beadle and Tatum (p. 212)
Neurospora crassa (p. 212)

biotin (p. 212)
nutritional mutants (p. 212)
"one gene, one enzyme"
  hypothesis (p. 213)
HbS, sickle-cell hemoglobin (p. 213)
HbA, normal hemoglobin (p. 213)
beta-chain (p. 213)
electrophoresis (p. 214)
net surface charge (p. 214)
Ingram (p. 214)
"one-gene, one polypeptide"
  hypothesis (p. 214)
sticky patch (p. 215)

Objectives

1. List two of the inherited metabolic disorders studied by Garrod, state what Garrod thought caused the disorders, and define *inborn error of metabolism*.

2. Consult Fig. 13.6 and state what the longest threads are and what the shorter threads that project at right angles from the longer threads are.

3. Describe Beadle and Tatum's experiments with *Neurospora crassa* and explain how their experiments led to the one gene, one enzyme concept.

4. State the one gene, one enzyme concept and explain how it has been changed since Beadle and Tatum first suggested it in the early 1940s.

5. State how RNA differs from DNA in structure and function and what features RNA has in common with DNA. Tell what RNA code would be formed from the following DNA code: TAC-CTC-GTT-CCC-GAA.

6. Explain how the process of electrophoresis is used to study proteins.

Self-Quiz Questions

Fill-in-the-Blanks

In (1) _____, excess amounts of an oxidation product of phenylalanine appear in

the urine; in (2) _____, urine samples exposed to air turn a dark color because

of concentrations of the intermediate alkapton. (3) _____ linked each of these

two diseases to the absence or deficiency of an essential (4) _____, which is

specified by a single (5) _____ _____ _____, which, if defective,

will code for a defective (6) _____. (7) _____ and (8) _____

worked with *Neurospora crassa*, which is a kind of (9) _____ _____ _____.

N. crassa was a useful organism because it is (10) _____: it has only one gene

for each trait, it can reproduce (11) _____ through spore formation and give rise to many offspring, and it can grow on a simple chemically defined medium contain-

ing (12) _____ and biotin (one of the B vitamins). From their work with *N.*

crassa mutants, they stated the one (13) _____, one (14) _____ concept,

which has since been modified to recognize that there are intermediary (15) _____

_____ between the genes and the enzyme products for which they contain assembly

instructions. Today, we know that (16) _____ molecules serve as the message

carriers. Unlike DNA, (17) _____ contains (18) _____ instead of deoxy-

ribose, and, in place of the base thymine, it has (19) _____, which can form

hydrogen bonds with the purine (20) _____.

13–II THE GENETIC CODE

214-216

Summary

Genes are carried on DNA, but the products they specify are not assembled on DNA. The
DNA message is carried by RNA molecules to ribosomes in the cytoplasm where amino acid
sequences are assembled into linear strings. A string of amino acids represents a protein's primary structure, which, in turn, dictates the protein's final three-dimensional form. This form determines the substances the protein will interact with in helping
to build and maintain the cell. There are only four different kinds of nucleotides in
DNA and RNA, but there are twenty different amino acids commonly found in the proteins
of living things. If each nucleotide called forth a single amino acid, that would give
us only four different amino acids. Two nucleotide letters in a row would give us only
sixteen different amino acids, not twenty. The genetic code consists of nucleotide
bases that are read linearly, three at a time, with the sequence of each triplet sig-
nifying an amino acid. Sixty-one of these *triplets* each specify an amino acid; some
different triplets specify the same amino acid. Three other triplets serve to terminate the protein chain. Any nucleotide triplet in messenger RNA (RNA that carries protein assembly instructions from the nuclear region to the ribosomes in the cytoplasm)
is now known as a *codon*. In all living organisms the same genetic code is the basic
language of protein synthesis—compelling evidence for the fundamental unity of life
at the molecular level.
     Mitochondrial DNA, with its slightly variant codons, is genetically isolated
from the rest of the eukaryotic cell. Many biologists suspect that free-living microorganisms that resembled some modern bacteria were the original ancestors of mitochondria. Somehow the ancestral organisms established an intracellular symbiosis with
their host cell. The descendants lost the ability to live independently.

Key Terms

sequestered (p. 214)
Crick and Brenner (p. 214)
base insertion
  experiments (p. 215)
defective enzymes (p. 215)
genetic code (p. 215)

nucleotide bases (p. 215)
triplet (p. 215)
reading frame (p. 215)
nonoverlapping (p. 215)
Khorana (p. 215)
Nirenberg (p. 216)

chain termination (p. 216)
codon (p. 216)
tryptophan (p. 216)
methionine (p. 216)
proline (p. 216)
"synonyms" (p. 216)

1. State the relationship between the DNA genetic code and a protein's primary structure.

2. Explain why singlets or doublets of nucleotide bases are insufficient to specify all of the amino acids in the proteins of living forms.

3. Describe Brenner's base-insertion experiments and explain how they led Crick to suggest that the code was read in triplets.

4. Scrutinize Fig. 13.6 and decide whether the "genetic code" in this instance applies to DNA, messenger RNA, or some other type of RNA. Write your answer in big letters above Fig. 13.6 _____ only, and explain how you determined your answer.

5. Explain how the DNA message TAC-CTC-GTT-CCC-GAA would be used to code for a segment of protein and state what its amino acid sequence would be.

Self-Quiz Questions

Fill-in-the-Blanks

A protein's (1) _____ structure is a string of amino acids, each of which is

specified by a sequence of (2) _____ nucleotide bases. If the code were read

in units of two bases instead of three, only (3) _____ amino acids could be

coded for. In the table that showed which triplet specified a particular amino acid,

the triplet code was in (4) _____ molecules. Each of these triplets is referred

to as a(n) (5) _____. The (6) (choose one) ○ 1st ○ 2nd ○ 3rd base in a codon

is much less critical in specifying the intended amino acid than are either of the

first two.

Summary

Three kinds of RNA govern the assembly of amino acids into proteins. All three are assembled in a complementary fashion from the template of DNA by the process of *transcription*. *Messenger RNA* (mRNA) molecules are linear, unbranched chains of at least 60 nucleotides with leader sequences (promoters) that control the function of mRNA. Messenger RNA alone carries the instructions for assembling a specific protein from the DNA in the nuclear region to the ribosomes. *Ribosomal RNA* (*rRNA*) becomes complexed with proteins to form ribosomes, which are the actual protein construction sites in the cytoplasm. rRNA is synthesized largely in the nucleolus of eukaryotic cells.

Transfer RNA (*tRNA*) is a shuttle molecule that picks up whatever amino acid is dictated by the anticodon at one end, attaches the amino acid to the opposite end, and moves over to the ribosome, where it waits its chance to dump its load into the growing amino acid chain. A *codon* (base triplet on mRNA) is a hydrogen-bonding site for an *anticodon* (complementary base triplet of one or more kinds of tRNA). Codon-anticodon recognition is precise between their first two base pairs. At the third base, some tRNA molecules can form hydrogen bonds with more than one kind of base; hence they can recognize more than one codon. When the slot for its particular amino acid load clicks into place on the ribosome, the complementary anticodon hydrogen bonds to the codon of mRNA, and this temporary bonding positions the amino acid perfectly in order for the appropriate enzymes to form peptide bonds by dehydration synthesis and incorporate the amino acid into the growing protein. That accomplished, the tRNA detaches from the codon and chugs off, searching for another amino acid like the first to bring back again. The mRNA strand continues to advance its codons through the groove in the ribosome until protein assembly is complete. In this way, protein chains of as many as 3,000 amino acid subunits are assembled from a collection of twenty different amino acids. In summary, DNA transcribes its message into three forms of RNA, which cooperate to translate the message into a three-dimensional protein.

Key Terms

transcription (p. 216)   translation (p. 217)   peptide linkages (p. 219)
DNA template (p. 216)   codon (p. 218)   A site (p. 219)
complementary (p. 216)   anticodon (p. 218)   dipeptide (p. 219)
chain elongation (p. 217)   "wobble" effect (p. 218)   P site (p. 220)
RNA polymerases (p. 217)   ribosome (p. 219)   chain termination (p. 220)
RNA transcripts (p. 217)   polysome (p. 219)   "stop" codons (p. 220)
promoters (p. 217)   initiation (p. 219)   release factors (p. 220)

Objectives

1. Distinguish the three types of RNA in terms of structure, function, where each is made, and where each carries out its function.

2. Define *anticodon* and tell the type of RNA it is associated with.

3. Describe the process of transcription and indicate how it differs from replication.

4. Describe how the three types of RNA participate in the process of translation.

5. Summarize the steps involved in the transformation of genetic messages into proteins, using a diagram that shows the relationships among DNA, RNA, and proteins.

6. Define *RNA polymerase* and state its particular function.

Self-Quiz Questions

Fill-in-the-Blanks

(1) _____ RNA alone carries the instructions for assembling a particular sequence of amino acids from the DNA to the ribosomes, where (2) _____ of the polypeptide occurs. (3) _____ RNA makes up much of the body of a ribosome, and (4) _____ RNA acts as a shuttle molecule as each type brings its particular (5)

_____ _____ to the ribosome where it is to be incorporated into the growing (6) _____. A(n) (7) _____ is a triplet on mRNA that forms hydrogen bonds with a(n) (8) _____, which is a triplet on tRNA. During the process of (9) _____, the two strands of the (10) _____ double helix are separated; (11) _____ _____, an enzyme, attaches to and moves along (12) (choose one) ○ one of the strands ○ both strands, all the while speeding up the assembly of RNA nucleotide subunits onto (13) _____ regions of the exposed DNA. Then the (14) _____ strand detaches from the DNA and moves into the (15) _____.

INTEGRATING AND APPLYING KEY CONCEPTS

Genes code for specific polypeptide sequences. Not every substance in living cells is a polypeptide. Explain how genes might be involved in the production of a storage starch (such as glycogen) that is constructed from simple sugars.

14
CONTROLS OVER GENE EXPRESSION

Summary

Cells depend on controls over which proteins are assembled at any given moment of the life cycle. These controls govern transcription, translation, and enzyme activity. *Feedback inhibition* is accomplished by means of allosteric enzymes shutting down their function as end-product molecules accumulate in the cell; in other words, control is proportional to the amount of end product present. Both prokaryotes and eukaryotes depend on allosteric enzyme inhibition as a means to control protein synthesis, but there are major differences in the way these groups regulate the expression of their genes. In prokaryotes, controls over gene expression are basically short-range responses to changing environmental conditions; in eukaryotes, both short-range and long-range controls over gene expression govern development and differentiation.

    Prokaryotes can respond to changes in the environment by activating or inactivating an *operon*—a set of genes that transcribes the appropriate pieces of mRNA, which can translate the code into a series of enzymes comprising a particular metabolic pathway. The lactose operon of *E. coli* examined by Lwoff, Jacob, and Monod is an example. When they switched the bacterial culture from a medium lacking lactose to a medium that contained only lactose as an energy source, they found the bacteria able, within minutes, to metabolize lactose. Jacob and Monod identified all the genes in the gene system that regulated the lactose pathway. *Structural genes* code for enzymes and other proteins, and transcriptional controls, of which there are three types, regulate the behavior of the structural genes. The term *operon* was coined to signify any set of structural genes that operate as a coordinated unit under the direction of DNA control elements. The lactose operon is shut down by *repressing* transcription; it is reactivated by changing the shape of the repressor molecule so that it can no longer repress transcription and transcription begins anew.

Repression, induction, and corepression are all operon controls over transcription that work by changing the way RNA polymerase interacts with its binding site (the promoter). The regulation of operons costs many ATP molecules, so it is used by the cell only where adaptation to changes in resource availability promotes survival in a particular environment. The *rate* of RNA transcription is generally controlled by varying the affinity of different promoters to RNA polymerase. Controls over translation regulate the rate at which protein products are assembled by varying the accessibility of certain parts of the mRNA molecules to translation enzymes, and by varying the amounts of the different tRNA molecules in the cell.

Key Terms

antibodies (p. 223)
transcriptional controls (p. 223)
structural gene (p. 223)
post-transcriptional controls (p. 223)
modification (p. 223)
translational controls (p. 223)
RNA transcripts (p. 223)
nuclear envelope (p. 223)
post-translational controls (p. 223)
controls over enzyme activity (p. 223)
substrate concentrations (p. 224)
feedback inhibition (p. 224)
short-range controls (p. 224)
Escherichia coli (p. 224)
Lwoff, Jacob, and Monod (p. 224)

regulator (p. 225)
repressor protein (p. 225)
promoter (p. 225)
operator (p. 225)
operon model (p. 225)
induction, induces (p. 225)
repression, represses (p. 225)
inducer molecule (p. 225)
RNA polymerase binding site (p. 225)
conformation changes (p. 225)
negative control (p. 225)
corepression (p. 225)
negative feedback control (p. 225)
tryptophan (p. 225)
lactose binding site (p. 226)

Objectives

1. Describe the studies of Lwoff, Jacob, and Monod with *E. coli* nutrition, and explain how their findings led them to develop the operon model.

2. Diagram an operon, locating the regulator, the promoter, the operator, and the structural genes. Indicate where the repressor protein and RNA polymerase bind to the operon.

3. Define the function of each of the components listed in Objective 2.

4. Describe how repression occurs in an operon. Then describe how induction undoes the events of repression in prokaryotes.

5. Show how corepression differs from simple repression.

6. State what repression, corepression, and induction have in common and explain why all operons are not controlled by such mechanisms.

7. Describe how the rate of transcription can change from time to time.

8. Define what a translational control is, indicate how it differs from a transcriptional control, and give two examples of translational controls.

9. Explain why it is advantageous to a cell to be able to activate and inactivate its enzymes in response to changes in the environment.

Self-Quiz Questions

Fill-in-the-Blanks

Cells depend on (1) _____ that govern transcription, translation, and enzyme activity. (2) _____ _____ is a regulatory process in which an end product or an intermediate of a metabolic pathway combines with and alters an (3)_____ enzyme that is usually the first in the pathway; the altered enzyme can no longer catalyze the reaction, so the pathway's activities are stopped. In prokaryotes, controls over gene expression are basically (4) _____-range responses to changing environmental conditions; in eukaryotes, both types of controls govern the events of (5) _____ and (6) _____.

(7) _____ is the process by which genes are prevented from being transcribed and gene function becomes increasingly restricted. If not blocked by a (8) _____ protein, (9) _____ _____ attaches to the promoter, and transcription begins. A(n) (10) _____ is any set of structural genes that operate as a coordinated unit under the direction of DNA control elements. Transcription is reactivated by inducing a change in the shape of the (11) _____ molecule so that it can no longer inhibit transcription.

Sometimes a (12) _____ molecule binds with the repressor molecule so that their combined shape is able to bind to the (13) _____ and thereby block transcription. The (14) _____ codes for the formation of mRNA, which assembles a repressor protein. The affinity of the (15) _____ for RNA polymerase dictates the *rate* at which a particular operon will be transcribed. Some specific tRNA types are less abundant than others in the cell environment; because this tends to slow or inhibit the assembly of that protein, it is an example of a (16) _____ control.

Summary

Prokaryotic *genomes* are largely single-copy DNA sequences, most of which are protein-coding structural genes. Each structural gene in prokaryotes is an uninterrupted stretch of nucleotides. Eukaryotic genomes contain single-copy, moderately repetitive, and highly repetitive DNA sequences. Seventy percent of your own genome consists of single-copy DNA nucleotide sequences, but, of this, protein-coding genes make up only a small part of the genome. Nearly all protein-coding genes (*exons*) are fragmented by intervening nucleotide sequences (*introns*) whose functions are unknown. Highly repetitive DNA generally is not transcribed, but might align chromosomes during mitosis and meiosis. Moderately repetitive DNA contains multiple copies of genes that transcribe rRNA and tRNA.

Most of the DNA and RNA in the nucleus is used to direct the activities of the control elements. Proteins called *histones*, which make up about half of all proteins in the nucleus, are tightly linked to DNA and form a nucleoprotein fiber which resembles a chain of beads. Each bead is a *nucleosome*, a complex packing arrangement that seems to be involved in controlling when transcription occurs.

During mitosis or meiosis, all of the chromatin network of nucleoprotein fibers with their nucleosomes condense into chromosomes. Transcription apparently never occurs on any chromatin that has condensed into the chromosome state. In mammalian females, one of each pair of X chromosomes always remains condensed so that, of any pair of X chromosomes, only one X chromosome is being transcribed in a given cell. The condensed X chromosome is a *Barr body*. The process of differential transcription is known as *Lyonization*, which ensures that any XX female is a "mosaic" of X-linked traits.

In addition to these means of gene regulation in eukaryotes, the nuclear membrane acts as a selective barrier that retains some mRNA and lets other mRNA pass into the cytoplasm. Sometimes mRNA is changed once it reaches the cytoplasm. Sometimes the different amounts of enzymes and types of tRNA influence which mRNA is translated into protein. And even after translation, proteins can be activated, inactivated, or destroyed in a selective manner that can vary from cell to cell or throughout the life of any given cell.

Key Terms

genome (p. 227)
single-copy DNA (p. 227)
moderately repetitive DNA (p. 227)
highly repetitive DNA (p. 227)
intervening DNA, introns (p. 227)
exons (p. 227)
RNA processing (p. 227)
β^+ thalassemia (p. 228)
globin (p. 228)
histones (p. 229)

nonhistones (p. 229)
nucelosome (p. 229)
TATA box (p. 230)
CAT box (p. 230)
Weintraub (p. 230)
X chromosome inactivation (p. 230)
Barr body (p. 230)
Lyonization (p. 230)
mosaic (p. 230)
calico cats (p. 230)

Objectives

1. Describe how eukaryotic DNA is organized in the nucleus. Distinguish between single-copy DNA and the two types of repetitive DNA, and comment on their probable uses by the cell.

2. Describe the process of transcript processing as it involves exons and introns. Discuss how the cutting and splicing processes are controlled.

3. Define *histones* and explain their relationship to DNA and nucleosomes. Comment on the supposed function of histones.

4. Define what is meant by *Barr body* and discuss its relationship to Lyonization.

5. State how human females or calico cats that did not undergo Lyonization would differ from those that did undergo Lyonization.

Self-Quiz Questions

Fill-in-the-Blanks

The RNA assembled on most (1) _____-_____ DNA never leaves the nucleus, so not much is known about its function. (2) _____ _____ DNA, however, occurs as multiple copies of genes that code for (3) _____ and (4) _____.

(5) _____ are snipped out of RNA transcripts that are destined to leave the nucleus; the remaining (6) _____ regions are nucleotide sequences that will be spliced together into a mature mRNA message and translated out in the cytoplasm. (7) _____ and DNA are tightly linked into a (8) _____ fiber, which looks like a beaded chain. Each bead is called a(n) (9) _____, which is thought to be involved in shutting down transcription. A (10) _____ _____ is a condensed X chromosome; the process by which an XX embryonic cell "chooses" which X chromosome will remain active is called (11) _____.

Short-Answer Essay

(12) If human males (XY) have no Barr bodies and human females (XX) have one Barr body per somatic cell, how many Barr bodies would you expect to find in each somatic cell of an XXX individual?

Summary

Genetic controls direct the unfolding of a single-celled zygote into all of the specialized cells and structures of the adult form; in eukaryotes, the controls span
longer periods of time and involve more intricate cellular arrangements in space than
in prokaryotes. During *development* of all complex eukaryotes, cells come to differ in
their positions in the body, in developmental potential, and in appearance, composition
and function. These differences are brought about by mitosis and *differentiation*,
during which cells of the same genetic makeup become more abundant and structurally and
functionally different from one another according to the prescribed program for the
species. Both environmental change and interactions between cells influence the activity of genes necessary for differentiation, as demonstrated by several studies using
the slime mold, *Dictyostelium discoideum*. Communication networks with hormones, and
other signaling agents become established and maintain cell interactions. Intercellular
junctions and the recognition factors on cell surfaces also help in maintaining intercellular communications. Evidence suggests, however, that, the DNA content remains the
same in all differentiated cells except the gametes of sexually reproducing organisms.

Hämmerling's *Acetabularia* graft experiments suggest that, though the DNA content
remains the same, some regions are selectively activated or inactivated in different
cells during the developmental program of the offspring.

Evidence that gene activity varies among cells within the same embryo comes from
studies of the giant chromosomes in the salivary glands of fruit flies; the transcription rate in cells with these chromosomes has been correlated with how large and diffuse the regional puffs appear. Puffing varies from one DNA region to the next during
different developmental stages.

Key Terms

development (p. 230)
genome (p. 230)
differentiation (p. 231)
selective gene expression (p. 231)
Dictyostelium discoideum (p. 231)
cellular slime molds (p. 231)
slug (p. 231)
sporocarp (p. 231)
self-assembly (p. 232)
cyclic adenosine monophosphate
  (cAMP) (p. 232)
chemotaxis (p. 232)
Acetabularia (p. 233)
rhizoid (p. 233)

A. crenulata (p. 233)
A. mediterranea (p. 233)
Hämmerling (p. 233)
Drosophila (p. 233)
polytene chromosomes (p. 233)
chromosome puff (p. 233)
precursors of uracil (p. 234)
ecdysone (p. 234)
molting (p. 234)
actinomycin (p. 234)
lampbrush chromosomes (p. 234)
oogenesis (p. 234)
RNA transcripts (p. 234)

Objectives

1. Define and distinguish the terms *development* and *differentiation*. State which sorts of organisms are highly differentiated and which show very little differentiation.

2. Describe the life cycle of *Dictyostelium discoideum* and summarize it by means of a diagram with arrows.

3. Give an example of differentiation occurring in *D. discoideum*. Then state why its life cycle leads biologists to say that environmental change *and* interactions between cells influence the activity of genes necessary for differentiation.

4. Explain why the cytoplasmic constituents of embryos differ even though, as a result of mitotic activity, their nuclei contain identical assortments of DNA.

5. Outline briefly how Hämmerling conducted his grafting experiments with two species of *Acetabularia* and how these experiments shed light on nuclear-cytoplasmic relationships.

Self-Quiz Questions

Fill-in-the-Blanks

Development in all complex eukaryotes consists of cell division and (1) _____, in which genetically identical cells become structurally and functionally distinct from one another. When food is scarce, *Dictyostelium discoideum* amoebas swarm together to form a (2) _____ that crawls around for a while, then differentiates to form a (3) _____ that eventually discharges spores. During much of this activity, some cells start producing the nucleotide (4) _____ _____ _____, which serves as a signaling agent that eventually stimulates differentiation. The life cycle of *D. discoideum* tells us that both (5) _____ _____ and intercellular interactions influence genes to transcribe their products in a differential manner.

True-False If false, explain why.

____ (6) Experiments with *Dictyostelium discoideum* demonstrated that both environmental change and interactions between cells influence gene activity involved in differentiation.

____ (7) In the polytene chromosomes of *Drosophila*, each band represents the location of one or more genes.

____ (8) Polytene chromosomes develop "puffing" patterns that vary at different times during development; specific puff patterns correlate strongly with the transcription and translation of specific assemblages of proteins.

____ (9) Hämmerling's grafting experiments with two species of *Acetabularia* demonstrated that cytoplasmic substances regulate nuclear functioning.

____ (10) Hämmerling's grafting experiments with two species of *Acetabularia* demonstrated that the nucleus is the control center that governs the regeneration of cellular structures.

Summary

Nuclear-cytoplasmic interactions that maintain variable gene expression do break down occasionally. Sometimes a single cell loses the controls that indicate when to stop cell division; it divides endlessly, crowding surrounding cells and interfering with vital cell functions. The mass that results with be either *benign*, with only its controls over cell division and growth gone awry, or *malignant,* with the recognition factors on its cell surfaces also disturbed. Benign tumors generally can be removed surgically. Malignant growth can invade and destroy surrounding tissues. Some malignancies do not stay as one solid tumor, but *metastasize*, or disperse to other sites in the body. Such cancers are generally treated with specific chemical agents, radiation, or both to selectively destroy uncontrollably dividing cells wherever they are in the body. Such traumatic changes in cellular controls seem to have a variety of causes; specific viruses, chemicals, continuous exposure to sunlight, and constant physical irritation of some tissue all may increase susceptibility to some forms of cancer. Although developmental biologists have elucidated several principles that govern eukaryotic development, they have yet to discover how all of the diverse cells in a complex body read the same genetic library in their own selective ways and thereby differentiate.

Key Terms

progeny (p. 234)   membrane surface   cancer (p. 235)
tumor (p. 235)   recognition factors (p. 235)   metastasis (p. 235)
benign tumor (p. 235)   malignant cells (p. 235)   leukemia (p. 235)
HeLa cell (p. 235)

Objectives

1. Give an example of cells that normally cease dividing. Then state how one such cell could be made to begin dividing again.

2. Define *tumor* and distinguish between benign and malignant tumors.

3. Explain what happens when a tumor metastasizes. Use leukemia as an example. Then describe how the treatment of a metastasized cancer differs from the treatment of benign tumors.

Self-Quiz Questions

   True-False If false, explain why.

____ (1) In humans, brain cells stop dividing when a child is about two years old.

____ (2) All abnormal growths and massings of new tissue in any region of the body are called tumors.

____ (3) Malignant tumors have lost only their normal surface markers, but benign tumors have lost both their surface markers and the mechanism that stops mitosis.

____ (4) The term *metastasis* means the acquiring of the shape or form typical of that species.

Fill-in-the-Blanks

Each differentiated cell (except human red blood cells and gametes) contains its complete allotment of (5) _____; hence it contains all the genes necessary to produce a complete individual. It follows that differentiated cells must be selectively (6) _____ different parts of the same set of genes. The nature of cytoplasmic changes induced by the (7) _____ and by interactions with other (8) _____ dictates which genes will be expressed.

INTEGRATING AND APPLYING KEY CONCEPTS

Suppose that you have been restricting yourself to a completely vegetarian diet for the past six months. Quite unexpectedly, you find yourself in a social situation that requires you to eat a half-pound sirloin steak. Would you expect to be able to digest the steak as easily as you digest soy bean "burgers"? Explain your yes or no answer in terms of transcriptional controls or feedback inhibition.

15
HUMAN GENETICS

Summary

For some human genetic diseases such as phenylketonuria, galactosemia, and some forms of diabetes, stopgap measures called *phenotypic cures* can preserve the lives of many afflicted individuals if the disorder is detected early enough. Galactosemia is an example of a disorder caused by a recessive allele carried on an autosome (a nonsex chromosome). The disorder appears to be transmitted according to the rules of simple Mendelian inheritance because it is expressed in homozygotes of both sexes.

    Hemophilia and the green weakness are examples of *sex-linked recessive* inheritance: the female almost always carries the recessive allele, and the male almost always is the one afflicted with the disorder.

    Chromosomal mutations occur in all organisms, even humans. Improper separation of chromosomes during gamete formation causes Down's syndrome, trisomic XXY, trisomic XYY, and Turner's syndrome.

pedigrees (p. 237)
genetic disorders (p. 237)
phenotypic cures (p. 237)
genotypic cures (p. 237)
recombinant DNA (p. 238)
galactosemia (p. 238)
autosomal recessive
   inheritance (p. 238)
gene locus (p. 238)
autosome (p. 238)
sex chromosome (p. 238)

simple Mendelian
   inheritance (p. 238)
hemophilia (p. 238)
sex-linked recessive
   inheritance (p. 239)
mutant allele (p. 239)
green weakness (p. 240)
retinal pigment (p. 240)
nondisjunction (p. 240)
chromosomal mutations
   (p. 240)

Down's syndrome (p. 240)
trisomic, trisomy 21 (p. 240)
   (2n + 1)
miscarriage (p. 240)
Turner's syndrome (p. 240)
sex chromosome
   abnormality (p. 240)
monosomic (2n - 1) (p. 240)
trisomic XXY (p. 241)
trisomic XYY (p. 241)

Objectives

1. Distinguish phenotypic cures from genotypic cures and state which kind of cure is used most at this time.

2. Explain how a genotypic cure is achieved.

3. List and describe the causes and phenotypic features of two human disorders that result from improper separation of chromosomes during gamete formation.

4. List and describe the causes and phenotypic features of two human disorders that are due to point mutations.

5. List three features of a genetic disorder that would suggest that it is transmitted as a form of *autosomal* recessive inheritance.

6. State the feature(s) of a genetic disorder that would lead you to believe that it is a form of *sex-linked* inheritance.

Self-Quiz Questions

Matching Choose the single most appropriate letter.

(1) ____ Down's syndrome

(2) ____ galactosemia

(3) ____ green weakness

(4) ____ hemophilia

(5) ____ trisomic XYY

(6) ____ Turner's syndrome

A. autosomal recessive inheritance

B. sex-linked recessive inheritance

C. nondisjunctive inheritance

Summary

A variety of *therapeutic measures* (diet modification, environmental adjustment, surgical correction, and chemotherapy) are used in treating genetically-based disorders. *Preventive* measures include mutagen reduction, genetic screening, genetic counseling, and prenatal diagnosis; gene replacement therapy also shows promise but is still in the research stage. Normal alleles would be substituted for mutant alleles in sperm or eggs. Through *amniocentesis* (the sampling of the fluid in the mother's uterus), it is possible to determine whether or not the developing fetus has a severe genetic disorder. If such a severe disorder is discovered, the parents may elect to induce the fetus to be expelled from the uterus and thus circumvent years of suffering brought about by such tragedies.

Key Terms

therapeutic measures (p. 242)	palate (p. 242)	genetic counseling (p. 243)
preventive measures (p. 242)	Wilson's disease (p. 242)	prenatal diagnosis (p. 243)
diabetes (p. 242)	phenylketonuria,	amniocentesis (p. 243)
chemotherapy (p. 242)	PKU (p. 242)	karyotype analysis (p. 243)
genetic screening (p. 242)	phenylalanine (p. 242)	gene replacement therapy
albino (p. 242)	tyrosine (p. 242)	(p. 244)
sickle-cell anemia (p. 242)	phenylpyruvic acid (p. 242)	abortion (p. 244)
harelip (p. 242)	mutagen reduction (p. 243)	pyloric stenosis (p. 245)

Objectives

1. Discuss some of the ethical predicaments that have arisen with our increased knowledge of human genetics and increased technical abilities.

2. Explain why gene replacement therapy is still in the experimental stage. What is involved in making the procedure work?

3. Describe the phenotypic cures in current use for any of the human disorders known; tell how the "cure" is administered.

4. Define *amniocentesis* and state one potential benefit and one potential drawback of the technique.

Matching Choose the one most appropriate answer for each.

(1) ____ amniocentesis

(2) ____ diabetes

(3) ____ Down's syndrome

(4) ____ galactosemia

(5) ____ hemophilia

(6) ____ trisomic XXY

(7) ____ phenylketonuria

(8) ____ Turner's syndrome

A. females with only one X chromosome, ovaries nonfunctional

B. XXY → sterile males with breast enlargement

C. trisomy 21 → mental retardation; large, misshapen head

D. blood fails to clot; Victoria was a carrier

E. should be placed on a milk-free diet if he or she has this

F. mutant allele codes for a defective enzyme; a toxic substance in the blood damages the nervous system

G. too little sugar in cells, too much in the blood

H. a diagnostic technique that samples uterine fluid

INTEGRATING AND APPLYING KEY CONCEPTS

Suppose that you are a genetics researcher who discovers in humans a sequence of genes that code for the production of several proteins that bring about aging and cell death. How might you use this information in writing a proposal to the National Science Foundation for a grant that would finance research investigating ways to prevent cells, tissues, and organisms from aging and dying? What specific research program would you propose? What equipment and personnel do you envision that you would need? How long would you be doing this research, where would you propose to conduct this study, and how much money would you ask for? You may wish to consult appropriate previous chapters and their bibliographic references in answering this question. Don't assume that famous researchers wouldn't want to be included in your research team!

16
PLANT CELLS, TISSUES, AND SYSTEMS

16–I   THE PLANT BODY: AN OVERVIEW

On "Typical" Plants

How Plant Tissues Arise

Summary

Nonvascular plants (certain algae, mosses, and liverworts) lack internal transport
systems; *vascular* plants (ferns, gymnosperms, and angiosperms) have well-developed internal tissues that conduct food and water through the multicellular plant body. *Gymnosperms* include pines, junipers, and redwoods; *angiosperms* are represented by roses,
cherry trees, corn, and dandelions. The plant body differentiates first into three
kinds of *vegetative* organs: root, stem, and leaf. Later in the life cycles of flowering
plants, flowers composed of several organs that serve as a *reproductive* system also appear. Many parts of the mature plant body contain clusters of cells that are undifferentiated; these "embryonic" zones (where mitosis and differentiation occur) are known
as *meristems*. Newly formed cells differentiate first into three more specialized embryonic tissues: (1) *protoderm*, which develops into the plant's surface layers; (2) the
ground meristem, which forms the bulk of the plant body; and (3) *procambium*, which differentiates still further to become the primary vascular tissues. The vegetative body
of vascular plants grows continuously throughout its lifetime by producing the same
tissue systems that are common to all of its parts from the same kinds of embryonic
tissues that were derived from apical meristems. The structure and function of these
tissue systems depend on their locations in the mature plant body.

woody (p. 248)
vascular plants (p. 248)
nonvascular
    plants (p. 249)
rudimentary (p. 249)
gymnosperms (p. 249)
angiosperms (p. 249)
Monocotyledonae
    (monocots) (p. 249)
Dicotyledonae
    (dicots) (p. 249)
cotyledons (p. 249)
seed leaves (p. 249)
shoot system (p. 249)
root system (p. 249)

reproductive system (p. 249)
open growth (p. 249)
maturity (p. 249)
undifferentiated
    tissue regions (p. 249)
meristems (p. 249)
shoot apical meristem (p. 249)
protoderm (p. 249)
ground meristem (p. 249)
procambium (p. 249)
primary plant body (p. 249)
vegetative body (p. 249)
epidermis (p. 249)
vascular tissue (p. 249)
pith (p. 249)

cortex (p. 249)
leaf mesophyll (p. 249)
primary phloem (p. 249)
primary xylem (p. 249)
shoot apex (p. 250)
root tip (p. 250)
node (p. 250)
internode (p. 250)
dermal system (p. 250)
ground system (p. 250)
vascular system (p. 250)
vascular bundles (p. 250)
ground tissue (p. 250)
secondary growth (p. 250)
vascular cambium (p. 250)
cork cambium (p. 250)

Objectives

1. Distinguish vascular plants from nonvascular plants, gymnosperms from angiosperms, vegetative organs from reproductive organs.

2. Compare the growth pattern of animals with that of plants.

3. (a Identify the three primary embryonic tissues that are derived from apical meristems.
    (b) Then identify the tissue(s) in the mature plant body formed from each of the three primary embryonic tissues.

4. Identify the three main tissue systems that compose the mature plant body and tell how each is structurally related to the others.

Self-Quiz Questions

Fill-in-the-Blanks

Root and shoot tips have dome-shaped (1) _____ meristems where new cells form

rapidly through mitotic divisions. These meristems give rise to three other kinds of

embryonic tissues: (2) _____, which will form the epidermis; (3) _____,

which will form the vascular tissue; and the (4) _____ _____, which will

form most of the plant body's substance. (5) _____ is an example of ground tis-

sue that lies centrally inside the rings of vascular bundles in plant stems. (6)

_____ is ground tissue that lies between the vascular ring and the dermal sur-

face layer of plant stems and roots. The main photosynthetic area of a leaf is composed

of (7) _____, which lies between the leaf's upper and lower (8) _____.

Vascular tissues consist of primary (9) _____, which conducts water and ions from the roots to the photosynthetic areas, and primary (10) _____, which conducts the products of photosynthesis away to storage areas and helps to support the plant.

16–II PLANT TISSUES AND THEIR COMPONENT CELLS

Summary

Three tissue systems—dermal, vascular, and ground—extend continuously through the plant body. Tubes and fibers of *vascular* tissue are embedded in *ground* tissue, which in turn is covered and protected by *dermal* tissue. The *ground* system comprises most of the plant body. It consists mainly of (1) *parenchyma*, which participates in photosynthesis, food storage, wound healing, and regenerating plant parts; (2) *collenchyma*, a thin-walled flexible tissue that helps support young plant parts; and (3) *sclerenchyma*, a rigidly walled strengthening tissue that consists mainly of dead cells.

Key Terms

parenchyma (p. 251)	collenchyma (p. 251)	sclerenchyma (p. 252)
ground tissue system (p. 251)	cellulose (p. 251)	sclereids (p. 252)
mesophyll (p. 251)	pectin (p. 251)	stone cells (p. 252)
regenerating (p. 251)	hydrophilic (p. 251)	lignin (p. 252)
graft (p. 251)	pliable (p. 251)	fibers (p. 252)

Objectives

1. Define and distinguish from each other the three principal tissue types that compose most of the ground system of the plant body.

2. State the function(s) of the three types of ground tissue and tell where each is located in the plant body.

3. Explain why pectin and cellulose make collenchyma cells flexible, but lignin promotes thick, rigid cell walls in sclerenchyma.

Self-Quiz Questions

   Fill-in-the-Blanks

(1) _____ cells form continuous storage tissues in the stem and root cortex and in stem (2) _____. The photosynthetic tissue of a leaf lies between the leaf's upper and lower (3) _____ and is known as (4) _____, which translates to mean "middle leaf." (5) _____ cells are responsible for healing wounds,

regenerating plant parts, and ensuring successful grafts. (6) _____ cells contain cellulose and (7) _____, which are hydrophilic substances promoting pliable cell walls that support young plant parts. (8) _____ tissues strengthen mature plant parts with (9) _____ and fibers.

16–III   Xylem: Water Transport, Storage, and Support
Phloem: Food Transport, Storage, and Support

Summary

The *vascular system* contains food-conducting tissues (*phloem*) and water-conducting tissues (*xylem*) that form a network through the entire plant. *Xylem* contains parenchyma cells that store water and food, sclereids and fibers that provide mechanical support, and *tracheids* and *vessel* members that passively conduct water and dissolved salts. Like xylem, *phloem* also contains parenchyma, sclereids, and fibers. Unlike the tracheids and vessel members of xylem, which die as soon as they reach maturity, the food-conducting cells of phloem, called *sieve-tube members*, are alive when mature, as are their adjacent parenchyma cells, called *companion cells*, which help load and unload the sieve element pipelines.

Key Terms

xylem (p. 252)
wood (p. 253)
vessel members (p. 253)
tracheids (p. 253)
parenchyma (p. 253)
sclereids (p. 253)
fibers (p. 253)
lignin-impregnated (p. 253)
secondary walls (p. 253)

passively (p. 253)
primary cell walls (p. 253)
pits (p. 253)
pit-pair (p. 253)
perforation plate (p. 253)
vessels (p. 253)
phloem (p. 254)
sieve elements (p. 254)
parenchyma (p. 254)

fibers (p. 254)
sclereids (p. 254)
sieve-tube members (p. 254)
sieve plates (p. 254)
companion cells (p. 254)
phloem pipeline (p. 254)
concentration
 gradient (p. 254)

Objectives

1. Identify the three main functions of xylem and name the types of cells that participate in each function.

2. Distingish the structural features of tracheids from those of vessel members.

3. List the principal differences in xylem and phloem tisses and state the undifferentiated tissue that both arose from.

4. Identify the three main functions of phloem and name the types of cells that participate in each function.

5. Explain how sieve elements differ from companion cells.

Fill-in-the-Blanks

The highly specialized cells of xylem that passively conduct water and (1) _____

are (2) _____ _____, which are perforated at one end, and (3) _____,

which are not perforated but have (4) _____ that allow water to move from cell

to cell. The specialized cells of phloem that conduct and store (5) _____ are

called (6) _____-_____ _____; the end walls of these cells are

called (7) _____ _____ because they have larger pores than do the side

walls. (8) _____ _____ are thought to help load and unload the phloem

pipelines.

16–IV   Epidermis and Periderm: Interfaces With the Environment   255-256

Summary

In general, *epidermis*, the outermost layer of cells, covers the plant as a single,
compact, and continuous layer of cells. Root hairs in root epidermis promote water
absorption from the surrounding soil. Cuticle, the surface coat on aerial plant parts,
restricts water loss and discourages microbial attack. *Stomata* are gates to the interior of leaves and stems that may be opened or closed to regulate the inward movement
of carbon dioxide and the outward movement of water. A variety of secretory structures
that aid in attracting insects, discouraging predators, and ridding the plant body of
excess salts are also associated with the epidermis. Periderm is a protective covering
that replaces the epidermis of gymnosperms and angiosperms undergoing secondary growth.

Key Terms

epidermis (p. 255)
root hairs (p. 255)
aerial (p. 255)
impregnated (p. 255)
wax (p. 255)
cutin (p. 255)
cuticle (p. 255)

guard cells (p. 255)
stoma, -ata (p. 255)
mucopolysaccharides (p. 255)
saltbush (p. 255)
tamarisk (p. 255)
nectaries (p. 255)

periderm (p. 256)
cork cambium (p. 256)
cork (p. 256)
bark (p. 256)
vascular cambium (p. 256)
suberin (p. 256)

Objectives

1. Distinguish between epidermis and periderm in terms of structure.

2. Describe how root epidermis differs in function and in structure from shoot
    epidermis.

3. Show how guard cells are related to a stoma and tell what general roles are per-
    formed by guard cells.

4. List some of the structures commonly found in epidermis that change its function
    from protection to some other role.

Fill-in-the-Blanks

In a young plant, (1) _____, the outermost layer of cells, is a single, continuous protective covering of cells. Root epidermis tends to promote (2) _____ _____ from the surrounding soil; shoot epidermis tends to restrict (3) _____ _____ and discourage (4) _____ _____. (5) _____ are gates in stem and leaf epidermis that are flanked by (6) _____ _____. (7) _____ replaces epidermis in older gymnosperms and angiosperms.

16-V   THE ROOT SYSTEM   256-258

Summary

In gymnosperms and flowering plants, the first (primary) root develops from the apical meristem at the tip of the embryo's root. Both taproot and fibrous root systems anchor the plant and store photosynthetically derived food; their most important function is absorption of water and dissolved ions from the soil. The meristematic tip of the primary root, shielded by its root cap as it pushes its way through the soil, gives rise to cells that differentiate into the vascular, dermal, and ground systems. In a mature root, epidermis with root hairs surrounds the cortex, which surrounds a vascular cylinder. Immediately outside the vascular cylinder lies the *endodermis*: a specialized, innermost layer of cortex; this single layer of cells, through active and passive transport mechanisms, helps to control the movement of water and dissolved salts into the xylem pipeline to the stem and leaves.

Secondary growth in many gymnosperms and flowering plants increases the diameter of roots and stems. *Procambium* that lies between primary xylem and phloem in each vascular strand develops into *vascular cambium*, which gives rise to secondary xylem and phloem. As the volume of new phloem increases, it pushes the vascular cambium toward the root's periphery. Part of the ground tissue between the phloem and endodermis begins dividing soon afterward and becomes the pericycle; pericycle cells line up and merge with the vascular cambium to produce layer after layer of secondary xylem and phloem. Cork cambium arises on the phloem tissue's outer ridges and produces periderm, the corky covering that replaces the epidermis.

Key Terms

primary root (p. 256)	root hairs (p. 256)	Casparian strip (p. 257)
lateral (p. 256)	cortex (p. 256)	active transport (p. 257)
taproot system (p. 256)	parenchyma (p. 256)	passive transport (p. 257)
adventitious roots (p. 256)	endodermis (p. 256)	solute (p. 257)
fibrous root system (p. 256)	vascular column (p. 257)	procambium (p. 257)
root apical	mineral salts (p. 257)	pericycle (p. 257)
meristem (p. 256)	xylem pipelines (p. 257)	pith (p. 257)
root cap (p. 256)	suberin (p. 257)	primary phloem (p. 258)

primary xylem (p. 258)   vascular cambium (p. 258)   secondary phloem (p. 258)
secondary growth (p. 258)   secondary xylem (p. 258)   periderm (p. 258)
lateral meristem (p. 258)

Objectives

1. Distinguish between taproot and fibrous root systems in terms of structure and developmental origins.

2. State the principal functions of roots and explain how the structural components of roots enable those functions to be achieved.

3. Describe how secondary growth increases the diameter of a root. Define all the structural features that participate in the secondary growth of roots.

Self-Quiz Questions

   Fill-in-the-Blanks

(1) _____ systems are based on a primary root and its lateral branches; (2)

_____ root systems arise from the young stem. The root (3) _____ protects

the delicate young root from being torn apart during growth through the soil; (4)

_____ _____ gives rise to cells that differentiate into the three basic

tissue systems. (5) _____ with root hairs surrounds the (6) _____, which

surrounds the (7) _____ _____. Immediately outside the vascular column

lies the innermost layer of cortex, the (8) _____, which regulates water and so-

lute movements into the xylem. During secondary growth, the procambium develops into

(9) _____ _____, a thin layer of lateral meristem that gives rise to sec-

ondary xylem and phloem. Divisions in the (10) _____ _____ produce peri-

derm, which replaces the epidermis.

16–VI   THE SHOOT SYSTEM

Summary

Stems provide a structure that (1) exposes leaves to light; (2) provides routes for food and water movement between roots and leaves; and (3) stores food in the parenchyma cells of cortex, pith, xylem, and phloem. The stem and its leaves constitute the shoot. At specific points along the stem (nodes), leaves are differentiated and buds develop.

Flowering plants have two general patterns of stem structure. *Monocot* stems usually are uniformly thick along their length, they generally do not undergo secondary growth, and cross-sections of their stems reveal that strands of vascular tissue are arranged in many bundles scattered throughout the ground tissue. *Dicot* stems are usually tapered, they usually undergo secondary growth, and their vascular bundles are arranged in a cylinder that separates the ground tissue into pith and cortex. Most leaves are sites for photosynthesis; their structure promotes the absorption of sunlight and the rapid uptake of carbon dioxide.

Secondary growth in stems occurs much as it does in roots. Two kinds of cells in the vascular cambium (the fusiform initials and the ray initials) produce the vertical and ray systems, respectively, in woody stems. The *vertical system* conducts food and water up and down the stem; the *ray system* consists mostly of parenchyma cells that act as lateral conduits and food storage centers. Through the ray system, the vascular cambium and phloem are fed water, and food from the secondary phloem enters the vascular cambium and still-living cells of secondary xylem. Growth rings are produced by alternating periods of activity and inactivity occurring in the vascular cambium.

Key Terms

stems (p. 258)
nodes (p. 259)
internodes (p. 259)
bud primordia (p. 259)
axillary buds (p. 259)
terminal buds (p. 259)
monocot (p. 259)
dicot (p. 259)
leaf primordia (p. 260)
apical meristem (p. 260)
leaf (p. 260)

veins (p. 260)
stem vascular
   bundle (p. 260)
cuticle (p. 260)
palisade mesophyll (p. 260)
spongy mesophyll (p. 260)
succulent (p. 260)
stomata (p. 260)
simple leaves (p. 261)
blade (p. 261)
leaflets (p. 261)

compound leaves (p. 261)
secondary growth in
   stems (p. 261)
vascular cambium (p. 261)
fusiform initials (p. 261)
vertical system (p. 261)
ray initials (p. 261)
ray system (p. 261)
early wood (p. 261)
late wood (p. 261)
"tree rings" (p. 262)

Objectives

1. List the principal functions of stems.

2. Describe the primary structure of a generalized stem. Distinguish nodes from internodes, tell where buds are located, and indicate how cross-sections of monocot stems differ from those of dicot stems.

3. Describe the primary structure of a leaf. List the principal functions of leaves and tell how leaf structure enables these functions to be carried out.

4. Describe some modifications in leaf structure that enable plants to exploit exceedingly dry or exceedingly wet environments.

5. Explain how some stems increase their diameters each year. Contrast the functions of the vertical and ray systems of xylem.

6. Tell how the xylem cells of woody plants indicate the relative amounts of water available to the plant from season to season.

Self-Quiz Questions

<u>Fill-in-the-Blanks</u>

Leaves are differentiated and buds develop at specific points along the stem called

(1) _____. (2) _____ stems generally do *not* undergo secondary growth,

and cross-sections of their stems reveal (3) _____ _____ scattered

throughout the ground tissue. Most leaves absorb (4) _____, take up (5)

_____ _____ from the air, and use (6) _____ delivered by the xylem

in carrying out the complex series of chemical reactions known as (7) _____.

Two kinds of cells in the vascular cambium produce the (8) _____ and (9)

_____ systems in woody stems; the (10) _____ system consists mostly of

(11) _____ cells that act as lateral conduits and food storage centers. Growth

rings are produced by fluctuating levels of activity in the (12) _____

_____.

INTEGRATING AND APPLYING KEY CONCEPTS

Try to imagine the specific behavioral restrictions that might be imposed if the human
body resembled the plant body in having (1) open growth with apical meristematic regions, (2) stomata in the epidermis, (3) cells with chloroplasts, (4) excess carbohydrates stored as starch rather than as fat, and (5) to be dependent on the soil as a
source of water and inorganic compounds.

17

WATER, SOLUTES, AND PLANT FUNCTIONING

17–I INTRODUCTION

264-265

Summary

Unlike most animals, land plants generally absorb water, carbon dioxide, and necessary ions from sources in which concentrations of these substances are low. About ninety percent of the volume of a mature plant cell is occupied by the vacuole, which is mostly water; thus the vacuole requires minimal energy outlay to be assembled and maintained. Nutrients can be stockpiled in vacuoles as they become available and dispensed as they become scarce. Extensive systems of cylindrically shaped roots enable efficient absorption of dilute substances and efficient burrowing through soil. Broad, thin leaves present a large exterior surface area for sunlight absorption and a large interior surface area in the spongy parenchyma for gas absorption.

Key Terms

central vacuole (p. 264)
vascular systems (p. 265)

plant physiology (p. 265)

Objectives

1. List four problems that had to be solved before plants could adapt to life on land.

2. Explain why a central vacuole that contains mostly water is useful for plant cells but is not an advantage for animal cells.

3. Explain why cylindrically shaped roots are better suited to carry out the functions of roots, and broad, thin leaves are better suited to carry out the functions of leaves, than vice versa.

Self-Quiz Questions

Fill-in-the-Blanks

(1) _____, which composes most of the contents of the central (2) _____

in a plant cell, is energetically (3) ○ more expensive ○ less expensive to accumulate compared with other cytoplasmic substances that might otherwise be stored.

Summary

Most plants require sixteen essential elements to grow and reproduce. *Oxygen, hydrogen,* and *carbon* account for about ninety-six percent of the dry weight of plants. Oxygen is incorporated into organic compounds that make up the plants' dry weight; it comes from water, gaseous oxygen (O_2), and carbon dioxide (CO_2) in the air. Nitrogen is a necessary constituent of amino acids and nitrogen bases from which plants synthesize their proteins and nucleic acids. Ions such as potassium, calcium, magnesium, phosphorus, chlorine, iron, boron, manganese, zinc, copper, and molybdenum are necessary to form and activate plant enzymes and other substances and are absorbed from the soil dissolved in water. They are also important in establishing osmotic gradients that distribute water necessary for plant growth and for sustaining a nonwilted plant form.

Key Terms

essential elements (p. 265)
macronutrients (p. 266)
micronutrients (p. 266)
trace elements (p. 266)
molybdenum (p. 266)

nitrogen-fixing
  bacteria (p. 267)
nodules (p. 267)
legumes (p. 267)
ammonia (p. 267)

potassium (p. 267)
enzymes (p. 267)
aerobic respiration (p. 267)
osmotic gradient (p. 267)
turgor pressure (p. 267)
mineral ions (p. 267)

Objectives

1. List the sources of oxygen, hydrogen, carbon, and nitrogen that make up ninety-six percent of the dry weight of plants.

2. Describe how plants obtain the important ions they need for enzyme formation and growth.

3. Explain how turgor pressure participates in the growth of cells.

Fill-in-the-Blanks

In plants, nitrogen is a necessary component of amino acids and (1) _____

_____, from which they synthesize proteins and (2) _____ _____.

Nitrogen is available to plants in the form of either (3) _____ or (4)

_____ ions. (5) _____ enters the plant via water, O_2, and CO_2. (6)

_____ also enters the plant via water. (7) _____ _____ is the

pressure applied to cell walls as water is absorbed into the cell; it allows the plant

to sustain a(n) (8) _____ form.

Matching Choose at least one and no more than two letters per blank.

(9) ___ boron	A. macronutrient
(10) ___ calcium	B. micronutrient
(11) ___ chlorine	C. component of phospholipids
	D. at center of chlorophyll molecule
(12) ___ copper	E. helps to establish osmotic gradients; as CO_2 dwindles in guard cells, this is pumped in
(13) ___ iron	
(14) ___ magnesium	F. part of a coenzyme needed in enzyme-mediated reactions that reduce nitrate
(15) ___ manganese	G. in ammonium, nitrate, and nitrite ions; needed for protein and nucleotide synthesis
(16) ___ molybdenum	H. involved in electron transport during photosynthesis and aerobic respiration
(17) ___ nitrogen	I. needed in cell walls, in cell growth, and division
(18) ___ phosphorus	
(19) ___ potassium	
(20) ___ sulfur	
(21) ___ zinc	

Summary

Water is absorbed into the roots through the root hairs of the thin-walled epidermal cells and then is passed along the highly permeable cortex cells. At the endodermis water is funnelled directly into the endodermal cytoplasm and into the vascular columns.

Evaporation from stems and leaves, known as *transpiration*, is caused by the drying power of air. According to the cohesion theory of water transport, as long as water molecules vacate transpiration sites, replacement water is sucked up through the xylem from the roots in continuous columns as a result of hydrogen bonding between the water molecules. The cuticle prevents extensive water loss from plants; stomata in the epidermis open and close according to the amounts of water and carbon dioxide in their guard cells. High levels of potassium in guard cells are established by active transport as carbon dioxide levels dwindle during photosynthesis; high levels of potassium cause water to move in osmotically, and the increased turgor pressure causes stomata to open. During heat stress, the plant produces abscisic acid, which causes potassium ions to leave the guard cells and causes turgor pressure to drop. As a result, stomata close, photosynthesis slows and stops, and so must growth.

Key Terms

uptake (p. 267)
fibrous root systems (p. 267)
taproot system (p. 267)
mycorrhizae (p. 268)
vegetative body (p. 268)
tensile forces (p. 268)

transpiration (p. 268)
mesophyll cell walls (p. 268)
negative pressure (p. 268)
cohesion theory of water transport (p. 269)
cohere (p. 270)

hydrogen bonds (p. 270)
cuticle (p. 270)
guard cells (p. 270)
stoma, -ata (p. 270)
abscisic acid (p. 272)

Objectives

1. Explain how water passes from dry soil into the vascular column of the root.
2. Define *transpiration* and list the main points of the cohesion theory of water transport.
3. Compare the amount of water that gets stored or used in metabolism with the amount that is transpired.
4. Explain how land plants regulate water loss as environmental conditions change. List the plant structures that participate in regulating water loss.

Fill-in-the-Blanks

(1) _____ is evaporation of water from stems and leaves. The (2) _____ theory of water transport suggests that (3) _____ _____ allows water molecules to cohere tightly enough to keep from breaking apart as they are pulled up through the plant body. High levels of (4) _____ in guard cells are established by (5) _____ _____ as carbon dioxide levels dwindle during photosynthesis. During (6) _____ stress, the plant produces (7) _____ _____, which causes (8) _____ ions to leave the guard cells, the turgor pressure to drop, and the stomata to close.

Summary

Solute absorption and accumulation occur as energy from ATP drives the membrane pumps involved in active transport to move substances into cells against concentration gradients. Mineral ion absorption and accumulation are coordinated throughout the plant body in ways that affect growth. Insoluble storage starches and fats are converted to soluble forms (most often sucrose) for transport from one plant organ to another via the sieve-tube system of phloem tissue (translocation). According to the pressure flow theory, translocation is driven by differences in water pressure from one region of the phloem pipeline to another. Sucrose is actively transported into sieve tube cells in the leaf, water follows osmotically due to the increased solute concentration, and pressure builds up in the leaf sieve tube regions. Stems, fruits, and roots are all regions of lower pressure because cells there convert the food molecules delivered to them to substances they need and use them in cellular respiration. Food (sucrose) therefore tends to flow from the leaves (the source) to the fruits, seeds, and roots (sink regions).

Key Terms

passive process (p. 272)   active transport (p. 272)   hydrolyzed (p. 273)
solute (p. 272)   concentration   translocation
membrane pumps (p. 272)   gradient (p. 272)   (p. 273)

sieve elements (p. 273)   source regions (p. 274)   pressure gradients (p. 274)
aphids (p. 273)   sink regions (p. 274)   loading process (p. 274)
stylet (p. 273)   pressure flow theory (p. 274) osmotic gradient (p. 274)

Objectives

1. Explain how plants absorb and accumulate solutes. Then explain how the rate at
    which a plant absorbs and accumulates solutes affects rates of photosynthesis,
    storage of foods, rates of cellular respiration, and rate of growth.

2. Discuss the evidence that leads us to believe in the pressure flow theory.

3. Describe the process of translocation and list the key points of the pressure flow
    theory of translocation. Explain what causes sucrose to be transported from one
    plant organ to another.

Self-Quiz Questions

   Fill-in-the-Blanks

When soil is sufficiently moist, dissolved ions are carried rapidly into roots, where

(1) _____ _____ moves them through the cortex and into the (2) _____

for transport upwards. (3) _____, the dominant food storage product in plants,

generally is converted to (4) _____ for transport through the plant body to an-

other plant organ; this transport process is known as (5) _____. The (6)

_____ _____ theory suggests a mechanism by which food can be transported

from one plant region to another. Sucrose is actively transported into (7) _____-

_____ cells, (8) _____ follows osmotically, and (9) _____ builds

up in the leaves. Sucrose flows from the leaves to the stems, fruits, and (10)

_____ because all three are regions of lower pressure.

INTEGRATING AND APPLYING KEY CONCEPTS

How do you think maple syrup is made from maple trees? Which *specific* systems of the
plant are involved and why do you think that maple trees are only tapped at certain
times of the year?

18

PLANT REPRODUCTION AND EMBRYONIC DEVELOPMENT

Summary

Flowering plants (angiosperms) can reproduce asexually (sex cells and fertilization are
not involved) as well as sexually. In flowering plant sporophytes, the vegetative
growth phase gives way to a reproductive phase that begins with flower production. A
flower is a cluster of reproductive and nonreproductive organs that are thought to
originate from highly modified leaves. The outermost green *sepals* protect the buds;
petals attract bird and insect pollinators and enclose the stamen and one or more carpels in a perfect flower. *Stamens* are male reproductive organs, each of which consists
of an anther (in which microspores develop into pollen grains) perched on its supportive stalk (the filament). *Carpels* are female reproductive organs, each of which consists of a swollen base (the ovary) that produces at least one *ovule*, within which
megaspores are produced; megaspores develop into egg-bearing gametophytes.

    Angiosperms form flowers that have microspore- and megaspore-producing structures,
as well as accessory parts that have both direct and indirect roles in sexual reproduction. Sexual reproduction depends on pollen grains being transferred from anthers
(male reproductive structures) to stigmas (female reproductive structures). Once a
pollen grain has been deposited on the stigma, a pollen tube forms and grows down to
the ovarian chamber below. In flowering plants, double fertilization occurs: One sperm
nucleus fuses with the nucleus of the egg, and the other sperm nucleus fuses with two

other nuclei in the ovule, forming a single triploid (3n) nucleus. Following double fertilization, the ovule expands and develops into a seed. The ovarian wall expands and ripens into a fruit, which may be an edible structure, a small capsule, or one of a variety of other forms.

Key Terms

meiosis (p. 276)	pollination (p. 277)	stamen (p. 278)
fertilization (p. 276)	pollen tube growth (p. 277)	anther (p. 278)
asexual reproduction (p. 276)	embryo development (p. 277)	filament (p. 278)
somatic cells (p. 276)	seed (p. 277)	carpel (p. 278)
pollination agent (p. 276)	fruit (p. 277)	ovary (p. 278)
vegetative organs (p. 276)	germination (p. 277)	style (p. 278)
sporophyte (p. 276)	flower (p. 277)	stigma (p. 278)
gametophyte (p. 276)	fertile (p. 277)	perfect flower (p. 278)
annuals (p. 277)	sterile (p. 277)	imperfect flower (p. 278)
perennials (p. 277)	receptacle (p. 277)	ovule (p. 279)
microspore formation (p. 277)	sepals (p. 277)	pollen grain (p. 279)
megaspore formation (p. 277)	petals (p. 277)	

Objectives

1. List in sequence the major events that constitute the reproductive portion of the flowering plant life cycle.

2. Explain how the inclusion of a reproductive phase in their life cycles is advantageous to flowering plants. Explain why reproduction by budding or other sexual means is not as advantageous to plant populations as reproduction by sexual means.

3. List the parts of a flower and state the functions of each part.

4. Distinguish perfect and imperfect flowers.

Self-Quiz Questions

Fill-in-the-Blanks

In a flower, the outermost (1) _____ protect the buds; (2) _____ attract bird and insect pollinators. (3) _____ are male reproductive organs, each of which consists of a(n) (4) _____, in which (5) _____ develop into pollen grains, and its supportive stalk, the filament. (6) _____ are female reproductive organs, each of which consists of a swollen base, the (7) _____, that produces at least one (8) _____; inside this (8), (9) _____ develop into egg-bearing gametophytes.

Microspores to Pollen Grains   278

Megaspores to Eggs   279-280

Summary

Reproductive cells develop and undergo meiosis, forming *microspores* in the male reproductive organs (anthers). Microspores develop into *pollen grains*, which are immature
male gametophytes. Pollen grains later develop into mature, sperm-bearing gametophytes.
Meanwhile, in the ovary lie one or more ovules, within which reproductive cells develop and undergo meiosis, forming *megaspores*. Megaspores develop into egg-bearing female gametophytes.

Key Terms

microspore mother   pollen grain (p. 279)   megaspore mother
   cells (p. 278)   tube cell (p. 279)   cell (p. 280)
meiospore (p. 278)   generative cell (p. 279)   megaspores (p. 280)
microspores (p. 278)   ovule (p. 279)   female gametophyte (p. 280)
pollen sac (p. 278)   seed (p. 279)   embryo sac (p. 280)
pollination (p. 278)   integuments (p. 280)   endosperm mother
pollen tube (p. 278)   micropyle (p. 280)   cell (p. 280)

Objectives

1. Explain where male sex organs form in a flowering plant. Then describe, beginning
    with microspore production, all of the steps and structures formed along the way
    to producing sperm nuclei.

2. Distinguish between pollination and fertilization.

3. State the physical basis that prevents one species of pollen grain from polli-
    nating the stigmas of another species.

4. Explain where female sex organs form in a flowering plant. Then describe, begin-
    ning with megaspore production, all of the steps and structures formed along the
    way to producing egg and endosperm nuclei.

Self-Quiz Questions

  Matching

(1) ____ carpel   A. a tiny opening through the integuments

(2) ____ generative cell   B. will become a seed especially if fertilization occurs

                                C. female sex organs

(3) ____ micropyle   D. male sex organs

(4) ____ microspore mother   E. undergo meiosis
         cell
                                F. will divide mitotically to form two sperm nuclei

(5) ____ ovule   G. immature male gametophyte

(6) ___ pollen grain

(7) ___ stamen

18–III   POLLINATION AND FERTILIZATION

Summary

Pollen grains (the immature male gametophytes) are transferred from the anthers to the
female gametophytes, where they deliver their mature sperms. One sperm nucleus fuses
with the egg to form a *zygote*, which, through mitotic division, eventually develops
into the plant *embryo*. The other sperm nucleus fuses with nuclear material in the ovule
and develops into *endosperm*, a mass of tissue that surrounds and nurtures the developing embryo.

Many successful plants (grasses, oaks, birches, and maples) are wind pollinated;
their flowers typically don't have nectar or perfume and tend to lack colorful petals.
Many flowering plants, however, depend on insects, birds, or bats to disperse their
pollen. Colors, patterns, and odors are important attractants for many pollinators.
Once a pollinator locates a flower, color patterns and petal shapes guide it to the
nectar and channel the pollinator's movements in ways that aid pollination; for example, petals of hummingbird-pollinated flowers often form a long tube, which correspond
to bill length and shape. The more refined the "fit" between plant and pollinator, the
more efficient pollination can be. But the more specialized the plant-pollinator relationship, the greater the chance that the plant may face extinction if its pollinator
should happen to disappear. However these interactions developed, diverse species of
pollinators and flowering plants have been interdependent since about the dawn of the
Cenozoic, about 65 million years ago.

Key Terms

pollination (p. 280)	primary endosperm	Cenozoic (p. 282)
generative cell (p. 280)	cell (p. 280)	hybridization (p. 282)
micropyle (p. 280)	triploid (p. 280)	propagation (p. 282)
double fertilization (p. 280)	Cretaceous Period (p. 280)	
endosperm mother cell (p. 280)	Mesozoic (p. 280)	

Objectives

1. Describe fertilization as it occurs in a flowering plant.

2. Identify the parts of the flower in Figure 18.3 and locate the sites of the (a)
   megaspore mother cell, (b) megaspore, (c) female gametophyte, (d) egg cell, (e)
   pollen grain, (f) pollen tube, (g) microspore, (h) immature male gametophyte, and
   (i) mature male gametophyte.

3. Define *double fertilization* and *endosperm*.

4. Explain how seed formation and fruit development are related.

5. State when during geologic history the flowering plants became established and tell which other groups of organisms paralleled their expansion.

6. Explain how some insect (or bird or bat) populations helped some plant populations to compete more effectively for nutrients, living space, and water. Then tell how those same plant populations helped those insect (or bird or bat) populations.

7. Summarize the nature of the plant-pollinator coevolutionary relationship.

Self-Quiz Questions

Fill-in-the-Blanks

Pollen grains are transferred from (1) _____ to (2) _____. The mature male gametophyte is the (3) _____ _____, which grows into the ovule and delivers two (4) _____ _____. One of the (4) _____ fuses with the egg and forms the (5) _____; the other fuses with nuclear material and develops into (6) _____, which surrounds and nourishes the developing embryo. Fully mature ovules are (7) _____; the ripened ovary of one or more carpels constitutes most or all of a (8) _____.

Flowering plants appeared on Earth about (9) _____ million years ago; by the close of the (10) _____ era, about 65 million years ago, they had become abundant and were found in diverse environments. Bird-pollinated flowers tend to be (11) _____. Bee-pollinated flowers tend to be blue or (12) _____, with prominent (13) _____ components.

18–IV EMBRYONIC DEVELOPMENT

Summary

Following double fertilization, the *ovule* of an angiosperm expands. Integuments thicken and harden, forming the *seed coat*. Inside, the zygote develops into an embryo that consists of diploid cells. Nutritive endosperm is formed by repeated mitotic divisions and consists of triploid cells. Fully mature ovules are *seeds*. The ripened ovary of one or more carpels constitutes most or all of a *fruit*. The largest mature fruit weighs 20 kilograms. Under appropriate conditions, seeds germinate and the reproductive phase is completed. Examination of a variety of embryonic plant tissues shows that they contain different amounts of certain organelles; hence transcription and translation of the genes coding for these proteins, as well as enzyme activity, vary in different parts of the developing plant.

Key Terms

zygote (p. 284)
polarization (p. 284)
Capsella, shepherd's
  purse (p. 284)
suspensor (p. 284)
embryo (p. 284)
ovule (p. 285)
integuments (p. 285)
endosperm (p. 285)

seed (p. 285)
seed coat (p. 285)
cotyledons (p. 285)
germinate (p. 285)
dicot (p. 285)
monocot (p. 285)
fruit (p. 285)
Coco-de-mer (p. 285)
coleoptile (p. 285)

aggregate fruit (p. 285)
multiple fruit (p. 285)
sepals (p. 285)
stamen (p. 285)
dry fruits (p. 286)
fleshy fruit (p. 286)
simple fruit (p. 286)
receptacle (p. 286)

Objectives

1. Define *polarization* and explain how it affects the fate of the two daughter cells that result from a flowering plant zygote undergoing cell division. State in your own words how cytoplasmic constituents of daughter cells can affect differentiation.

2. Explain how a flower is related to a fruit and how an ovule is related to a seed.

3. Describe the role(s) played by the cotyledon(s) of flowering plants.

4. Distinguish simple, aggregate, and multiple fruits. Provide an example of each.

5. Describe some of the strategies that have evolved in plants that decrease the competition between parent plants and offspring.

Self-Quiz Questions

Fill-in-the-Blanks

The (1) _____ seed is contained within the ovary as it develops. Palms, lilies, and orchids are examples of (2) _____; roses, beans, oaks, and squash are examples of (3) _____, so-called according to the number of (4) _____ in each seed that secrete (5) _____ _____ into the endosperm and absorb digested products in return. A fully matured ovule is a seed; the (6) _____ surrounding it develops into most or all of a fruit. (7) _____, which is triploid tissue, results when a single sperm fuses with two other nuclei in the (8) _____ during double fertilization in flowering plants. Grasses, oaks, birches, and maples are (9) _____-pollinated; their (10) _____ typically don't have perfume or nectar and are not colorful. (11) _____ is the unequal distribution of cytoplasmic substances in the zygote; when this happens, the two daughter cells generally develop into radically different types of tissues.

Summary

Asexual reproduction in plants occurs by several mechanisms: (1) reproduction on modified stems (runners, rhizomes, corms, tubers, and bulbs); (2) *parthenogenesis*, (3)
vegetative reproduction, and (4) tissue culture propagation. By means of *tissue culture*,
thousands of identical plants can be propagated from that one specimen.

Key Terms

runners (p. 286)   rhizome (p. 287)   tissue culture
parthenogenesis (p. 286)   corm (p. 287)   propagation (p. 287)
vegetative   tuber (p. 287)   axillary bud (p. 287)
  reproduction (p. 286)   bulb (p. 287)
clone (p. 286)   adventitious (p. 287)

Objectives

1. Distinguish between parthenogenesis and vegetative reproduction.

2. Name and give one example each of four asexual modes of reproduction by modified
    stems. Describe each of the four modified stems.

3. Outline the steps involved in tissue culture and the production of plant clones.

Self-Quiz Questions

   Fill-in-the-Blanks

An embryo that develops from an unfertilized egg is an example of (1) _____ .

(2) _____ along the stems of many plants will develop new roots and shoots if

cultivated properly. Onions and lilies can grow from (3) _____ , which are exam-

ples of underground stems; new strawberry plants can grow from (4) _____ which

are above-ground horizontal stems. Asexually-produced offspring are genetically identi-

cal with their parents and are called (5) _____ .

INTEGRATING AND APPLYING KEY CONCEPTS

Unlike the growth pattern of animals, the vegetative body of vascular plants continues
growing throughout life by means of meristematic activity. Explain why most representatives of the plant kingdom probably wouldn't have survived and grown as well as they do
if they had followed the growth pattern of animals.

19
PLANT GROWTH AND DEVELOPMENT

19–I PLANT GROWTH: ITS NATURE AND DIRECTION

Genetic Controls Over Development

Summary

In plants, development involves controlled activation and deactivation of genes. Specific enzymes required in the formation of particular cell structures are produced in
response to signals from controlling genes at the appropriate times. Transcription and
translation of the genes coding for these proteins, as well as gene activity, vary in
different parts of the developing plant. Polarizing influences, unequal cell divisions,
differing orientations to their environment, and differing receptor sites cause the
different cells to interact in different ways. Some *hormones* (chemical messengers that
are synthesized and released from certain cells) either stimulate or inhibit protein
synthesis within *target cells*. Many of the proteins synthesized are *enzymes*, which
are required in the formation of cellular structures and substances and are produced
at particular times according to the developmental program.

Key Terms

polarization (p. 289)
germinate (p. 289)
specialized (p. 289)

microtubules (p. 289)
transcription (p. 289)
translation (p. 289)

hormones (p. 289)
target cell (289)

Objectives

1. Explain how, even though all plant embryo cells contain the same DNA code, plant cells can become specialized into different kinds of cells.
2. List three factors that cause different cells to interact in different ways.
3. Describe how hormones are related functionally to enzymes.

Self-Quiz Questions

Fill-in-the Blanks

After (1) ____________ of a cell, various organelles are positioned at one end of the cell and a vacuole occupies most of the cell volume at the other end. (2) ____________ is a process in which a plant embryo divides and daughter cells grow to form first roots, then a stem and leaves. (3) ____________ are assembled from protein subunits with the aid of enzymes. Specific (4) ____________ are chemical messengers that inhibit or stimulate protein synthesis within certain types of (5) ____________ ____________.

19–II Patterns of Cell Enlargement 289-291

Summary

The whole seed swells with incoming water molecules, which hydrogen bonding has attracted to proteins and polysaccharides stored in the seed. In the seeds of most species, the first cells to grow are in the embryonic root (the radicle), and they grow longitudinally. A slender primary root results; when it protrudes from the seed, germination is complete. Plants grow because their cells grow. On the average, about half of the daughter cells formed enlarge—often by twenty times. The other half remain meristematic; they grow only as large as the mother cell, then probably divide again. Cell growth is driven by water uptake and the ensuing turgor pressure against the cell wall. To a large extent, the arrangement of polysaccharides in cell walls dictates the direction in which the wall yields under turgor pressure. Therefore, how the wall polysaccharides are oriented governs whether a cell becomes long and slender or short and broad. If most cells in a plant organ are elongated, so is the organ; this is true of roots and stems. If most are spherical, so is the organ; this is true of many fruits. The arrangement of cellulose molecules in a cell wall controls the direction of cell growth, and cell shapes resulting from directional growth help determine plant form. The expansive properties of primary cell walls permit cell enlargement.

Key Terms

coleoptile (p. 289)
adventitious (p. 289)
radicle (p. 290)
primary root (p. 290)
lateral roots (p. 290)
cotyledons (p. 290)

hypocotyl (p. 290)
node (p. 290)
turgor pressure (p. 291)
polysaccharides (p. 291)
cellulose microfibrils (p. 291)
meristematic cells (p. 291)

parenchyma (p. 291)
pith (p. 291)
cortex (p. 291)
microtubular growth (p. 291)
elastic (p. 291)
plastic (p. 291)

Objectives

1. Compare the growth of corn (a monocot) with that of a soybean plant (a dicot).

2. Explain how the arrangement of cellulose molecules in a cell wall controls the nature and direction of plant cell growth.

3. State the mechanism by which some cells remain meristematic and others enlarge and differentiate.

4. Distinguish elastic from plastic.

Self-Quiz Questions

Fill-in-the-Blanks

Corn plants develop (1) _____ roots, but each soybean plant develops a single

embryonic root called the (2) _____, which develops into the primary root that

later sends out lateral branches. On the average, about half of the daughter cells (3)

_____, often by twenty times; the other half remain (4) _____. Cell

growth is driven by (5) _____ uptake and the ensuing (6) _____ _____

against the cell wall. To a large extent, the arrangement of (7) _____ in cell

walls dictates the direction in which the wall yields under pressure. (8) _____

growth is thought to somehow govern the direction in which cellulose microfibrils be-

come deposited during the formation of cell walls.

19–III PLANT HORMONES

Types of Plant Hormones

Examples of Hormonal Action

Summary

Plant hormones are information-carrying messenger molecules that cause certain cell types to change their activities in response to environmental conditions. Plant hormones are less specific than animal hormones with respect to the physiological responses they activate, and the target tissues themselves may be less specific. Five hormones (or groups of hormones) have been identified in vascular plants. *Auxins* are best known for their ability to stimulate cell elongation in coleoptiles and stems. 2, 4-D is a synthetic auxin that is used to kill dicot weeds. *Gibberellins*, like auxins, promote stem elongation in plants; they have been identified in fungi. *Cytokinins* stimulate cell division, promote leaf cell expansion, and retard leaf aging. *Abscisic acid* stimulates stomate closure, is probably involved in root gravitropism, may bring about seed and bud dormancy, and may promote leaves, flowers, and fruits to drop from the parent plant. *Ethylene* is a simple hydrocarbon gas that stimulates fruit ripening. The growth of a coleoptile (a hollow, cylindrical organ that protects tender young leaves growing within it) is controlled by IAA (an auxin) synthesized in its tip. The hormone moves down the coleoptile and increases cell wall plasticity, which encourages elongation. Auxin synthesized at a dicot stem tip is essential for the growth of cells below. The growth of stems that elongate slowly is especially influenced by gibberellin. In contrast to coleoptiles and dicot stems, root and leaf cells themselves synthesize most or all of the hormones they require.

Key Terms

plant hormones (p. 291)
target tissue (p. 291)
auxins (p. 292)
gibberellins (p. 292)
cytokinins (p. 292)
abscisic acid, ABA (p. 292)
ethylene (p. 292)

indoleacetic acid, IAA (p. 292)
phenylacetic acid (p. 292)
horticulturist (p. 292)
2, 4-D (p. 292)
gibberellic acid (p. 292)
zeatin (p. 292)
adenine (p. 292)

abscission (p. 292)
stomate (p. 292)
dormancy (p. 292)
hydrocarbon (p. 292)
emanation (p. 292)
coleoptile (p. 292)

Objectives

1. State what plant hormones are and, in general, what they do.

2. Identify the five principal plant hormones (or hormone groups) and state the known or suspected effects of each.

3. List three examples of specific plant hormones and describe how each hormone affects the different parts of a plant.

Self-Quiz Questions

Fill-in-the-Blanks

Plant (1) _____ are information-carrying messenger molecules that cause certain cell types to change their activities in response to environmental conditions. (2) _____ and (3) _____ both promote elongation of cells in stems. (4) _____ stimulate cells to divide. (5) _____ _____ promotes the closure of stomata and is probably involved in root (6) _____. (7) _____ promotes the ripening of fruit. The growth of a coleoptile is controlled by (8) _____ synthesized in its tip.

19-IV HORMONES, THE ENVIRONMENT, AND PLANT DEVELOPMENT

Summary

Both temperature and sunlight influence the growth of coleoptiles, stems, roots, and leaves. Environmental temperature affects cell metabolism and transport of water, ions, and sucrose. Especially for dicots, exposure to sunlight promotes leaf expansion and stem branching but inhibits stem elongation; *phytochrome*, a blue-green pigment, receives light energy and seems to be an on-off switch for expansion, branching, and elongating activities, and, in many plants, for seed germination and flowering as well. *Plant tropisms* are familiar but as yet unexplained growth responses in which an environmental stimulus causes the plant to respond with a faster rate of cell elongation on one side or the other. In gravitropism (a response to Earth's gravitational force), the root cap synthesizes the growth inhibitor abscisic acid; transport of ABA to lower growing portions of a horizontally positioned root causes the downward curvature of the growing root tip. *Phototropism* is caused by light coming in mainly from one side. IAA moves to the shaded side and downward, where it causes stems to curve toward the light, and the flat surface of a leaf blade becomes perpendicular to the light. *Thigmotropism* is a positive and *thigmomorphogenesis* a negative growth response to contact.

Key Terms

phytochrome (p. 294)
inactive phytochrome, Pr (p. 294)
red light (p. 294)
active phytochrome, Pfr (p. 294)
far-red light (p. 294)
carotenoid pigments (p. 295)

tropisms (p. 295)
stimulus (p. 295)
phototropism (p. 295)
gravitropism (p. 296)
IAA (p. 296)
flavoprotein (p. 296)
riboflavin (p. 296)
primary wall (p. 296)

middle lamellae (p. 296)
growth inhibitor (p. 296)
thigmotropism (p. 297)
thigmomorphogenesis (p. 297)
auxins (p. 297)
gibberellins (p. 297)
abscisic acid (p. 297)

Objectives

1. Explain how temperature influences the growth of coleoptiles, stems, roots, and leaves.

2. Describe how sunlight influences plant growth by explaining how phytochrome responds to specific oolors of light and how seed germination, stem elongation, leaf expansion, stem branching, and flowering are affected by the activated form of phytochrome.

3. Define *geotropism*, state the cellular and hormonal events that bring it about, and indicate how geotropism promotes the survival of a plant.

4. Define *phototropism*, state the cellular and hormonal events that bring it about, and indicate how phototropism promotes the survival of the plant.

5. Distinguish thigmotropism from thigmomorphogenesis and indicate how each promotes the survival of specific plant types.

Fill-in-the Blanks

For dicots, exposure to sunlight (1) _____ leaf expansion and stem branching

but (2) _____ stem elongation. (3) _____ seems to be an on-off switch

for hormone activities that govern leaf expansion, stem branching, stem length, and,

in many plants, seed (4) _____ and flowering. Before the sun rises, phytochrome

exists mainly in an (5) ○ active ○ inactive form that absorbs (6) ○ red ○ far-red

light and is converted to the (7) ○ active ○ inactive form (Pfr). In the shade, at

sunset, or at night, (8) ○ Pr is converted to Pfr ○ Pfr is converted to Pr.

Matching Choose the one most appropriate answer for each.

____ (9) gravitropism

____ (10) phototropism

____ (11) thigmotropism

____ (12) thigmomorphogenesis

A. response mechanism is not yet known

B. thought to be controlled by abscisic acid

C. thought to be controlled by an auxin

D. known to be controlled by indoleacetic acid and blue light

19–V   THE FLOWERING PROCESS

SENESCENCE

DORMANCY

Summary

As a flowering plant matures, its physiological processes become directed toward flower, fruit, and seed production. Plants such as corn, soybeans, and peas live only one
growing season; they are known as *annuals*. Plants that live for many growing seasons
are *perennials*. Still other species produce only roots, stems, and leaves the first
growing season, die back to soil level in autumn, then grow a new flower-forming stem
from a bud. These plants that typically live for two growing seasons are *biennials*;
they include cabbages, carrots, and turnips.

Daylength and low temperature are the strongest stimuli for flowering of land
plants. Low-temperature stimulation of flowering is called *vernalization*. Daylength
cues flowering and reproduction in many species. *Long-day* plants are adapted to long
days and reproduce early in summer. *Short-day* plants are attuned to shorter day lengths
of late summer and flower in autumn. *Day-neutral* plants flower almost independently of
daylength. A relationship between the pigment phytochrome and an elusive hormone,
florigen, is supposed to exist but has never been demonstrated conclusively.

All of the processes that lead to the death of a plant or any of its organs are
called *senescence*. One stimulus for senescence could be the drain of nutrients during
the growth of reproductive organs, but some sort of "death signal" that forms during
short days is also suspected. When any plant stops growing under physical conditions
that are actually quite suitable for growth, it is said to have entered a period of
dormancy. Abscisic acid movement from leaves to buds in late summer may trigger

dormancy, and abscisic acid breakdown along with gibberellin accumulation in buds during late autumn and winter may end dormancy.

Key Terms

annual (p. 297)
perennial (p. 297)
biennial (p. 297)
vernalization (p. 297)
long-day plants (p. 298)
short-day plants (p. 298)
day-neutral plants (p. 298)
Pfr (p. 298)

florigen (p. 298)
reproductive (p. 299)
vegetative (p. 299)
parenchyma (p. 299)
abscission (p. 299)
ethylene (p. 299)
abscisic acid (p. 299)
senescence (p. 299)

cytokinins (p. 299)
dormancy (p. 300)
Pr (p. 301)
dormancy-breaking
  mechanism (p. 301)
gibberellin (p. 301)
abraded (p. 301)
germination (p. 301)

Objectives

1. Define *annual, perennial,* and *biennial.*

2. Identify factors that stimulate flowering in land plants. Present the reasons some botanists believe there must be a flowering hormone named *florigen.*

3. Distinguish long-day, short-day, and day-neutral plants and give an example of each.

4. Identify factors that bring about senescence in plants.

5. Distinguish between senescence and dormancy, and identify factors that promote dormancy.

6. Explain how dormancy can promote the survival of plants.

7. Identify factors that help to break dormancy.

Self-Quiz Questions

Fill-in-the-Blanks

Plants that typically live two growing seasons are called (1) _____, whereas

plants that live year after year are (2) _____. (3) _____ is a highly

reliable cue for flowering and reproduction of many species. (4) _____ organs

withdraw nutrients from vegetative organs through connecting (5) _____ _____.

In deciduous trees, nutrients are transported to reproductive organs and to (6)

_____ cells in twigs, stems, and roots prior to abscission. (7) _____

is the principal substance that stimulates abscission. Any plant that stops growing

under physical conditions that are actually quite suitable for growth is said to have

entered a period of dormancy. (8) _____ _____ movement from leaves to

buds in late summer may trigger dormancy; its breakdown, along with gibberellin accumu-

lation in buds in late autumn and winter, may end dormancy.

Case Study: From Embryogenesis to the Mature Oak

Summary

Time-measuring devices that allow all eukaryotes, including plants, to anticipate and
adjust to environmental changes are called biological clocks. These clocks control
rhythmic activities in the absence of any apparent external cues (*endogenous* rhythms).
The clocks also operate in response to environmental cues (*exogenous* rhythms). Circadian rhythms occur daily; *Lunar* rhythms every 28 days; *Annual* rhythms every 365 days.
An organism such as the Coast Live Oak develops and grows in an ecosystem according to
interactions between environmental cues and its own genotype. Humans, in their ignorance,
of the requirements of other organisms, often sentence their intended neighbors to
death.

Key Terms

endogenous (p. 302)   pericycle (p. 302)
exogenous (p. 302)   mycorrhizae (p. 303)
circadian (p. 302)   phytochrome (p. 303)

Objectives

1. Distinguish endogenous from exogenous rhythms.

2. Describe annual, circadian, and lunar rhythms.

3. List three things that humans unwittingly did to hasten the death of the live oak.

Self-Quiz Questions

   Fill-in-the-Blanks

(1) _____ rhythms are not cued by environmental factors, but (2) _____

rhythms are. (3) _____ is a pigment that triggers hormonally-induced branching

and leaf expansion by monitoring day lengths. (4) _____ are fungal strands

symbiotically intermingled with tree roots that enhance the absorption of water, es-

sential ions, and gases.

INTEGRATING AND APPLYING KEY CONCEPTS

An oak tree has grown up in the middle of a forest. A lumber company has just cut down
all of the surrounding trees except for a narrow strip of woods that includes the oak.
How will the oak be likely to respond as it adjusts to its changed environment? To what
new stresses will it be exposed? Which hormones will most probably be involved in the
adjustment?

CROSSWORD NUMBER THREE

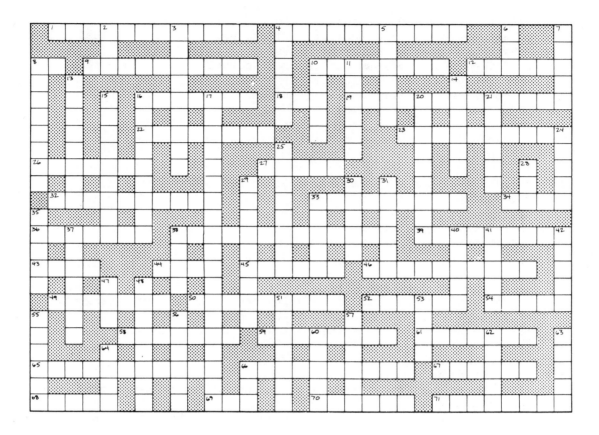

Across

1. not vegetative
4. the transfer of the immature male gametophyte
8. inactive phytochrome
9. chiefly conifers in this group
10. slope
12. ground tissue located between the vascular ring and the dermal surface
16. a reversible form of growth
18. the most widely encountered auxin
19. evaporation from stems and leaves
22. fusiform ___ produce xylem and phloem cells that are arranged parallel with the stem's long axis
23. the portions of the stem between the nodes
26. embryonic root
27. one of the two groups of angiosperms
32. gives rise to vascular tissue
33. the exterior region of any body
34. a straight line about which a body or geometrical object rotates
36. elements of xylem
38. low-temperature stimulation of flowering

39. components of xylem
43. the below-ground portion of a plant
44. ground tissue located within a ring of vascular bundles
45. functions in food conduction, storage, and support
46. stimulate cell division; promote leaf cell expansion and retard leaf aging
49. meristem that runs parallel with the sides of roots and stems
50. ___ fruit: formed from numerous but separate carpels of a single flower
52. female reproductive organ in a flower
54. stalk that connects the ovary with the stigma
58. male reproductive organs in flowers
59. circulatory
61. a ___ system has a primary root and its various lateral branchings
65. flowering plants
66. seed leaves
67. at the peak of development; generally able to reproduce
68. the above-ground part of a plant
69. plant fluid
70. a group of interacting elements
71. sticking together

Down

2. the ___ system acts as a system of conduits and food storage centers in wood
3. the ___ strip is a continuous band of fatty suberin deposits
4. a hydrophilic substance found in the cell walls of collenchyma
5. of, or related to, the apex
6. active phytochrome
7. a class of plant growth hormones
8. the ___ flow theory says that translocation depends on pressure gradients in the sieve-tube system
10. a small, hard seed or fruit, especially that produced by a grass such as rice, corn, or wheat
11. a pollen-bearing structure
13. the vascular ___ is surrounded by cortex in a plant root
14. long cells that have somewhat flexible walls
15. ground tissue between the phloem and endodermis in roots and stems that becomes part of the active cambium
16. surface layer of cells
17. generally, overall plant growth is inhibited as a response to contact
20. leading to the death of a plant or any of its organs
21. a plant that lives for one growing season
24. fertilized ovules
25. plant that typically lives for two growing seasons
28. any of a particular class of fatty derivatives that are insoluble in water
29. will develop into an egg-bearing female gametophyte
30. the ripened ovary of one or more carpels
31. ___ parenchyma lies between the palisade tissue and the lower cuticle-covered epidermis
33. a sieve ___ has larger pores than do the side walls
35. a multicellular organ that produces eggs
37. gates across the epidermis of leaves and stems
38. a vascular structure
40. the ___ system of secondary vascular tissues conducts food and water up and down the stem
41. root ___ are absorptive structures
42. ___ elements and companion cells are components of phloem

47. abscisic acid
48. a ___ fruit is formed from the carpels of several associated flowers
51. a part of one plant is joined to another at an incision made in a stem or root
53. ___ are on the side walls of vessel elements
55. conspicuously colored leaves
56. an organism that is in an undifferentiated state before it has achieved a distinctive recognizable form
57. the flattened portions of a leaf
60. the simplest complete living systems
62. structure within which megaspores are produced
63. a waxy waterproofing substance
64. a seed ___ protects against desiccation
66. the root ___ protects against abrasion during growth

20
SYSTEMS OF CELLS AND HOMEOSTASIS

SOME CHARACTERISTICS OF
ANIMAL CELLS AND TISSUES

KINDS OF ANIMAL TISSUES

    Epithelial Tissues
    Connective Tissues
    Nerve Tissues
    Muscle Tissues

OVERVIEW OF ORGAN SYSTEMS
AND THEIR FUNCTION

HOMEOSTASIS AND SYSTEMS CONTROL

    The Internal Environment
    Homeostatic Control Mechanisms

20–I SOME CHARACTERISTICS OF ANIMAL CELLS AND TISSUES 306-308

Summary

Individual cells can recognize and adhere to each other because patterns of surface receptors on the plasma membrane serve as adhesion sites. Chemical and structural bridges established at the molecular level unite cells in form and function as a cohesive *tissue*. Tissues that work together to perform a task form an *organ*. Organs acquire their particular character from the tissues of which they are composed and the ways they are joined together. All the diverse body parts found in different animals may be assembled from a few tissue types simply through variations in how those tissues are combined and arranged. *Somatic cells* constitute the physical structure of the animal body, and *germ cells* develop into gametes. Somatic cells become differentiated into four main tissue types: epithelial, connective, muscular, and nervous tissues. In animal embryos, ectoderm, mesoderm, and endododerm are unspecialized layers of tissues that will differentiate into the various specialized tissues. Ectoderm forms the tissues of the nervous system and outer skin; mesoderm forms muscle, skeletal, and circulatory tissue, in addition to other tissues. Endoderm becomes the gut lining and digestive glands.

Key Terms

sponge (p. 306)
tissue (p. 306)
Humphreys and Moscona (p. 306)
rudimentary (p. 306)
adhesion (p. 306)
somatic cells (p. 306)

germ cells (p. 306)
embryo (p. 306)
primordial (p. 307)
ectoderm (p. 307)
mesoderm (p. 307)
endoderm (p. 307)

surface receptors (p. 307)
tight junctions (p. 307)
adhering junctions (p. 307)
gap junctions (p. 307)
epithelial tissues (p. 307)
organ (p. 307)
organ system (p. 308)

Objectives

1. Describe Humphrey's and Moscona's sponge research in the 1960s and indicate the significance of the results.
2. Explain the nature of three different cell-to-cell junctions and state the types of tissues in which these junctions occur.
3. Distinguish *tissues* from *organs* and give an example of each structure.
4. List the various types of tissues that are assembled into animal organs, organ systems, and organisms.

Self-Quiz Questions

Fill-in-the-Blanks

Groups of *like* cells that work together to perform a task are known as a(n) (1)_____.

Groups of *different* types of tissues that interact to carry out a task are known as

a(n) (2) _____. All of the diverse body parts found in different animals

may be assembled from a few (3) _____ types through variations in the way

they are combined and arranged. (4) _____ cells constitute the physical struc-

ture of the animal body; they become differentiated into the components of four main

types of tissue: (5) _____, (6) _____, (7) _____ and (8)

_____.

20–II KINDS OF ANIMAL TISSUES   308–313

Epithelial Tissues   308–309

Summary

Epithelial tissues cover the external body surface of all animals, and they also line internal organs from gut cavities to vertebrate lungs. These tissues are sheets of densely-packed cells with little space or intercellular material at intercellular junctions. They form a continuous barrier between the body parts they cover and the surrounding medium. There may be one flat layer of cells, or the tissue may be strati-fied into different layers. Cells in epithelial tissues may resemble flat floor tiles, cubes, or columns. Functionally, epithelium may serve as a covering tissue or glandu-lar tissue. Simple epithelium blocks out some substances and facilitates passage of some other substances across the tissue. Stratified epithelium protects underlying tissues. Glandular epithelium includes endocrine glands that secrete hormones into the bloodstream and exocrine glands that secrete their various products (digestive enzymes, mucus, saliva, wax, oil, milk) through ducts that open out onto an epithelial surface.

Objectives

1. Define what is meant by *epithelial tissue*.

2. Describe how the general structure of epithelial tissues permits it to carry out its principal functions.

3. Identify and distinguish each of the three structural types of epithelial tissue.

4. Characterize both functional types of epithelial tissue and state the general location of each type.

5. Explain the meaning of the term *gland*, cite three examples of glands, and state the extracellular products secreted by each.

Self-Quiz Questions

True-False If false, explain why.

____ (1) Mammalian skin contains squamous epithelium.

____ (2) The more tight junctions there are in a tissue, the more permeable the tissue will be.

____ (3) Endocrine glands secrete their products through ducts that empty onto an epithelial surface.

____ (4) Endocrine cell products include digestive enzymes, saliva, and mucus.

____ (5) Goblet cells are multicelled flask-shaped exocrine glands.

20–III Connective Tissues   310-313

Summary

Connective tissue cells are usually scattered through an extensive extracellular ground substance; these tissues bind together and generally support other animal tissues. Loose connective tissue functions as a packing material composed of strong, flexible collagen fibers and a few highly elastic *elastin* (a protein) fibers. Dense fibrous tissue attaches and holds movable body parts together; it consists generally of parallel collagen fibers with little ground substance between them. *Tendons* and *ligaments* are made of dense fibrous tissue. *Adipose* tissue consists of large cells, each containing a single fat-filled vacuole. Fat contains stored energy, cushions parts of the body against shocks and blows, and helps to retain body heat.

Cartilage and *bone* are supportive connective tissues. Cartilage has small clusters of living cartilage cells suspended in the firm, rubbery ground substance they secreted. *Bone* consists of living bone cells embedded in a dense collagen-rich ground substance they secreted; calcium salts strengthen and make the collagen matrix rigid. The long bones of mammals contain red marrow, a major site of blood cell formation, and yellow marrow, a reserve tissue.

Blood consists of a fluid ground substance (*plasma*) and free cells; it transports substances to and from body cells, regulates pH, and helps to distribute body heat, nutrients, water, wastes, hormones, and infection-fighting cells.

Key Terms

connective tissues (p. 310)
connective tissue proper
  (p. 310)
ground substance (p. 310)
fibroblast (p. 310)
loose connective tissue
  (p. 310)
collagen (p. 310)
elastin (p. 310)
blood (p. 310)
plasma (p. 310)

platelets (p. 310)
dense connective tissue
  (p. 311)
tendons (p. 311)
ligaments (p. 311)
adipose tissue (p. 311)
cartilage (p. 311)
intervertebral disks
  (p. 311)
bone (p. 311)
osseous tissue (p. 311)

calcium phosphate (p. 311)
calcium carbonate (p. 311)
spongy bone tissue (p. 311)
latticelike (p. 311)
red marrow (p. 311)
yellow marrow (p. 312)
compact bone tissue (p. 312)
lamellae (p. 312)
Haversian canal (p. 312)
osteoblasts (p. 312)
osteocytes (p. 312)
osteoclasts (p. 312)

Objectives

1. Describe the basic features of connective tissue and explain how they enable connective tissue to carry out its various tasks.

2. List each type of connective tissue and characterize each in terms of its component materials, its location in the body, and its function.

3. Explain how bone is shaped and reshaped through the lifespan of a large mammal.

4. State the functions of marrow and adipose tissue.

5. List the components of blood and three of its functions.

Self-Quiz Questions

Fill-in-the-Blanks

Cells of connective tissues are scattered through an extensive extracellular (1) _____ _____. (2) _____ connective tissue contains a weblike scattering of strong, flexible protein fibers ((3) _____) and a few highly elastic protein fibers ((4) _____) and serves as a packing material that holds in place blood vessels, nerves, and internal organs. (5) _____ and (6) _____ are examples of dense fibrous tissue that helps to attach together elements of the skeletal and muscular systems. (7) _____ and (8) _____ are examples of supportive connective tissue. (9) _____ systems are found in the long bones of mammals and contain living bone cells that receive their nutrients from the blood. (10) _____ _____ is a major site of blood cell formation. The fluid matrix of blood is called (11) _____; suspended in it are the three types of formed elements: (12)

_____ _____ _____, which carry oxygen; (13) _____ _____

_____, which fight against infections; and (14) _____, which are involved

in the formation of blood clots. (15) _____ tissue serves to store fat as a food

reserve.

20–IV   Nerve Tissues 313
Muscle Tissues 313

Summary

Nerve cells have highly excitable plasma membranes. In most animals, nerve cells
(neurons) are arranged in organized networks concerned with receiving signals and
conducting nerve impulses to other neurons, muscles, or glands, causing them to respond by contracting or secreting substances.

   Muscle cells enable an animal to move in place or through the environment; contractile threadlike structures shorten upon chemical stimulation and move body parts
to which the muscle cells are attached. *Striated* (striped) muscle forms skeletal and
cardiac (heart) muscle tissues. *Smooth* muscle tissue is not banded, contracts slowly,
and is responsible for involuntary movements such as those associated with digestion,
circulation, and respiration.

Key Terms

nerve tissue (p. 313)	skeletal muscle tissue (p. 313)	involuntary (p. 313)
neurons (p. 313)	cardiac muscle tissue (p. 313)	muscle fibers (p. 313)
muscle tissue (p. 313)	contractile (p. 313)	muscle bundle (p. 313)
smooth muscle	striated (p. 313)	muscle (p. 313)
tissue (p. 313)	spindle-shaped (p. 313)	intercalated disks (p. 313)

Objectives

1. Explain how changes in plasma membranes provide the basis to receive and respond
   to signals in the environment.

2. Define neuron and describe the basic features of a true nervous system.

3. Describe the relationships among muscle fibers, muscle cells, muscle bundles,
   and muscles.

4. Distinguish skeletal, cardiac, and smooth muscle tissues in terms of location,
   structure, and function.

Self-Quiz Questions

   True-False   If false, explain why.

____ (1) Muscle fibers are skeletal muscle cells.

____ (2) Both skeletal and cardiac muscle tissues are striated.

___ (3) Cardiac muscle cells are fused, end to end, at regions called muscle bundles.

___ (4) Smooth muscle tissue contracts more slowly than striated muscle tissue.

___ (5) Smooth muscle tissue contracts voluntarily.

___ (6) Neurons conduct messages to another neuron(s) or to muscles or glands.

20–V OVERVIEW OF ORGAN SYSTEMS AND THEIR FUNCTION   314-316

Summary

Human skin is an example of an organ system composed of organs such as sebaceous glands, blood vessels, sweat glands, and so on. All four basic types of tissues are included in the skin, which is arranged in three principal layers: the epidermis, the dermis, and the subcutaneous layer. All of the specialized organs of the animal body are composed of two or more types of tissues, and all developed as adaptations to a specific kind of environment. Most animal groups have characteristic body plans with their own particular assortment of the eleven organ systems: body covering, nervous, endocrine, skeletal, muscular, circulatory, defense, respiratory, digestive, excretory, and reproductive.

Key Terms

skin (p. 314)	sebaceous (p. 315)	excretory (p. 315)
epidermis (p. 314)	differentiated (p. 315)	cranial cavity (p. 316)
keratin (p. 314)	follicle (p. 315)	thoracic cavity (p. 316)
subcutaneous (p. 314)	evagination (p. 315)	
dermis (p. 315)	integumentary system (p. 315)	

Objectives

1. Define *epidermis, dermis,* and *subcutaneous layer.*

2. List all of the various components found in each of the three skin layers.

3. Define and contrast *anatomy* and *physiology.*

4. List each of the eleven principal organ systems and match each with its main tasks.

5. State why the elimination of fecal material is not regarded as part of the excretory system.

Self-Quiz Questions

Fill-in-the-Blanks

Human skin is an organ system that consists of two layers: the outermost (1) _____,

which contains mostly dead cells and the middle (2) _____, which contains hair fol-

licles, nerves, tiny muscles associated with the hairs, and various types of glands.

The (3) _____ layer, with its loose connective tissue and stores of fat in (4)

_____ tissue lies beneath the skin. (5) _____ is the study of the major structural features—organs and tissues—of an organism; (6) _____ is the study

of how the organism functions. The excretory system of an animal is concerned with the

disposal of (7) _____ wastes; fecal material is not considered in this category.

The endocrine system is generally concerned with internal (8) _____ control; to-

gether with the (9) _____ system, it integrates physiological processes.

20–VI HOMEOSTASIS AND SYSTEMS CONTROL 316-318

The Internal Environment 316

Homeostatic Control Mechanisms 316-318

Summary

Individual cells carry out the same general kinds of activities as multicellular
organisms. However, because most cells in the multicellular animal body lack direct
access to the resources of the external environment, the complex of organ systems work
together to maintain a stable environment within the body that supplies resources and
ensures favorable conditions. Each cell must carry out its own tasks even as it per-
forms some special activity that stabilizes the internal environment of the whole
organism. Extracellular fluid bathes the cells and exchanges substances with them,
much as the ocean washes over sponges and keeps their cells alive. In vertebrates,
most extracellular fluid is interstitial (occupying spaces between cells and tissues).
The remainder is blood plasma (which is contained in blood vessels and the heart)
and lymph (which is contained in lymph vessels and lymph nodes). Organisms maintain
internal conditions through systems of homeostatic control, which try to preserve
physical and chemical aspects of the body's interior within specific ranges of tolerance. In *negative feedback mechanisms*, deviations from stable conditions activate reactions that work to counteract or reverse the change. *Positive feedback mechanisms*,
in which some disturbance to the homeostatic state activates a chain of events that
throws things even further off balance, are also called into play in such situations
as the formation of a blood clot or the birth process. Homeostatic feedback mechanisms have three basic components: receptors, integrators, and effectors. Receptors
detect specific types of stimuli (energy changes in the environment). Integrators are
control points where responses to the stimulus are selected and signals are sent to
the organism's effectors (the muscles and glands), which respond to the stimuli.
Feedforward mechanisms detect small environmental changes at their surface receptors
and send messages to organs in the body that can begin corrective measures before
the change in external environment significantly alters the internal environment.

Key Terms

Bernard (p. 316)
extracellular fluid (p. 316)
interstitial fluid (p. 316)
plasma (p. 316)
liver (p. 316)
glycogen (p. 316)
Cannon (p. 316)

homeostatic systems
  (p. 317)
negative feedback
  mechanisms (p. 317)
thermostat (p. 317)
receptors (p. 317)
integrators (p. 317)

effectors (p. 317)
integrative centers (p. 317)
positive feedback
  mechanisms (p. 317)
oxytocin (p. 317)
feedforward
  mechanisms (p. 317)

Objectives

1. Explain how, if each cell can perform all of its basic activities, organ systems contribute to cell survival.

2. Define *extracellular fluid* and *interstitial fluid*.

3. Explain how extracellular fluid helps cells to survive.

4. Draw a diagram that illustrates the mechanism of homeostatic control.

5. Distinguish between negative and positive feedback mechanisms. Give an example of each in human physiology.

6. Describe the relationships among receptors, integrators, and effectors in a negative feedback system.

7. Explain how feedforward mechanisms refine homeostasis.

Self-Quiz Questions

Fill-in-the-Blanks

(1) _____ fluid fills the spaces between cells and tissues in vertebrates.

Bernard discovered that the (2) _____ absorbs many of the nutrients carried

to it by blood and converts them to complex storage forms; he also showed that the

amount of blood being supplied to different body regions can be varied by the (3)

_____ and (4) _____ of small blood vessels. Blood, with its cargo of

(5) _____ and oxygen, can be diverted to those regions where it is needed

most.

True-False If false, explain why.

___ (6) The process of childbirth is an example of a negative feedback mechanism.

___ (7) The human body's effectors are, for the most part, muscles and glands.

___ (8) An integrator is constructed in such a way that it detects specific energy changes in the environment and relays them to a receptor.

INTEGRATING AND APPLYING KEY CONCEPTS

(1) Explain why, of all places in the body, marrow is located on the interior of long bones.

(2) Explain why your bones are remodeled after you reach maturity. Why doesn't your body keep the mature skeleton throughout life?

21

INTEGRATION AND CONTROL: NERVOUS SYSTEMS

21–I   THE NEURON

Summary

Throughout the life cycle, each body part is constantly monitored and evaluated, not only for its own sake but for its contribution to overall working patterns. Whole constellations of *neurons* (nerve cells arranged in precise message-conducting and information-processing pathways) and *endocrine cells* (which secrete chemical messages that affect target cells some distance away) form the framework of animal *integration*. Most *neurons* have three message-handling zones that occur on the cell body and on its cytoplasmic extensions (collectively called *processes*): (1) an *input* zone (the dendrites), where the neuron receives most incoming messages; (2) a *conducting* zone (the axon), along which information travels rapidly, without alteration, from one region to another; and (3) an *output* zone (the axon terminals), where chemical or electrical signals are released that affect another neuron, muscle cells, or gland cells. In some animals, billions of neurons are organized into an interconnected system. Some are *sensory* neurons, which are receptors for environmental stimuli. *Interneurons* connect and integrate different neurons in the system. *Motor* neurons connect with muscles or glands (effectors) that increase or decrease their activity according to incoming signals. *Neuroglial* cells, which make up at least half the cells in a nervous system, nurture the neurons in various ways.

neuron (p. 319)   integrating centers (p. 320)   dendrites (p. 320)
excitability (p. 319)   interneurons (p. 320)   axon (p. 320)
integrate (p. 319)   motor neurons (p. 320)   cell body (p. 320)
propagate (p. 319)   Schwann cells (p. 320)   input zones (p. 320)
transmit (p. 319)   neuroglia (p. 320)   conducting zones (p. 320)
sensory neurons (p. 320)   cytoplasmic extensions (p. 320)   output zones (p. 320)
                           processes (p. 320)

Objectives

1. Describe the basic two-pronged strategy that complex animals use to integrate diverse functions.

2. Draw a neuron and label it according to its three general zones, its specific structures, and the specific function(s) of each structure.

3. Define *neuroglia* and state how they are related to neurons and the nervous system.

Self-Quiz Questions

  Fill-in-the-Blanks

Nerve cells that conduct messages are called (1) _____. (2) _____ cells, which support and nurture the activities of neurons, make up about half the volume of the nervous system. Most neurons have three zones: an input zone consisting of one or more (3) _____; the (4) _____ _____, which consists of a single threadlike process called the (5) _____; and an output zone, the axonal (6) _____, where chemical or electrical signals are released. (7) _____ neurons are receptors for environmental stimuli, (8) _____ connect different neurons in the nervous system, and (9) _____ neurons are linked with muscles or glands.

21–II   MESSAGE CONDUCTION   320–325

The Neuron "At Rest"   321–322
Changes in Membrane Potential   322–324

Summary

The neuronal plasma membrane is permeable in different ways to different ions. Two steep concentration gradients—one for potassium ions, the other for sodium ions—exist across the neuronal membrane. In addition, an electrical gradient also exists across the neuronal membrane; the interior of a neuron at rest has an overall negative charge relative to the exterior. The difference in charge represents a potential to do work called the *resting membrane potential*. The potential energy from these ion-concentration and electrical gradients can be used to conduct messages whenever the gradients exist. The gradients for potassium and sodium ions are reestablished by enzyme systems called *sodium-potassium pumps*. Energy from a stimulus changes the proteins that span the membrane width at a certain point such that the membrane permeability to first sodium and then potassium ions is increased, and the ions flow along their natural gradients: sodium in and potassium out. This ion flow increases the permeability of the adjacent regions of the membrane, and the disturbance (now called the nerve impulse, or message) radiates outward from the initial point of stimulation.

Key Terms

differentially permeable (p. 320)	polarity of charge (p. 321)	sodium-potassium pump (p. 321)
concentration gradient (p. 320)	at rest (p. 321)	cotransport mechanism (p. 321)
electric gradient (p. 320)	millivolts (p. 321)	graded potential (p. 322)
reversal (p. 320)	voltage difference (p. 321)	action potential (p. 322)
	resting membrane potential (p. 321)	

Objectives

1. Explain how the plasma membrane of a neuron differs in function from a bone cell.

2. Describe the types of gradients associated with the neuronal membrane at rest.

3. Define *resting membrane potential*; explain what establishes it, and how it is used by the cell neuron.

4. Describe what happens to the gradients associated with the neuronal membrane when it is disturbed.

5. Define *sodium-potassium pump* and state how it helps to maintain the resting membrane potential.

6. Explain how alterations in the electrical gradient across the neuron membrane have to do with messages carried by the nervous system.

7. State the characteristics of neurons that help a message to travel without changing along the neuronal membrane.

Self-Quiz Questions

Fill-in-the-Blanks

A plasma membrane is (1) _____ permeable; it passively permits some molecules and ions to move across the membrane along their (2) _____ gradient, and it passively prevents others from doing so. Sometimes it actively transports ions so that they move (3) _____ their gradient. A neuronal membrane has far more positively charged (4) _____ ions inside than out, and far more positively charged (5) _____ ions outside than in. An electrical gradient also exists across the neuronal membrane; compared with the outside, the inside of a neuron at rest has an overall (6) _____ charge. For most neurons in most animals, the difference in charge across the neuronal membrane is within the range of 60-90 (7) _____, which represents an ability to do work; it is the (8) _____ _____ _____. A mechanism built into the neuronal membrane, the (9) _____ - _____ _____, prevents net diffusion equalizing concentrations across the membrane. Alterations in the electrical gradient across the membrane constitute the (10) _____ that travels over a neuronal membrane surface.

21–III   Action Potentials
Summary of Message Conduction Mechanisms

Summary

Some messages travel without changing along the neuronal membrane, and some don't. Fast-responding neurons usually have axons with broad diameters and effective insulation. Messages that travel unaltered and with great speed are encoded as *action potentials*. When a *threshold stimulus* (the minimum change in membrane potential needed to generate a nerve impulse) is applied to a resting neuron, the electrical disturbance associated with ion flow across the membrane causes all of the "sodium gates" to open at once, which causes the membrane potential to reverse abruptly. The sodium gates then close, and the potassium gates open and then partly close when the membrane potential has returned to its resting value. The time during which potassium ions are flowing out of the neuron is known as its *refractory period*—the time it is insensitive to stimulation. By the time the resting membrane potential is restored and the refractory period ends, the electrical disturbance has moved far enough away that it no longer can cause sodium gates to open in the region of the original impulse. Some fast-conducting cells have very thin axons that are wrapped in the electrically resistant myelin membranes of Schwann cells. *Graded potentials* are short-range messages that can change and vary in size as they travel along short sections of dendrites, sensory receptors, and axonal terminals. Graded potentials arriving at a neuron from different pathways can be added up (summed) to determine the course of further transmissions.

Key Terms

transmembrane
  channels (p. 322)
positive feedback (p. 322)
threshold value (p. 322)
all-or-nothing events (p. 322)
millisecond (p. 322)
nerve impulse (p. 323)

innervate (p. 323)
electrode (p. 323)
oscilloscope (p. 323)
self-propagating (p. 323)
refractory period (p. 323)
Schwann cells (p. 323)
myelin sheath (p. 324)

node of Ranvier (p. 324)
saltatory conduction
  (p. 324)
graded potential (p. 325)
action potential (p. 325)
summed, summation (p. 325)

Objectives

1. Describe how gated channels through resting membranes participate in moving sodium ions.

2. Define *action potential* by stating its three main characteristics.

3. Explain the chemical basis of the action potential. Look at Fig. 21.5d and determine which part of the curve represents:

(a) the point at which the stimulus was applied,
(b) the events prior to achieving the threshold value,
(c) the opening of the ion gates and the diffusing of the ions,
(d) the change from net negative charge inside the neuron to net positive charge and back again to net negative charge, and
(e) the active transport of sodium ions out of and potassium ions into the neuron.

4. Define *refractory period* and state what causes it.

5. Define *Schwann cell, nodes of Ranvier,* and *myelin sheath* and explain how each helps narrow-diameter neurons conduct nerve impulses quickly.

6. Describe *graded potentials* and contrast them with action potentials.

7. Explain what is meant by summing graded potentials "to determine the course of further transmissions."

Self-Quiz Questions

Fill-in-the-Blanks

A(n) (1) _____ _____ is an all-or-none, brief reversal in membrane potential; it is also known as a(n) (2) _____ _____. Once an action potential has been achieved, it is (3) ___-_____-_____; its amplitude won't change even if the strength of the stimulus changes. The minimum change in membrane potential needed to achieve an action potential is the (4) _____ value. Each action potential is followed by a (5) _____ period: a time of insensitivity to stimulation. Some narrow-diameter neurons are wrapped in the lipid-rich (6) _____ _____ of Schwann cells; each of these is separated from the next by a (7) _____ _____ _____: a small gap where the axon is exposed to extracellular fluid. An action potential jumps from one node to the next in line and is called (8) _____ conduction. Short-range neural messages that decay in strength within a short distance from the point of stimulation are called (9) _____ _____; those arriving at a neuron from different pathways can be (10) _____ to determine the course of further transmissions.

Synaptic Transmission Between Neurons   325-326

Summary

The junction between two neurons, which is specialized for transmission, is called a *synapse*. Transmitter substances are chemical transmitters that diffuse across the synaptic cleft from the axonal terminals of the transmitting neuron to the plasma membrane of the receiving cell. Acetylcholine is one such neurotransmitter; enzymes on the receiving membrane rapidly break it down, so it doesn't accumulate.

At each synapse excitatory (+) and inhibitory signals (-) generate graded potentials, which are added together. If the net effect at the synapse is excitatory, the information may be reinforced and sent on; if the net effect is inhibitory, the information is repressed. *Integration* is the moment-by-moment summation of all excitatory and inhibitory signals acting on a neuron.

Different messages about stimulus type, intensity, and location that converge simultaneously on a neuron are summed to provide information of even more precise content. A receiving neuron may have hundreds or thousands of synapses acting on it; how the receiving neuron will handle the incoming messages depends on whether its activity is enhanced or inhibited. At *excitatory synapses*, neurotransmitters increase membrane permeability to both sodium and potassium, which drives the membrane potential *toward* threshold. At *inhibitory synapses*, neurotransmitters increase membrane permeability to potassium and/or chloride, which drives the membrane potential *away* from threshold, decreasing the likelihood that the receiving neuron will fire action potentials.

Key Terms

message transmission (p. 325)	exocytosis (p. 325)	epinephrine (p. 325)
synapse (p. 325)	transmitter	excitatory synapse (p. 325)
synaptic knob (p. 325)	substances (p. 325)	inhibitory synapse (p. 325)
synaptic cleft (p. 325)	acetylcholine (p. 325)	excitatory input (p. 326)
vesicles (p. 325)	dopamine (p. 325)	integration (p. 326)

Objectives

1. Name and describe the three functional types of neurons.

2. Name all of the structures associated with a synapse and explain the role of each structure in conducting a message across a synapse to a receiving neuron.

3. State the functional features shared by muscle and nerve cells.

4. Distinguish how excitatory synapses function from inhibitory synapses.

5. State what we mean by *integration* used in conjunction with the nervous system.

Fill-in-the-Blanks

The junction specialized for transmission between a neuron and another cell is called

a (1) _____. Usually, the signal being sent to the receiving cell is carried by

chemical messengers called (2) _____ _____. (3) _____ is an exam-

ple of such a chemical messenger that diffuses across the synaptic cleft, combines with

protein receptor molecules on the muscle cell membrane, and soon thereafter is rapidly

broken down by enzymes. (4) _____ at the cellular level is the moment-by-

moment summation of all excitatory and inhibitory signals acting on a neuron. Incoming

information is (5) _____ by cell bodies, and the charge differences across the

membranes are either enhanced or inhibited; this summation of messages at the synapses

is referred to as (6) _____. At an (7) _____ synapse, the membrane poten-

tial is driven toward the threshold value and increases the likelihood that an action

potential will occur. At an (8) _____ synapse, the membrane potential is driven

away from the threshold value, and the receiving neuron is less likely to achieve an

action potential.

21–V   Junctional Transmission Between Neurons and Muscle Cells   326–327

THE REFLEX ARC: FROM STIMULUS TO RESPONSE   327–328

Summary

A *reflex* is a sequence of events elicited by a stimulus. In a monosynaptic reflex pathway, a sensory receptor is stimulated, resulting in an impulse being generated in the
sensory axon. The sensory neuron synapses in the spinal cord or brain with a motor
neuron, which generates an impulse in the motor neuron that is carried to a muscle or
gland (effectors), where the response will be contraction or secretion if the impulse
was sufficiently strong.

Key Terms

neuromuscular junction (p. 326)   receptors (p. 327)   stretch reflex (p. 327)
reflex arc (p. 327)   effector cells (p. 327)   muscle spindles (p. 328)

1. Explain what a reflex is by drawing and labeling a diagram and telling how it functions.

2. Explain what the stretch reflex is and how it helps an animal to survive.

Self-Quiz Questions

Fill-in-the-Blanks

A (1) _____ is an involuntary sequence of events elicited by a stimulus. During

a (2) _____ _____, a muscle contracts involuntarily whenever conditions

cause a stretch in length; many of these help you maintain an upright posture despite

small shifts in balance. Located within skeletal muscles are length-sensitive organs

called (3) _____ _____, which generate action potentials when stretched

beyond a critical point; these potentials are conducted rapidly to the (4) _____

_____, where they are communicated to motor neurons leading right back to the

muscle that was stretched.

21-VI NERVOUS SYSTEMS: INCREASING THE OPTIONS FOR RESPONSE   328-335

Evolution of Nervous Systems   328-331

Summary

Nervous systems presumably evolved from simple systems such as those seen in cnidarians (radially arranged multidirectional networks dispersed through the body) to a bilaterally symmetrical, centralized system with interneurons generally arranged along the body axis and the major sense organs in the head. A shift from radial to bilateral symmetry could have led, in some evolutionary lines of animals, to paired nerves and muscles, paired sensory structures such as eyes, and paired brain regions. Differences that do exist among complex animals relate largely to brain size and its degree of control over the rest of the nervous system.

No matter how specialized a nervous system has become, the same basic principles apply: (1) Information flows by way of *graded* potentials (in receptors and at synapses) and *action* potentials (along axons). (2) Pathways of divergence allow a message being conducted along a neuron to travel through its branched endings and activate many other neurons. (3) Pathways of convergence enable each neuron to receive both excitatory and inhibitory signals coming from many other neurons. (4) For all bilaterally symmetrical animals, each through-conducting pathway leads from sensory axons to the central nervous system and then to motor axons leading away to muscles, glands, or other effectors. (5) Excitatory and inhibitory signals are summed at each synaptic transfer. (6) Signals of all sorts are sent to the brain, which sends out command signals when a change in response is required. The brain's action represents the highest level of integration yet developed.

cnidarian (p. 328)
radial symmetry (p. 328)
nerve nets (p. 328)
statocyst (p. 329)
ganglion (p. 329)
ganglia P. 329)
integrative
  center (p. 329)

planula (p. 329)
polyp (p. 329)
cephalization (p. 329)
bilateral
  symmetry (p. 329)
central nervous system
  (p. 330)

peripheral nervous
  system (p. 330)
segmental patterning (p. 330)
divergence (p. 330)
convergence (p. 330)
synaptic transfer (p. 330)

Objectives

1. Contrast the sensitivity and motor abilities of sponges and cnidarians.

2. List some of the changes that might have occurred in promoting the change from a
    radially arranged, noncephalized, multidirectional nervous system to a system
    that is bilaterally arranged, cephalized, and with a one-directional flow of in-
    formation.

3. Describe how the shift from radial to bilateral symmetry influenced the number of
    organs and their location in the body.

4. Define and contrast *central* and *peripheral* nervous systems.

5. List, in your own words, six basic principles that govern the organization of ner-
    vous systems.

Self-Quiz Questions

Fill-in-the-Blanks

The symmetrical arrangement of constituents, especially of radiating parts, about a

central point is (1) _____ symmetry. Organized knots of neurons encased in con-

nective tissue that form integrative centers are known as (2) _____. (3)

_____ increases the concentration of nervous structures and coordinative func-

tions in the head; organisms with this body pattern tend to have (4) _____

symmetry. A shift from radial to bilateral symmetry could have led, in some evolution-

ary lines of animals, to paired (5) _____ and (6) _____, paired sensory

structures such as eyes, and paired (7) _____ regions. The (8) _____ ner-

vous system consists of the brain and nerve cord (or paired cords). The (9) _____

nervous system includes cell bodies of sensory neurons and all nerves (bundles of

axons).

Summary

In all vertebrates, the central nervous system is enclosed in protective walls and cushioned and bathed with fluid. Bony fishes, amphibians, reptiles, birds, and mammals all have bony plates joined together forming a *skull* that protects the *brain* and another series of bony segments linked into a *vertebral column* that protects the *spinal cord*. In the more primitive fishlike vertebrates, cartilage replaces bone in this protection. The spinal cord is a region of local integration and reflex connections and serves as the nerve pathway leading to and from the brain.

At the head end of the spinal cord are three distinct masses of neurons: the hindbrain, midbrain, and forebrain. The *hindbrain* contains the *medulla oblongata*, a center that controls breathing, heart rate, and blood pressure reflexes; and the *cerebellum*, which helps coordinate limb movements concerned with maintaining balance, posture, and spatial orientation. The *midbrain* is the primary "association center" in fishes and amphibians. The optic lobes, which are visual centers associated with the optic nerves, are located dorsally. In reptiles, birds, and mammals, some integration occurs in the midbrain, but messages are sent on to the forebrain for further interpretation. The *forebrain* is divided into two halves—the cerebral hemispheres—which overlie forebrain regions called the thalamus, hypothalamus, and pituitary.

The skeletal muscles are the effectors of the *somatic* nervous system, which in humans, is under conscious control. The *autonomic* nervous system, on the other hand, generally is not under conscious control; parts of the peripheral and central nervous systems cooperate to adjust the body's organs and organ systems to changing conditions. The autonomic nervous system is subdivided into two networks that act antagonistically to each other. *Parasympathetic* nerves generally dominate internal events when environmental conditions permit normal body functioning, and *sympathetic* nerves dominate internal events in times of stress and danger, mobilizing the whole body for rapid response to change.

Key Terms

skull (p. 331)
vertebral column (p. 331)
spinal cord (p. 331)
white matter (p. 331)
gray matter (p. 331)
hindbrain (p. 332)
medulla oblongata (p. 332)
cerebellum (p. 322)
spatial orientation (p. 332)
tremor (p. 332)
midbrain (p. 332)
optic lobes (p. 332)

association center (p. 332)
forebrain (p. 332)
cerebral hemispheres (p. 332)
thalamus (p. 332)
hypothalamus (p. 332)
effectors (p. 332)
somatic nervous system (p. 333)
conscious control (p. 333)
autonomic nervous system (p. 333)
sympathetic nerves (p. 333)
parasympathetic nerves (p. 333)
fight-or-flight response (p. 333)

Objectives

1. Describe the basic structural and functional organization of the spinal cord. In your answer, define *white matter* and *gray matter* and distinguish spinal cord from vertebral column.

2. Describe the ways the brain is protected and list its three principal divisions.

3. List the parts of the brain found in the hindbrain, midbrain, and forebrain and tell the basic functions of each.

4. For each part of the brain in the list you developed for objective 3, state how the behavior of a normal person would change if he or she suffered a stroke in that part of the brain.

5. Tell what happened to the importance of the midbrain during the evolution of the vertebrates.

6. Distinguish the somatic and autonomic nervous systems with respect to location and chief activities.

7. Explain how parasympathetic nerve activity balances sympathetic nerve activity. List activities of the sympathetic and parasympathetic nerves in regulating pupil diameter, rate of heartbeat, activities of the gut, and elimination of urine.

8. Describe how your autonomic nervous system would act to promote your survival if you were in a serious car accident and were rendered unconscious.

Self-Quiz Questions

Fill-in-the-Blanks

The (1) _____ _____ is a region of local integration and reflex connections with nerve pathways leading to and from the brain; its (2) _____ _____, which contains myelinated sensory and motor axons, is the through-conducting zone. The (3) _____ _____ includes nerve cell bodies, dendrites, and nonmyelinated axon terminals; this is the (4) _____ zone. The hindbrain is an extension and enlargement of the upper spinal cord; it consists of the (5) _____ _____, which contains the control centers for the heartbeat rate, blood pressure, and breathing reflexes. The hindbrain also includes the (6) _____, which helps coordinate motor responses associated with refined limb movements, maintenance of posture, and spatial orientation. The optic lobes are part of the (7) _____.

The (8) _____ contains two cerebral hemispheres composed of gray matter. Underlying the cerebral hemispheres is the (9) _____, a region that links many of the activities of the nervous and (10) _____ systems, and the (11) _____, which is the principal gateway to the cerebral hemispheres. All motor-to-skeletal muscle pathways and all sensory pathways make up the (12) _____ nervous system. The remaining nerve tissue, which generally isn't under conscious control, is collectively known as the (13) _____ nervous system; it is subdivided into two

parts: (14) _____ nerves, which respond to emergency situations, and (15)

_____ nerves, which oversee normal body functioning.

21–VIII THE HUMAN BRAIN

Summary

The cerebral hemispheres are paired masses that structurally appear to be mirror images of each other. Functionally, however, they differ considerably. Inside each hemisphere is a core of *white matter* that contains axonal pathways connecting the rest of
the central nervous system with the brain's surface layer. An expanse of *gray matter*
about .6 cm thick, the *cerebral cortex*, covers the white matter. Some cortical regions
receive signals from receptors on the body's periphery; others coordinate and process
sensory input, and yet others coordinate instructions for motor responses.

Neural messages contain information about stimulus type; that is, about the activation of particular receptors, which carry information about one form of stimulus
only. Variations in stimulus strength are encoded as variations in the frequency of
action potentials traveling one after another, in series, and in the number of receptors activated. The body's receptors are distributed through different tissue regions
called *receptive fields*. Sensory neurons lead from receptive fields to the brain and/
or spinal cord, where they synapse with interneurons. The smaller the receptive field,
the more accurately a message can be traced back to the receptor(s) that were stimulated.

The two cerebral hemispheres are connected by the *corpus callosum*. Experiments
severing the corpus callosum have revealed how the cerebral hemispheres function in
relation to each other. Many of the lines leading into and out of one hemisphere deal
with the opposite side of the body. Each cerebral hemisphere can function separately,
but it functions in response to signals mainly from one side of the body. The main
association regions responsible for language and analytical skills generally reside in
the left hemisphere. The main association regions responsible for nonverbal perception
(intuition) generally reside in the right hemisphere. The fossilized remains of skulls
and sensory structures suggest that developments in the vertebrate forebrain enabled
conscious experiences approximately 3 or 4 million years ago.

Key Terms

white matter (p. 335)	temporal lobe (p. 336)	nerves (p. 338)
cerebral cortex (p. 335)	occipital lobe (p. 336)	cerebral hemisphere
gray matter (p. 335)	Broca's area (p. 336)	(p. 339)
thalamus (p. 335)	fissure of Rolando (p. 336)	corpus callosum (p. 339)
cortical (p. 336)	association cortex (p. 337)	Sperry and colleagues
motor cortex (p. 336)	stimulus type (p. 337)	(p. 339)
somatosensory cortex (p. 336)	receptor (p. 337)	epilepsy (p. 339)
visual cortex (p. 336)	receptive field (p. 337)	split-brain (p. 339)
auditory cortex (p. 336)	stimulus location (p. 337)	conscious experience
parietal lobe (p. 336)	stimulus intensity (p. 338	(p. 339)
frontal lobe (p. 336)	frequency (p. 338)	language (p. 339)

Objectives

1. Describe the basic layout of the cerebral hemispheres relative to the other parts of the forebrain.

2. List the four lobes of the cerebral cortex (see Fig. 14.20) and state the activities that occur in each.

3. Explain what is meant by *conscious experience*.

4. Name three aspects of stimuli that are detectable by the part of the nervous system responsible for integration.

5. Describe what constitutes a receptive field. Then tell what leads from the receptive field and where it goes.

6. State what the results of the "split-brain" experiments suggest about the functioning of the cerebral hemispheres.

7. Explain how Hubel and Wiesel's experiments suggest that the key to visual perception resides in the organization and synaptic connections between columns of neurons in the brain.

Self-Quiz Questions

Fill-in-the-Blanks

The surface layer of each hemisphere, the (1) _____ _____, is gray matter about a quarter of an inch thick. Neurons in the (2) _____ _____ act as direct channels from the brain to motor neurons, and stimulating differing points of this surface causes contractions of muscle groups in different body parts. The synaptic zone for nerves concerned with signals from somatic receptors is the (3) _____ cortex, located just behind the motor cortex. Communication pathways from the eyes terminate in a primary receiving center, the (4) _____ _____ in the (5) _____ lobe of each cerebral hemisphere. Pathways from the ears terminate in the (6) _____ _____, a small region of each temporal lobe. Regions of the (7) _____ _____ may give rise to conscious perception of changing events in the surrounding world. Information regarding (8) _____ _____ is conveyed by the activation of particular receptors, which carry information about one form of stimulus only. (9) _____ _____ is communicated as variations in the frequency of action potentials traveling one after another, in series, from a specific number of receptors. A (10) _____ is a nerve cell that conducts messages, but a (11) _____ consists of bundles of separate axons from more than one nerve cell. The body's receptors are distributed through different tissue regions called (12) _____ _____. (13) _____ _____ is communicated by the number of signals on specific communication lines from a given body tissue region. The two brain halves are connected by a thick tract of white matter, the (14) _____ _____; severing this changes the domain of (15) _____ experience perceived and executed by each cerebral hemisphere. (16) _____ _____ coordinates muscles required for speech.

Summary

Conscious experience entails a capacity for *memory* (the storage of individual bits of information somewhere in the brain). So far, experiments suggest that at least two stages are involved in establishing memory traces in the brain. One is a short-term formative period lasting only a few minutes; information then becomes spatially and temporally organized in neural pathways (long-term storage). Long-term memory depends on structural changes in the brain. There is also evidence that neuron fine structure is not static, but rather can be modified in several ways, most likely depending on electrical and chemical interactions with neighboring neurons.

Between the mindless drift of coma and total alertness are many levels of conscious experience: sleeping, dozing, meditating, and daydreaming. Neurons chattering among themselves show up as characteristic wavelike patterns in an *electroencephalogram* (EEG), an electrical recording of the frequency and strength of potentials from the brain's surface. Alpha waves predominate during meditation, a relaxed state of wakefulness. The slow-wave sleep pattern composes about 80 percent of the total sleeping time for adults. EEG arousal occurs when individuals make a conscious effort to focus on external stimuli or even on their own thoughts.

The REM sleep pattern coincides with vivid dreams. The *reticular formation* governs changing levels of consciousness; within this formation are neurons (the RAS) that connect to the thalamus and arouse the brain and maintain wakefulness. The *sleep centers* are also in the reticular formation. The alternating states of sleeping and wakefulness are thought to have a physiological basis in neurotransmitters produced in the brain stem and whose interactions influence other brain regions. People deprived of REM sleep and its accompanying dream state experience irritability and mental instability. Some individuals with brain damage never fall asleep, yet they survive.

Endorphins (which include enkephalins) are *analgesics* (pain relievers that the brain can produce); they bind to neural membranes on the spinal cord and limbic system and thereby inhibit neural activity. Emotional states are expressions of the balance of transmitter substances (such as acetylcholine, serotonin, dopamine, etc.) as they are produced in response to changing conditions in our complex world. Imbalances in the production of transmitter substances can provoke mental illness, the symptoms of which can often be brought under control by drugs that act to balance the imbalance. Tranquilizers, opiates, stimulants, and hallucinogens inhibit, modify, or enhance the behavior of chemical messengers throughout the brain.

thinking (p. 340)
memory (p. 340)
short-term formative
  period (p. 340)
long-term storage (p. 340)
retrograde amnesia (p. 340)
sleeping (p. 340)
meditating (p. 340)
electroencephalogram (p. 340)
alpha rhythm (p. 340)
coma (p. 34)
slow-wave sleep (p. 341)
EEG arousal (341)
REM sleep (p. 341)

reticular formation (p. 341)
reticular activating
  system (RAS) (p. 341)
thalamus (p. 341)
sleep centers (p. 341)
serotonin (p. 341)
pons (p. 341)
analgesics (p. 341)
limbic system (p. 341)
endorphins (p. 341)
enkephalins (p. 341)
transmitter
  subtances (p. 342)

norepinephrine (p. 342)
dopamine (p. 342)
serotonin (p. 342)
tranquilizers (p. 342)
opiates (p. 342)
stimulants (p. 342)
hallucinogens (p. 342)
ACTH 4-10 (p. 342)
amphetamines (p. 342)
antidepressants (p. 342)
phenothiazines (p. 342)
LSD (p. 342)

Objectives

1. Define *memory* and distinguish it from *memory trace*.

2. Explain why scientists believe that the formation of a permanent memory trace is a two-step process.

3. Describe current theories that attempt to explain how neurons might be the repositories of memory.

4. Tell what an electroencephalogram is and what EEGs can tell us about the levels of conscious experience. Describe three typical EEG patterns and tell which level of consciousness each characterizes.

5. Locate and identify the function of the reticular formation. Explain what happens if a normal RF is stimulated or is damaged.

6. Distinguish the RAS from the sleep centers. Tell which area releases serotonin, and state its effect.

7. Explain the relationship between transmitter substances and analgesics.

Self-Quiz Questions

  Fill-in-the-Blanks

The storage of individual bits of information somewhere in the brain is called (1)

_____; the neural representation of such bits is known as a (2) _____

_____. Experiments suggest that there are at least two stages involved in its

formation. One is a (3) _____-_____ _____ period, lasting only a

few minutes; then information becomes spatially and temporally organized in neural

pathways. The other is a (4) _____-_____ _____; then information

is put in a different neural representation and permanently filed in the brain. An

electrical recording of the frequency and strength of potentials from the brain's surface is known as a(n) (5) _____ . The principal wave pattern for someone who is relaxed, with eyes closed, is a(n) (6) _____ _____ . The (7) _____ - _____ _____ pattern occupies about 80 percent of the total sleeping time for adults. When individuals shift from sleep to a conscious focus on external stimuli or on their own thoughts, the pattern is called (8) _____ _____ . (9) _____ _____ accompanies vivid dreaming periods. Activities in the (10) _____ _____ determine whether you are awake or asleep. High (11) _____ levels in the brainstem's core bring about drowsiness and sleep. (12) _____ are analgesics produced by the brain that inhibit regions concerned with our emotions and perception of (13) _____ . Imbalances in (14) _____ _____ can produce emotional disturbances; (15) _____ affect these imbalances by depressing activity in neurons that use these transmitter substances.

INTEGRATING AND APPLYING KEY CONCEPTS

Suppose that anger eventually is determined to be caused by excessive amounts of specific transmitter substances in the brains of angry people. Suppose that an inexpensive antidote to anger that neutralizes these anger-producing transmitter substances is also readily available. Can violent murderers now argue that they have been wrongfully punished because they were victimized by their brain's transmitter substances and could not have acted in any other way? Suppose an antidote is prescribed to curb violent tempers in a person. Suppose also that the easily angered person forgets to take the pill and subsequently murders a family member. Can the murderer still claim to be "victimized" by transmitter substances?

22

INTEGRATION AND CONTROL: ENDOCRINE SYSTEMS

Summary

Neurotransmitters are secreted from nerve cell endings and travel only a short distance, across a synaptic cleft, to an adjacent cell, in effect saying, "It's time to change your pattern of activity." Some neurotransmitters are acetylcholine, epinephrine, norepinephrine, serotonin, and dopamine. They function, for a fleeting moment, as information carriers. *Neurohormones* are produced by neurosecretory cells; they travel slowly and farther by way of the bloodstream to many nonadjacent cells. In simple animals, neurohormones are the principal physiological controls over cells and tissues concerned with growth and reproduction.

In more complex animals, secretory *endocrine* cells function in a variety of ways; some operate individually, while others are organized into tissue patches or organs. All secrete true animal *hormones*, which are transported by the bloodstream and regulate specific cellular reactions in tissues and organs some distance away. Some hormones activate short-term adjustments (heart rate, blood chemical composition); others promote long-term adjustments, such as those involved in growth, differentiation, and reproduction. Generally, through hormonal action, specific enzyme activities (or enzyme formation) are accelerated or slowed down.

Steroid hormones pass easily through the plasma membrane of a target cell. Inside they bind to a receptor molecule in the cytoplasm and form a complex that "turns on" specific genes in the nucleus to make specific mRNA transcripts that alter the cell's activity. *Polypeptide hormones* do not pass through the plasma membrane, but they *do* bind to receptor sites on the membrane, which activates adenyl cyclase, an enzyme that converts ATP to cyclic AMP. cAMP activates a different enzyme that alters cellular activity.

Four factors determine the level of specific hormones in the blood: the availability of chemical precursors, the rate of hormone secretion, the rate of hormonal activation, and the rate that hormones are removed from the bloodstream. All help to determine the levels of specific hormones and their responses.

Key Terms

neurosecretory hormones (p. 344)
interstitial fluid (p. 344)
transmitter substances (p. 345)
true hormones (p. 345)
glands (p. 345)
target cells (p. 345)
membrane receptor sites (p. 345)
gastrin (p. 345)
steroid hormones (p. 345)
peptide (p. 345)
amine (p. 345)
cytoplasmic receptor molecule (p. 345)
adenyl cyclase (p. 345)
cyclic AMP, cAMP (p. 345)
polypeptide hormones (p. 345)

second messenger (p. 345)
mediate (p. 345)
protein kinases (p. 345)
chemical precursor (p. 346)
ACTH (p. 346)
adrenal cortex (p. 346)
cortisol (p. 346)
anterior lobe of pituitary (p. 346)
testosterone (p. 346)
dihydrotestosterone (p. 346)
pineal gland (p. 346)
thyroid gland (p. 346)
parathyroid glands (p. 346)
pancreatic islets (p. 346)
thymus gland (p. 346)

Objectives

1. Define *neurotransmitters, neurohormones,* and *hormones* and list their functions.

2. State what is secreted by nerve cell endings, neurosecretory cells, endocrine cells, and exocrine cells.

3. Contrast the proposed mechanisms of hormone action on target cell activities by (a) steroid hormones and (b) hormones that are proteins or derived from proteins.

4. Name four factors that influence hormone levels in the bloodstream.

5. Name six examples of endocrine glands.

Self-Quiz Questions

   Fill-in-the-Blanks

(1) _____ are substances that are secreted from nerve endings and travel

only a short distance across a synaptic cleft to an adjacent cell. (2) _____

cells release information carriers, (3) _____, that travel more slowly and

travel further, by the body fluids, to many nonadjacent cells. Animal (4) _____

are produced by endocrine cells and transported by the bloodstream to tissues and organs some distance away, where they regulate specific cellular reactions in (5)

_____ cells. Pheromones are produced by specialized (6) _____ glands,

which have ducts that lead out to the body surface; pheromones may activate behavioral

changes in other animals of the (7) _____ species. (8) _____ is one kind

of second messenger that communicates information from cell surfaces to specific inner

regions of the cell. The (9) _____ are the female gonads, and the (10) _____

are the male gonads.

22–II NEUROENDOCRINE CONTROL CENTER 347–351

On Neural and Endocrine Links 347
Interactions Between the Hypothalamus and Pituitary 347–351

Summary

Most endocrine elements are controlled either directly by a separate nerve supply or
indirectly by secretions from particular centers of the nervous system. Many physiolo-
gists believe endocrine tissue evolved from mutated neurosecretory cells that conferred
a survival advantage on their bearers. All vertebrates have a neuroendocrine control
center, which consists of the hypothalamus and pituitary.
    Because its secretions affect many other endocrine elements, the pituitary is
traditionally thought of as the "master gland." But the hypothalamus controls the pituitary. The hypothalamus itself is influenced by the cerebral cortex and by concentrations
of hormones, ions, and nutrients in the bloodstream.
    Information about changing conditions in the external and internal worlds flows
constantly to the hypothalamus. Some information arrives along neural pathways from the
cerebral cortex. Some endocrine elements arrive with the bloodstream. In the hypothalamus, incoming neural messages are summed, shifts in hormonal concentrations are detected, and responses are sent out in the form of *releasing factors* or *inhibiting factors*, which are routed to one of two pituitary regions. The anterior pituitary is mostly glandular tissue; the posterior lobe and its connecting stalk are nervous tissue. A
variety of regulatory hormones produced by the anterior pituitary participate in feedback loops.

Key Terms

neurosecretory cells (p. 347)
hypothalamus (p. 347)
pituitary (p. 347)
posterior lobe of the pituitary (p. 347)
anterior lobe of the pituitary (p. 348)
intermediate lobe (p. 348)
melanocyte-stimulating hormones (p. 348)
endorphins (p. 348)
brain analgesics (p. 348)
neuroendocrine control center (p. 348)
oxytocin (p. 349)

vasoconstriction (p. 349)
releasing factors (p. 349)
inhibiting factors (p. 349)
portal vessels (p. 349)
adrenocorticotropin, ACTH (p. 350)
thyroid-stimulating hormone, TSH (p. 350)
follicle-stimulating hormone, FSH (p. 350)
luteinizing hormone, LH (p. 350)
growth hormone, somatotropin (p. 350)
prolactin (p. 350)
exocrine glands (p. 350)

antidiuretic hormone, vasopressin,   <u>dwarfism</u> (p. 350)
   ADH (p. 349)   <u>gigantism</u> (p. 350)

Objectives

1. Explain how the endocrine system could have evolved and state any evidence that
    supports your ideas.

2. Explain why the hypothalamus and pituitary are regarded as the neuroendocrine
    control centers.

3. State how, even though the anterior and posterior lobes of the pituitary are com-
    pounded as one gland, the tissues of each part differ in character.

4. Identify the hormones produced by the anterior lobe of the pituitary and tell
    which target tissues or organs that each acts on.

5. Identify the hormones released from the posterior lobe of the pituitary and state
    their target tissues.

6. Provide an example of a negative feedback relationship in which the anterior lobe
    of the pituitary is involved.

Self-Quiz Questions

Fill-in-the-Blanks

In the (1) _____, incoming neural messages are summed, shifts in hormonal con-

centrations are detected, and responses are sent out in the form of (2) _____

_____ to one of two regions of the (3) _____. The tissues of the

anterior lobe are (4) _____ in nature; the tissues of the posterior lobe are

(5) _____. Vessels that connect two distinct capillary beds are called (6)

_____ vessels; such vessels link the capillary bed in the (7) _____

with appropriate capillary beds in either lobe of the pituitary. From the anterior

pituitary come several hormones: (8) _____, which stimulates the adrenal cor-

tex; (9) _____, which stimulates the thyroid to produce thyroxin, and (10)

_____, which stimulates milk production in mammary glands.

Summary

The outer layer of the adrenal gland, the *adrenal cortex*, secretes three types of hormones: glucocorticoids, which help regulate food metabolism; mineralocorticoids, which influence salt and water concentrations; and sex hormones (the androgens and estrogens). The inner region, the *adrenal medulla*, secretes epinephrine (adrenalin) and norepinephrine, which help regulate blood circulation and carbohydrate metabolism.

The thyroid gland stores and releases hormones—among them thyroxin and calcitonin—that help govern growth, development, and metabolic rates throughout the body. The four parathyroid glands counterbalance the effects of calcitonin.

The gonads (ovaries and testes) make gametes and secrete hormones that build and maintain ducts, glands, and tissues involved in reproduction.

Although the pancreas serves primarily as a secretor of digestive enzymes, about 2 million pancreatic islets secrete insulin and glucagon, which counterbalance each other in regulating aspects of glucose, fat, and protein metabolism.

The *kidneys* are paired organs lying along the spine that balance salt and water concentrations in body fluids; they also influence blood pressure and red blood cell production indirectly through the production of *angiotensin*, a powerful vasoconstrictor, and directly through the production of *erythropoietin*. The pineal gland plays a role in reproductive physiology by cueing reproductive cycles to the changing seasons. The thymus produces infection-fighting cells. Prostaglandins are a grab bag of hormones secreted by tissues throughout the body: lungs, gut, prostate gland in males and liver; they seem to act as mediators between membrane receptors and adenyl cyclase.

Key Terms

adrenal cortex (p. 352)
glucocorticoids (p. 352)
cortisol (p. 352)
mineralocorticoids (p. 352)
aldosterone (p. 352)
androgens (p. 352)
estrogens (p. 352)
adrenal medulla (p. 352)
epinephrine (p. 352)
norepinephrine (p. 352)
thyroid gland (p. 352)
thyroxin (p. 352)
calcitonin (p. 352)
parathyroid glands (p. 352)

parathyroid hormone (p. 352)
gonads (p. 352)
testes (p. 352)
ovaries (p. 352)
accessory reproductive
   structures (p. 352)
pancreatic islets (p. 352)
insulin (p. 352)
glucagon (p. 352)
glycogen (p. 352)
glandular
   epithelium (p. 352)

gastrin (p. 352)
cholecystokinin (p. 352)
secretin (p. 352)
kidneys (p. 352)
renin (p. 352)
angiotensin (p. 352)
vasoconstrictor (p.352)
erythropoietin (p. 352)
pineal gland (p. 353)
photoperiodism (p. 353)
thymus (p. 353)
lymphocytes (p. 353)
thymosin (p. 353)
protaglandins (p. 353)

Objectives

1. List all the major endocrine glands and patches of tissue, tell which substance(s) each secretes, tell what tissues or organs each substance affects, and state how the "target" tissue responds.

2. State which two glands seem to be the principal regulators of the amount of calcium deposited in the bones and distributed in the muscles. Then tell which of the anterior pituitary hormones interact with these two glands.

3. List all the various hormones (and their secretory tissues) involved in regulating carbohydrate metabolism.

4. Describe how the endocrine elements of the gonads and adrenal cortex affect aspects of reproduction.

5. List all of the hormones involved in regulating salt and water balance.

Self-Quiz Questions

Fill-in-the-Blanks

Renin is secreted by the (1) _____ and released into the bloodstream where it acts on two plasma proteins produced by the liver; the protein is converted to a hormone called (2) _____, which is a powerful vasoconstrictor that keeps blood flow to the kidneys under sufficiently (3) ○ high ○ low pressure. This hormone also acts on the (4) _____ _____, stimulating it to release aldosterone. In humans, the pineal gland takes part in (5) _____ physiology, although its precise role has not been identified. When calcium levels in the blood plasma rise, the (6) _____ secretes calcitonin, which (7) ○ promotes ○ inhibits the release of calcium from bone storage sites. Counterbalancing the effects of calcitonin, the (8) _____ secrete their hormone, which (9) ○ promotes ○ inhibits the release of calcium into the bloodstream. (10) _____ is a hormone that promotes the breakdown of glycogen (a storage starch) into glucose subunits; it counterbalances some aspects of (11) _____ action. Both hormones are produced by the pancreatic islets. The adrenal medulla produces (12) _____ and (13) _____, which help regulate blood circulation and carbohydrate metabolism. The (14) _____ is involved in the immune response by producing a hormone that takes part in the production of specific types of infection-fighting cells.

INTEGRATING AND APPLYING KEY CONCEPTS

Suppose that you suddenly quadruple your daily consumption of table salt, NaCl. State which body organs would be affected and tell how they would be affected. Name three hormones, the levels of which would most probably be affected, and tell whether your body's production of them would increase or decrease. Suppose that you continue this high rate of salt consumption for ten years. Can you predict the organs that would be stressed most?

23
RECEPTION AND MOTOR RESPONSE

```
SENSING ENVIRONMENTAL CHANGE        MOVEMENT IN RESPONSE TO CHANGE
    Primary Receptors                    Motor Systems
    Chemical Reception                   Muscle Contraction
    Mechanical Reception                 Summary of Muscle Contraction
    Light Reception                      Types of Muscle Responses
                                           to Action Potentials
                                         Neuromotor Basis of Behavior
```

Summary

The only windows between the nervous system and events going on within and around the animal body are *primary receptors*: cells or parts of cells that detect specific kinds of stimuli. A *stimulus* is any form of energy change in the environment that the body actually detects. Receptor cells translate stimulus energy into electrochemical messages that can be dealt with by the nervous system. *Chemoreceptors* detect impinging chemical energy; they include odor and taste receptors and internal receptors such as those sensitive to levels of specific chemicals in the bloodstream. Taste buds are examples of organs that contain chemoreceptors. *Mechanoreceptors* detect mechanical energy associated with changes in pressure, position, or acceleration; they include receptors of stretch, touch, equilibrium, and hearing. Sound is perceived as a form of vibration. Sound detectors pick up on differences in the rate and density of high- and low-pressure regions in sequences of compressional waves. The sound receptors in the human ear are hair cells in the organ of Corti.

Key Terms

echolocate (p. 355)
high-frequency (p. 355)
primary receptors (p. 355)
stimulus (p. 355)
wavelength (p. 355)
receptor (p. 355)
electrochemical
  message (p. 355)

sensory organ (p. 355)
stimulus
  amplification (p. 355)
stimulus
  direction (p. 355)
chemoreceptors (p. 355)
mechanoreceptors (p. 355)
photoreceptors (p. 355)
thermoreceptors (p. 355)

electroreceptors (p. 355)
nocireceptors (p. 355)
acceleration (p. 355)
olfaction (p. 356)
taste buds (p. 356)
pheromone (p. 356)
bombykol (p. 356)
tactile receptors (p. 356)
stretch receptors (p. 356)

gated sodium
  channels (p. 356)
vibration (p. 356)
longitudinal
  waves (p. 356)
sound (p. 357)
amplitude (p. 357)
perceived pitch (p. 357)
frequency (p. 357)

hair cells (p. 357)
equilibrium (p. 357)
ear canal (p. 358)
eardrum (p. 358)
middle earbones (p. 358)
oval window (p. 358)
cochlea (p. 358)
round window (p. 358)
basilar membrane (p. 358)

auditory nerve (p. 358)
organ of Corti (p. 358)
tectorial membrane (p. 358)
semicircular canals (p. 359)
cupula (p. 359)
otoliths (p. 359)
ear stones (p. 359)
statocyst (p. 360)
statolith (p. 360)

Objectives

1. Describe how the receptors and effectors of bats allow them to exploit a food supply that most other animals cannot utilize.

2. State the particular characteristics of light stimuli, sound stimuli, and chemical stimuli.

3. Define and distinguish *chemoreceptors, mechanoreceptors, photoreceptors, thermoreceptors, electroreceptors,* and *nocireceptors.* Name at least one example of each type that appears in an animal.

4. Explain how a taste bud works and distinguish the types of stimuli it detects from those detected by tactile or stretch receptors.

5. Distinguish the types of stimuli detected by tactile and stretch receptors from those detected by hearing and equilibrium receptors.

6. Explain what sound is and how animals are attuned to it.

7. Follow a sound wave from pinna to organ of Corti, mention the name of each structure it passes, and state where the sound wave is amplified and where the pattern of pressure waves is translated into electrochemical impulses.

8. State how low- and high-pitch sounds affect the organ of Corti.

9. State how low- and high-amplitude sounds affect the organ of Corti.

10. Explain how the three semicircular canals of the human ear detect changes of position and acceleration in a variety of directions. Compare semicircular canal function with statocyst function.

Self-Quiz Questions

Fill-in-the-Blanks

Cells (or parts of cells) that detect specific kinds of stimuli are (1) _____

_____. A (2) _____ is any form of energy change in the environment that

the body actually detects. (3) _____ detect impinging chemical energy;

(4) _____ detect mechanical energy associated with changes in pressure,

position, or acceleration; (5) _____ detect the energy of visible and

ultraviolet light; (6) _____ detect radiant energy associated with temperature changes; (7) _____ detect energy changes that are injurious or

painful to the body; and (8) _____ detect currents of electrical energy.
(9) _____ receptors sampling odors from food in the mouth are important for our
sense of taste. (10) _____ receptors occur in the muscles of all four-legged
animals. The (11) _____ (perceived loudness) of sound depends on the density
difference between areas of high and low pressure in wave trains. The (12) _____
(perceived pitch) of sound depends on how fast the wave changes occur. The faster the
vibrations, the (13) ◯ higher ◯ lower the sound. Hair cells are (14) ◯ nocireceptors
◯ mechanoreceptors ◯ electroreceptors that detect vibrations. The hammer, anvil, and
stirrup are located in the (15) ◯ inner ◯ middle ear. The (16) _____ is a coiled
tube that resembles a snail shell and contains the (17) _____ _____
_____, the organ that changes vibrations into electrochemical impulses. Struc-
tures that detect rotational acceleration in humans are (18) _____ _____.

23–II Light Reception

Summary

Light is a flow of discrete energy packets known as photons. Each photon is a bit of
energy that has escaped from an excited atom—not only atoms of an original light
source, such as the sun, but atoms of objects in the environment. Photons travel
through the air in a straight line. Photoreceptors detect photon energy of visible and
ultraviolet light. As the *pigment molecules* of photoreceptors selectively absorb photons, their molecular structures are altered; eventually, the alteration causes the
incoming energy of photons to be transformed into the electrochemical energy of a nerve
signal. *Vision* requires precise focusing of light onto a layer of photoreceptive cells
that is dense enough to sample details concerning the light stimulus, followed by interpretation of the signal pattern into an image, which occurs in the brain. In the
vertebrate eye, adjustments by a lens ensure that the focal point for a given batch of
light rays will fall on the retina (a tissue containing densely packed photoreceptors).

Key Terms

photons (p. 359)
photoreception (p. 360)
wavelength (p. 360)
phototaxis (p. 360)
visual system (p. 360)
lens (p. 360)
vision (p. 360)
eyespots (p. 360)
microvilli (p. 361)

eyes (p. 361)
cornea (p. 361)
retina (p. 361)
iris (p. 362)
pupil (p. 362)
compound eyes (p. 363)
ommatidium,
  -dia (p. 362)
rhabdomeres (p. 362)

optic lobe (p. 363)
mosaic theory (p. 363)
sclera (p. 363)
choroid (p. 363)
aqueous humor (p. 363)
vitreous body (p. 363)
focal point (p. 363)
accommodation (p. 363)
nearsighted (p. 364)

farsighted (p. 364)   visual acuity (p. 364)   amacrine cells (p. 365)
rod cell (p. 364)   optic nerve (p. 365)   on-center field (p. 365)
cone cell (p. 364)   bipolar cells (p. 365)   off-center field (p. 365)
fovea (p. 364)   ganglion cells (p. 365)   preliminary integration (p. 365)

Objectives

1. Define *light* and explain the general principles that affect how light is detected by photoreceptors and changed into electrochemical messages.

2. Give some examples demonstrating that many organisms lack eyes yet respond to light.

3. Explain what a visual system is and list four aspects of a visual stimulus that are detected by different components of a visual system.

4. Describe the main parts of an image-forming eye and contrast the way it functions with the way that an ocellus (eyespot) functions.

5. Contrast the structure of compound eyes with the structures of ocelli and the human eye.

6. Distinguish the way accommodation occurs in fish and in humans.

7. Define *nearsightedness* and *farsightedness*, and relate each to eyeball structure.

8. Describe how the human eye perceives color and black-and-white.

Self-Quiz Customers

Fill-in-the-Blanks

Light is a stream of (1) _____: discrete energy packets. (2) _____ is a process in which photons are absorbed by pigment molecules and photon energy is transformed into the electrochemical energy of a nerve signal. (3) _____ requires precise light focusing onto a layer of photoreceptive cells that is dense enough to sample details of the light stimulus, followed by image formation in the brain. (4) _____ are simple clusters of photosensitive cells, usually arranged in a cuplike depression in the epidermis. (5) _____ are well-developed photo-receptor organs that allow at least some degree of image formation. The (6) _____ is a transparent cover of the lens area, and the (7) _____ consists of tissue containing densely packed photoreceptors. Compound eyes contain several thousand photosensitive units known as (8) _____. In the vertebrate eye, lens adjustments assure that the (9) _____ _____ for a specific group of light rays lands on the retina. (10) _____ refers to the lens adjustments that bring about precise focusing onto the retina. (11) _____-sighted people focus light from nearby objects beyond (posterior to) the retina. (12) _____ cells are concerned with daytime vision and, usually, color perception. A (13) _____ is a funnel-shaped pit on the retina that provides the greatest visual acuity.

Summary

Motor systems, which are adapted to respond to a variety of stimuli, run parallel with sensory systems. Motor systems are based on the ability of certain cells to contract (shorten) and relax (extend in length); they also depend on the presence of some medium or structure against which the contractile force may be applied. Contractile cells exert force against a variety of resistant surfaces (cylinders of fluid encased in membrane, inner plates, exterior plates, inner rods). Connective tissue generally links the contractile cells to the resistant structure, against which they pull in an *antagonistic muscle system*, where the action of one motor element opposes the action of another. Cnidarians and annelids have hydraulic skeletons; insects, crabs, and other anthropods have exoskeletons, and vertebrates have endoskeletons.

In skeletal muscle systems, *reciprocal* innervation of reflexes between antagonistic muscle pairs is the usual basis of coordinated contractions. Contraction of a skeletal muscle cell is initiated by action potential(s) from neuron(s), causing an action potential along the muscle membrane, which in turn causes calcium ions to be released from the sarcoplasmic reticulum that surrounds the myofibrils. Each myofibril contains actin and myosin (protein filaments that lie parallel to each other). Calcium ions bind to the actin filaments, freeing them to interact with cross-bridges extending outward from the myosin filaments. The hydrolysis of ATP releases free energy that is used to slide myosin past the actin in an enzyme-controlled mechanism that telescopes the protein filaments and thus shortens the muscle. Calcium pumps eventually transport calcium ions back to the sarcoplasmic reticulum, actin no longer interacts with myosin, and the sarcomere returns to the relaxed state.

Key Terms

motor system (p. 366)
contractile
  proteins (p. 366)
relax (p. 366)
actin filaments (p. 366)
myosin filaments (p. 366)
antagonistic muscle
  system (p. 367)
segmented (p. 367)
longitudinal
  muscles (p. 367)
circular muscles (p. 367)
hydraulic skeleton (p. 367)

setae (p. 367)
cuticle (p. 367)
chitin (p. 367)
exoskeleton (p. 367)
endoskeleton (p. 367)
axial portion (p. 367)
appendicular
  portion (p. 368)
tendons (p. 368)
ligaments (p. 368)
skeletal muscles (p. 368)
biceps (p. 368)
triceps (p. 368)

flex (p. 368)
extend (p. 368)
reciprocal
  innervation (p. 368)
smooth muscle
  tissue (p. 368)
tendon of origin (p. 368)
tendon of
  insertion (p. 368)
breast bone (p. 368)
cardiac muscle
  tissue (p. 369)
motor end plate (p. 369)

neuromuscular
  junction (p. 369)
motor unit (p. 369)
myofibrils (p. 369)
A band (p. 369)
Z lines (p. 370)
sarcomere (p. 370)
muscle fiber (p. 370)

sliding filament
  model (p. 370)
cross-bridges (p. 370)
myosin heads (p. 370)
actin binding sites (p. 370)
lactate (p. 370)
creatine phosphate (p. 370)
rigor (p. 370)
rigor mortis (p. 370)

action potential (p. 371)
acetylcholine (p. 371)
calcium ions (p. 371)
sarcoplasmic
  reticulum (p. 371)
tropomyosin (p. 371)
troponin (p. 371)
sarcolemma (p. 371)

Objectives

1. Distinguish between contraction and relaxation of muscle cells.

2. Explain what is meant by "antagonistic muscle system."

3. Explain how segmentation, longitudinal muscles, circular muscles, and a hydraulic skeleton are functionally interrelated in earthworms.

4. Describe the general design of the human skeletal system. Name six bones (or groups of bones) that compose an arm. Name the parts of the axial skeleton.

5. Distinguish *tendons* from *ligaments,* and *origin* from *insertion.*

6. Explain how an endoskeleton and an antagonistic muscle system are functionally related in the human action of moving the thumb away from the index finger and in moving the two together. Draw the general shapes of the bones involved and then try to draw in muscles showing how shortening of the muscles causes the action mentioned.

7. Define reciprocal innervation and explain how it helps coordinate motor elements.

8. Describe the fine structure of a muscle fiber; use terms such as *myofibril, sarcoplasmic reticulum, sarcomere, neuromuscular junction, actin*, and *myosin*.

9. List, in sequence, the biochemical and fine structural events that occur during the contraction of a skeletal muscle fiber. Then explain how the fiber relaxes.

10. Refer to Fig. 23.16 and indicate (a) a muscle used in sit-ups, (b) another used in dorsally flexing and inverting the foot, and (c) another used in flexing the elbow joint.

Self-Quiz Questions

Fill-in-the-Blanks

All motor systems are based on effector cells able to (1) _____ and (2) _____, and on the presence of a medium against which the (3) _____ force may be applied. Longitudinal and circular muscle layers work as an (4) _____ muscle system, in which action of one motor element opposes the action of another. A membrane filled with fluid resists compression and can act as a (5) _____ skeleton. Arthropods have antagonistic muscles attached to an (6) ○ endoskeleton ○ exoskeleton. In skeletal muscle, (7) _____ _____ of reflexes between antagonistic muscle pairs is the basis for coordinated contractions. Each myofibril contains (8) _____

filaments, which have crossbridges, and (9) _____ filaments, which are thin and

lack crossbridges. The repetitive fundamental unit of muscle contraction is the (10)

_____. The sarcoplasmic reticulum releases (11) _____ ions as a con-

sequence of charge reversal.

23–IV   Types of Muscular Responses to Action Potentials 373
Neuromotor Basis of Behavior 373

Summary

Muscular contractions that are initiated by a single action potential are known as
twitches. Each twitch consists of a latent period, actual contraction, then relaxation;
twitches are not generally important in normal sustained muscle contractions. A muscle
may be maintained in a state of contraction called *tetanus* (in which frequent action
potentials do not allow the muscle to relax). Tetanic contractions are normally encountered in skeletal muscle contraction. Low levels of tetanic contraction are called
muscle tone, the maintenance of which is coordinated by the spinal cord.
     Animal behavior is coordinated neuromotor response to changes in external and in-
ternal states. It is produced by the integration of diverse sensory, genetic, neural,
and endocrine factors, and it may be modified through learning. The internal states of
an animal influences its responsiveness to external stimuli and thus the behaviors that
are activated. Behaviors associated with feeding, aggression, and mating are programmed
through specific centers in the brain and are influenced by both internal and external
stimuli.

Key Terms

twitch contraction (p. 373)   tetanic contraction (p. 373)   stretch reflex (p. 373)
latent period (p. 373)   summation (p. 373)   reflex arc (p. 373)
actual contraction (p. 373)   tetanus (p. 373)   integrative centers (p. 373)
relaxation (p. 373)   muscle tone (p. 373)   memory banks (p. 373)

Objectives

1. Distinguish *twitch* contractions from *tetanic* contractions.

2. Name three factors that influence neuromotor responses.

Self-Quiz Questions

  Fill-in-the-Blanks

(1) _____ ordinarily do not contribute to normal, sustained muscle contrac-

tions, but (2) _____ contractions are the normal mode of contraction. Low levels

of tetanic contractions are collectively referred to as (3) _____ _____.

INTEGRATING AND APPLYING KEY CONCEPTS

How might human behavior be changed if human eyes were compound eyes composed of ommatidia and if humans perceived only vibrations--as fish do--rather than sounds?

24
CIRCULATION

Summary

A *circulatory system*, and the *blood* it transports through the animal body, carries vital material to cells, carries products and wastes from them, and helps maintain an internal environment that is favorable for cell activities. Flatworms and less complex animals lack networks of blood transport vessels because none of the body cells is too far from the external world to be able to obtain the necessary oxygen and nutrients or to eliminate wastes. In more complex animals, various circulatory systems have evolved for exchanging materials between the internal and external environments. The closed circulatory systems of some invertebrates and all vertebrates contain blood enclosed within the walls of one or more hearts and blood vessels. The *open circulation systems* of many invertebrates (including insects) pump blood from the heart into one or more blood vessels that open directly into tissue spaces. The blood bathes nearby cells and then seeps sluggishly back to the heart. Animals with closed circulatory systems often have a supplementary *lymph vascular system*, which recovers and purifies interstitial fluid and returns it to the major blood vessels.

Vertebrate blood is composed of *plasma* (55%) and formed elements (45%). Plasma is mostly water (91.5%) and plasma proteins (7%) such as albumin, the globulins, and fibrinogen; however, nutrients, dissolved gases, hormones, and wastes are also dissolved or suspended. *Red blood cells* transport O_2, *leukocytes* defend against tissue damage, and infections and platelets are involved in forming blood clots.

Key Terms

blood (p. 374)
open circulation
   systems (p. 374)
closed circulation
   systems (p. 374)
interstitial fluid (p. 375)
constriction (p. 375)
dilation (p. 375)
lymph vascular
   system (p. 375
blood volume (p. 375)
erythrocytes, red
   blood cells (p. 375)
biconcave (p. 375)
hemoglobin (p. 375)
heme (p. 375)

oxyhemoglobin (p. 375)
carbamino-
   hemoglobin (p. 375)
red bone marrow (p. 375)
cell count (p. 375)
phagocytic (p. 376)
erythropoietin (p. 376)
leukocytes, white
   blood cells (p. 376)
stem cells (p. 376)
lymphocytes (p. 376)
T-cells (p. 376)
B-cells (p. 376)
plasma cells (p. 376)
antibodies (p. 376)
neutrophils (p. 376)
monocytes (p. 376)

basophils (p. 376)
eosinophils (p. 376)
clotting (p. 376)
"granular" (p. 376)
"agranular" (p. 376)
platelets,
   thrombocytes (p. 376)
megakaryocytes (p. 377)
plasma (p. 377)
plasma proteins (p. 377)
albumin (p. 377)
alpha globulin (p. 377)
beta globulin (p. 377)
gamma globulin (p. 377)
fibrinogen (p. 377)
lipoproteins (p. 377)

Objectives

1. Explain why some organisms are able to manage without a system of pumps and vessels to distribute body fluids.

2. Distinguish between an open and a closed circulation system.

3. State what the lymph vascular system does.

4. Describe, using percentages of volume, the composition of human blood.

5. Name the plasma components and their functions.

6. Explain how red blood cells manage to carry both O_2 and CO_2 at different times.

7. State where erythrocytes, leukocytes, and platelets are produced.

8. Distinguish each of the five types of leukocytes from each other in terms of structure and functions.

Self-Quiz Questions

Fill-in-the-Blanks

Blood is a highly specialized fluid (1) _____ tissue that helps to stabilize

internal (2) _____ and to equalize internal temperature throughout an animal's body. Organisms with (3) _____ circulation systems generally also have a supplementary (4) _____ _____ _____ that recovers and purifies interstitial fluid and returns it to the major blood vessels. In the hemoglobin molecule, a central (5) _____ _____ is associated with four (6) _____ _____. Oxygen binds with the (7) _____ atom of a heme group and forms the substance (8) _____. When oxygen levels in tissues are low (as they would be if you moved from Phoenix, Arizona to Santa Fe, New Mexico), the kidneys secrete a substance that combines with a plasma protein, converting it to the hormone known as (9) _____ that stimulates the production of erythrocytes by (10) _____ _____. (11) _____ cells are immature cells not yet fully differentiated. (12) _____ bind to extracellular invaders and mark them for destruction; they are secreted by (13) _____ _____. (14) _____ and monocytes are highly mobile and phagocytic; they chemically detect, ingest, and destroy bacteria, foreign matter, and dead cells. (15) _____ (thrombocytes) are cell fragments that aid in forming blood clots.

24–II STRUCTURE AND FUNCTION OF BLOOD CIRCULATION SYSTEMS 377–386

Summary

Arteries, which branch into smaller tubes, or arterioles, carry blood away from the heart. Through controls over their musculature, arterioles can vary the resistance to blood flow along different routes; hence they control the distribution of blood flow through the body. Many thin-walled capillaries diverge from the arterioles and represent an immense cross-sectional exchange area between blood and interstitial fluid. Capillaries converge into venules which converge into veins. Because of their great elasticity, veins and venules can function as blood volume reservoirs during times of low metabolic output. Terrestrial vertebrates have two major blood transport routes: *pulmonary* circulation to and from the lungs; and *systemic* circulation to and from the rest of the body.

Zones of the vertebrate heart that receive blood being returned from the body are called *atria*; other zones called *ventricles* pump the blood out. Fishes have two-chambered hearts, with one atrium and one ventricle, amphibians have two atria and one ventricle, and birds and mammals both have two atria and two ventricles. The pacemaker of the heart has the fastest inherent rate of firing. Each contraction period (systole) is followed by a rest period (diastole), in which the heart muscle cells relax and the heart fills; each systole and its diastole constitute one cardiac cycle. The pulse pressure is the difference between the systolic and diastolic readings. Blood pressure is not the same at the start and the end of the circuit. The pressure gradient that heart contractions establish is aided by pressure generated by muscle pumps and by respiration; these forces together work to return venous blood to the heart.

Key Terms

heart (p. 377)
capillary beds (p. 377)
pulmonary
   circulation (p. 377)
systemic
   circulation (p. 377)
pericardium (p. 378)
myocardium (p. 378)
endocardium (p. 378)
atrium, atria (p. 378)
ventricle (p. 378)
aorta (p. 378)
pulmonary
   artery (p. 378)
valves (p. 378)
atrioventricular
   valve (p. 378)
semilunar valve (p. 378)
cardiac cycle (p. 378)
systole (p. 378)
diastole (p. 378)
superior vena cava (p. 379)

inferior vena cava (p. 379)
septum (p. 379)
intercalated disk (p. 380)
gap junctions (p. 380)
self-excitatory (p. 380)
cardiac conduction
   system (p. 380)
sinoatrial node (p. 380)
cardiac pacemaker (p. 380)
artery, arteries (p. 381)
arterioles (p. 381)
atrioventricular node (p. 381)
ventricular conducting
   fibers (p. 381)
heart apex (p. 381)
vasoconstriction (p. 381)
vasodilation (p. 381)
basement membrane (p. 382)
endothelium (p. 382)
capillary (p. 382)
osmotic pressure (p. 382)
hydrostatic force (p. 382)

net outward filtration
   pressure (p. 382)
net inward osmotic
   pressure (p. 382)
filtration (p. 382)
coronary
   circulation (p. 382)
coronary arteries (p. 382)
venules (p. 383)
veins (p. 383)
sympathetic nerves (p. 383)
precapillary
   sphincters (p. 383)
lumen (p. 383)
venous valves (p. 383)
blood pressure (p. 383)
cardiac output (p. 383)
total peripheral
   resistance (p. 383)
pulse pressure (p. 384)
pressure reservoir (p. 384)

Objectives

1. Describe how the vertebrate heart has evolved from that of fishes to that of mammals.

2. Trace the path of blood in the human body. Begin with the aorta and name all major components of the circulatory system that the blood passes through before it returns to the aorta.

3. Explain what causes a heart to beat. Then describe how the rate of heartbeat can be slowed down or speeded up.

4. List the factors that cause blood to leave the heart and the factors that cooperate to return blood to the heart.

5. List the five types of vessels common to vertebrate circulatory systems. Distinguish arteries from veins and then distinguish each from capillaries.

6. Explain how veins and venules can act as reservoirs of blood volume.

7. Explain why exchange of gases, nutrients, and wastes occurs across capillaries rather than across arterial walls.

8. Distinguish pulmonary circulation from systemic circulation.

Fill-in-the-Blanks

A "receiving zone" of a vertebrate heart is called a(n) (1) _____; a "departure zone" is called a(n) (2) _____. Each contraction period is called (3) _____; each relaxation period is (4) _____. Label the organ parts shown on the following diagram.

(5) _____

(6) _____

(7) _____

(8) _____

(9) _____

(10) _____

(11) _____

(12) _____

A(n) (13) _____ carries blood away from the heart. A(n) (14) _____ is a blood vessel with such a small diameter that red blood cells must flow through it single-file; its wall consists of no more than a single layer of (15) _____ cells. In each (16) _____ _____, small molecules move between the bloodstream and the (17) _____ fluid. (18) _____ are in the walls of veins and prevent backwashing. Both (19) _____ and (20) _____ serve as a temporary reservoir for blood volume. The heart is a pumping station for two major blood transport routes: the (21) _____ circulation to and from the lungs and the (22) _____ circulation to and from the rest of the body.

True-False If false, explain why.

____ (23) The pulse rate is the difference between the systolic and the diastolic pressure readings.

____ (24) Since the total volume of blood remains constant in the human body, blood pressure must also remain constant throughout the circuit.

COMMENTARY: On Cardiovascular Disorders   386-387

Summary

The blood vessels of many invertebrates and all vertebrates constrict in a rapid reflex
response to hemorrhage. In vertebrates, this muscular reflex is followed by platelets
adhering and clumping, plugging the rupture and releasing vasoconstrictors that pro-
long the muscle spasm. A clot forms as the blood is converted to a solid gel; the clot
then retracts into a compact mass that draws ruptured walls back together. The formation of a blood clot involves a series of at least thirteen reactions; if one important
substance is missing, as in hemophilia, a clot will not form.

A variety of cardiovascular disorders occurs in humans. Two disorders of the arterial walls, *arteriosclerosis* and *atherosclerosis* promote high blood pressure, which can
lead to the rupture of small bood vessels. The rupture of such vessels in the brain can
cause a *stroke*. A *coronary occlusion* is a blockage of the heart's own blood supply.
Hypertension--a gradual increase in arterial blood pressure--is one of the most
prevalent killers in the United States.

Key Terms

hemostasis (p. 385)   prothrombin (p. 385)   hypertension (p. 386)
hemorrhage (p. 385)   thrombin (p. 385)   plaque (p. 387)
muscle spasm (p. 385)   fibrinogen (p. 385)   platelets (p. 387)
platelet plug   fibrin (p. 386)   clot (p. 387)
   formation (p. 385)   blood clot (p. 386)   coronary arteries (p. 387)
collagen fibers (p. 385)   extrinsic clotting   coronary occlusion (p. 387)
serotonin (p. 385)   mechanism (p. 386)   high-density lipoproteins (p. 387)
vasoconstrictor (p. 385)   arteriosclerosis (p. 386)   low-density lipoproteins (p. 387)
clot formation (p. 385)   stroke (p. 386)   Type A personalities (p. 387)
intrinsic clotting   atherosclerosis (p. 386)   Type B personalities (p. 387)
   mechanism (p. 385)

Objectives

1. List the homeostatic events that occur when small blood vessels are damaged.

2. List in sequence the events that occur in the formation of a blood clot.

3. Distinguish a stroke from atherosclerosis.

4. Distinguish a stroke from a coronary occlusion.

5. Describe how hypertension develops, how it is detected, and whether or not it can
    be corrected.

6. State the significance of high- and low-density lipoproteins to cardiovascular
    disorders.

Self-Quiz Questions

   Fill-in-the-Blanks

Bleeding is stopped by several mechanisms that are referred to as (1) _____ ;

the mechanisms include blood vessel spasm, (2) _____ _____ _____,
and blood (3) _____. Once the platelets reach a damaged vessel, through chemical recognition they adhere to exposed (4) _____ fibers in damaged vessel walls.
When attached, the platelets release ADP (which attracts still more platelets) and (5)
_____ ions, which promote clumping into a plug. Clumped platelets release a
phospholipid that reacts with calcium ions, and damaged tissues release thromboplastin.
Together, these substances activate prothrombin, which is converted to the enzyme (6)
_____. Cascade reactions act on a soluble plasma protein, fibrinogen, which is
then assembled into long, insoluble (7) _____ fibers. These trap blood cells and
components of plasma. Eventually, under normal conditions, a clot is formed at the
damaged site. One cause of (8) _____ is the rupture of one or more blood vessels in the brain. A (9) _____ _____ blocks a coronary artery. (10)
_____ is a term for a formation that can include cholesterol, calcium salts,
and fibrous tissue. It is not healthful to have a high concentration of (11) _____-
density lipoproteins in the bloodstream.

24–IV   LYMPHATIC SYSTEM

Summary

The *lymph vascular system* accumulates, purifies, and returns excess fluid and a small
amount of proteins in the bloodstream from whence they came. It also transports fats
absorbed from the digestive tract to the bloodstream. *Lymph* is the tissue fluid in the
vessels of the lymphatic system. Cells of the immune system are strategically dispersed
through the body, especially in tissues of the lymphoid system: bone marrow, thymus,
lymph nodes, spleen, appendix, tonsils, adenoids, and tissues associated with the small
intestine. Nonspecific immune responses oversee first-time contacts as well as later
encounters with foreign matter.

hydrostatic pressure (p. 386)   collecting ducts (p. 388)   plasma cells (p. 388)
osmotic pressure (p. 386)   right lymphatic   macrophages (p. 388)
lymphatic system (p. 386)   duct (p. 388)   red pulp (p. 388)
lymph (p. 386)   quadrant (p. 388)   white pulp (p. 388)
lymph vascular   thoracic duct (p. 388)   thymus (p. 389)
  system (p. 388)   lymphoid organs (p. 388)   exocrine glands (p. 389)
lymph nodes (p. 388)   spleen (p. 388)   inflammation (p. 389)
lymph vessels (p. 388)   tonsils (p. 388)   complement (p. 389)
lymph capillaries (p. 388)   lymphocytes (p. 388)

Objectives

1. List three functions of the lymph vascular system.

2. List as many tissues and organs of the lymphoid system as you can.

3. Distinguish nonspecific from specific immune responses by defining and giving an example of each.

4. Explain what happens during the inflammatory response.

Self-Quiz Questions

Fill-in-the-Blanks

Label each numbered part on the drawing.

(1) _____

(2) _____

(3) _____

(4) _____

(5) _____

(6) _____

(7) _____

(8) _____

(9) _____ vessels reclaim fluid lost from the bloodstream, purify the blood of

microorganisms, and transport (10)_____ from the (11) _____ _____

to the bloodstream. Platelets are produced in the (12) _____ _____; two-

thirds of them circulate in the blood, but the remainder are stored in the (13)

_____, which contracts when the nervous system detects (14) _____. (15)

_____, a group of about fifteen circulating proteins, helps in mobilizing the

immune response more effectively.

Evolution of Vertebrate Immune Systems   390

Killer T-Cells and the B-Cell Ways of Chemical Warfare   390–392, 395

Summary

The *vertebrate immune system* includes a variety of cells (granulocytes, monocytes, macrophages, lymphocytes, and plasma cells) and cell products (complement, antibodies, antitoxins). This system is concerned with recognizing microbial invaders as well as cells or substances that don't belong in a given tissue, and selectively eliminating or neutralizing them.

  Specific immune responses amplify the nonspecific ones by recognizing and disposing of invaders in a highly discriminatory way. Lymphocytes are divided into two groups: *T-cells*, which spend some time in the thymus, where they acquire specific substances on their cellular surfaces that enable them to battle with viruses, fungi, parasites, cancer cells, and graft cells; and *B-cells*, which differentiate into the plasma cells that secrete the *antibodies* that set up bacterial invaders for subsequent destruction by macrophages.

Key Terms

granulocytes (p. 390)
monocytes (p. 390)
macrophages (p. 390)
lymphocytes (p. 390)
plasma cells (p. 390)
vertebrate immune
  system (p. 390)
defense (p. 390)
extracellar
  housekeeping (p. 390)
immune surveillance (p. 390)
"self" (p. 390)

"nonself" (p. 390)
T-cells (p. 390)
B-cells (p. 390)
stem cells (p. 390)
precommitted (p. 390)
killer T-cells (p. 390)
"helper" T-cells (p. 390)
antibodies (p. 390)
antigen (p. 392)
primary immune
  pathway (p. 392)
primary immune
  response (p. 390)

secondary immune
  pathway (p. 392)
memory lymphocytes (p. 392)
secondary immune
  response (p. 392)
amplified (p. 392)
immunization (p. 392)
vaccine (p. 392)
Sabin polio vaccine
  (p. 392)
"booster" shot (p. 395)
passive immunity (p. 395)

Objectives

1. Describe the possible origins and evolution of the vertebrate immune system.

2. Define and discuss the three principal functions of the immune system: *defense, extracellular housekeeping,* and *immune surveillance.*

3. Explain how stem cells differ from those that differentiate earlier in development.

4. Distinguish the roles of T-cells and B-cells.

5. Explain what is meant by *primary immune pathway* as contrasted with *secondary immune pathway.*

6. Describe two ways that people can be immunized against specific diseases.

Fill-in-the-Blanks

Among single-celled organisms, (1) _____ was a means of ingesting food that probably proved adaptive in defense. Through events called (2) _____ _____, cells of the vertebrate immune system remove mutant and cancerous cells as well as graft cells surgically transplanted from other individuals. Two distinct (3) _____ populations carry out specific immune responses: (4) _____ and (5) _____. Both kinds arise from (6) _____ _____ in bone marrow. Some of these progeny move to the thymus where they acquire specific (7) _____ on their cell surfaces; in doing so, they become (8) _____. (9) _____ clones indirectly attack their targets through the production of antibodies. The (10) _____ _____ _____ is the route taken during a first-time contact with an antigen. Deliberately provoking the production of memory lymphocytes is known as (11) _____.

Identify each numbered blank in the accompanying diagram.

(12) _____

(13) _____

(14) _____

(15) _____

(16) _____

(17) _____

(18) _____

(19) _____

(20) _____

(21) _____

Stimulus:
TISSUE INVASION

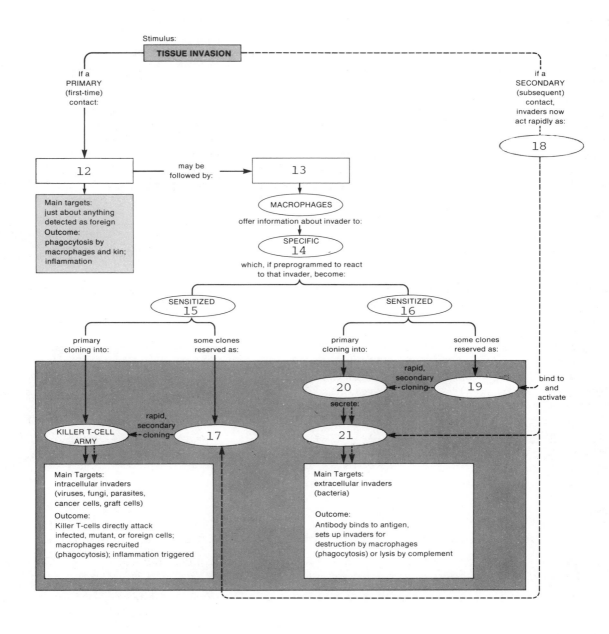

If a
PRIMARY
(first-time)
contact:

if a
SECONDARY
(subsequent)
contact,
invaders now
act rapidly as:

18

12 → may be followed by: → 13

Main targets:
just about anything
detected as foreign
Outcome:
phagocytosis by
macrophages and kin;
inflammation

MACROPHAGES

offer information about invader to:

SPECIFIC
14

which, if preprogrammed to react
to that invader, become:

SENSITIZED
15

SENSITIZED
16

primary
cloning into:

some clones
reserved as:

primary
cloning into:

some clones
reserved as:

rapid,
secondary
cloning

bind to
and
activate

20 ← rapid, secondary cloning → 19

secrete:

KILLER T-CELL
ARMY ← rapid, secondary cloning → 17

21

Main Targets:
intracellular invaders
(viruses, fungi, parasites,
cancer cells, graft cells)

Outcome:
Killer T-cells directly attack
infected, mutant, or foreign cells;
macrophages recruited
(phagocytosis); inflammation triggered

Main Targets:
extracellular invaders
(bacteria)

Outcome:
Antibody binds to antigen,
sets up invaders for
destruction by macrophages
(phagocytosis) or lysis by complement

Summary

The vertebrate body becomes able to distinguish self from nonself cells during embryonic development. In all mammals, a group of genes called the *major histocompatibility complex* controls the production of surface recognition factors present on all nucleated cells. Except for identical twins, the cells of different individuals have different patterns of surface recognition factors. T-cells recognize the differences and destroy the nonself invaders. Immune surveillance theory suggests that cancer cells may arise in your body every day as a result of mutations induced by events such as viral attack, chemical bombardment, or irradiation. As fast as cancer cells arise, circulating sensitized lymphocytes become activated and begin cloning themselves rapidly and secreting lymphokines, which mobilize voracious macrophages to destroy the mutant cells. *Allergies* and *autoimmune disease* are examples of the ways that the immune system can damage the body instead of protecting it.

Key Terms

malignant cells (p. 392)
immune surveillance
  theory (p. 392)
sensitized
  lymphocytes (p. 392)
activated
  lymphocytes (p. 392)
lymphokines (p. 392)
macrophage (p. 392)
chemotherapy (p. 393)
immune therapy (p. 394)
interferons (p. 394)

monoclonal
  antibodies (p. 394)
targeted drug
  therapy (p. 394)
IgE (p. 394)
allergy (p. 394)
histamine (p. 394)
serotonin (p. 394)
prostaglandins (p. 394)
embryo (p. 395)
split genes (p. 395)
major histocompatibility
  complex (MHC) (p. 395)

surface recognition
  factors (p. 395)
T-cells (p. 395)
"blocking"
  antibodies (p. 395)
autoimmune disease (p. 395)
insulin-dependent
  diabetes mellitus (p. 395)
thymus (p. 396)
complement (p. 396)
memory lymphocytes (p. 396)

Objectives

1. Describe the physical basis that establishes self cells and nonself cells.

2. Discuss the odds that transplants of organs or tissues between close family members will take hold.

3. Explain what immunosurveillance theory has to say about the development of cancer. Discuss the measures available when cancer cells slip through the immunosurveillance network.

4. Explain how allergy differs from autoimmune disease.

5. Trace the sequence of immunological events that would help you to defeat bacterial invaders that get past your first line of defense.

Fill-in-the-Blanks

The (1) _____ _____ _____ controls the production of surface

markers that identify specific nucleated cells. (2) _____ _____ is a

disorder in which the body mobilizes its forces against certain of its own tissues.

(3) _____ is an altered secondary response to a normally harmless substance

that may actually cause injury to tissues. The (4) _____ _____ _____

hypothesizes that cancer cells are normally destroyed before they can multiply suffi-

ciently for anyone to detect them. Mutant cells sensitize and activate lymphocytes;

some become (5) _____ _____ and others release (6) _____ — sub-

stances that arouse macrophages. Aroused macrophages war against any invaders, in-

cluding cancer cells. (7) _____ comprise a group of small proteins that display

antiviral activity and induce resistance to a wide range of viruses. Milstein and

Kohler developed a means of producing large amounts of (8) _____ _____ .

(9) _____ drug therapy hooks up anticancer drug molecules with monoclonal anti-

bodies, which locate precisely where the cancer cells are and deliver the drug to

them only. The nonspecific defense mechanism known as (10) _____ stimulates

phagocytes to follow the gradient of these small proteins to the offending invaders.

INTEGRATING AND APPLYING KEY CONCEPTS

You observe that some people appear as though fluid had accumulated in their lower
legs and feet. Their lower extremities resemble those of elephants. You inquire
about what is wrong and are told that the condition is caused by the bite of a
mosquito that is active at night. Construct a testable hypothesis that would
explain (1) why the fluid was not being returned to the torso, as normal, and
(2) what the mosquito did to its victims.

25
RESPIRATION

25–I   OXYGEN ACQUISITION: SOME ENVIRONMENTAL CONSIDERATIONS

RESPIRATION: AN OVERVIEW

Summary

The overall exchange of gases among cells, the blood, and the environment is known as respiration. Any respiratory membrane must be kept moist on both surfaces for gas exchange to occur. When moisture retention is not a problem (as it is not for water-dwelling animals), gas exchange is enhanced by *evagination* (the folding outward of a membrane, such as a gill) and by *countercurrent flow mechanisms*, in which water flows past the blood stream in the opposite direction. When moisture retention is a problem (as it can be for land-dwelling animals), gas exchange is enhanced by *invagination* (formation of an inward-directed pocket in a membrane, as occurs in air sacs and lungs.

Key Terms

aerobic
  respiration (p. 398)
viscous (p. 398)
gill (p. 398)
evagination (p. 398)
gill filaments (p. 398)

gill arch (p. 398)
epithelial layer (p. 399)
lamellae (p. 399)
countercurrent flow (p. 399)
invaginations (p. 399)

tracheas (p. 399)
chitin (p. 399)
spiracle (p. 399)
air capillaries (p. 399)
lungs (p. 399)
alveolus, alveoli (p. 400)

Objectives

1. Describe the ways that respiration differs in aquatic and terrestrial animals.

2. Define *countercurrent flow mechanism* and explain how it works. State where such a mechanism is found.

3. Describe how incoming oxygen is distributed to the tissues of insects and contrast it with the process that occurs in mammals.

4. List all the principal parts of the human respiratory system and explain how each structure contributes to transporting oxygen from the external world to the bloodstream.

Self-Quiz Questions

Fill-in-the-Blanks

A (1) _____ is an evaginated thin, moist membrane endowed with blood vessels.

Gas transfer is enhanced by a (2) _____ _____ _____, in which

water flows past the bloodstream in the opposite direction. Insects have (3)

_____: chitin-lined air tubes leading from the body surface to the interior.

One layer of a fragile membrane, the (4) _____, covers the surface of each lung;

the other membrane layer adheres to the wall of the chest cavity. Gas exchange occurs

across the epithelium of the (5) _____.

Summary

In the human respiratory system, oxygen moves inward and carbon dioxide moves outward through the same branched tubes. Gas exchange in the *same* tubes is possible because of rhythmic changes in pressure gradients between the lungs and the atmosphere. The diaphragm, a muscular partition that separates the chest and abdominal cavities, moves downward and flattens during normal *inhalation*; the rib cage moves outward and upward, and the volume of the chest cavity increases and its internal pressure drops. Since atmospheric pressure remains the same, air rushes into the respiratory tract. During *exhalation*, the diaphragm and rib cage return to their resting position, the volume of the chest cavity decreases, and the resultant rise in pressure on the lungs forces air outward.

A respiratory center in the *brainstem* governs the rate of breathing in accord with information detected by chemoreceptors in arterial walls and in the brainstem itself. In the human respiratory system, air normally enters the body via the nose and nasal cavity, where it is cleaned, warmed, and moistened. The air then moves through the *pharynx* and down through the cartilage-reinforced tubes called the *larynx* (voice-box) and *trachea* (windpipe). The trachea diverges into two *bronchi,* which diverge further into *bronchioles.* Air finally ends up in tiny pouches called *alveoli.* It is across the epithelium of these alveoli that gas exchange occurs.

Key Terms

epiglottis (p. 400)
true vocal cords (p. 400)
glottis (p. 400)
bronchus, bronchi (p. 400)
bronchioles (p. 400)
alveolar ducts (p. 400)
thoracic cavity (p. 400)
ventilation (p. 400)
respiration (p. 400)
respiratory system (p. 400)
nasal cavities (p. 400)
pharynx (p. 400)
larynx (p. 400)
pleural sac (p. 401)

inhalation (p. 402)
inspiration (p. 402)
bronchial tree (p. 402)
exhalation (p. 402)
expiration (p. 402)
diaphragm (p. 402)
intercostal
  muscles (p. 402)
intrapleural
  pressure (p. 402)
alveolar air
  pressure (p. 402)
parietal pleura (p. 402)
pulmonary pleura (p. 402)

intrapleural fluid (p. 402)
phrenic nerves (p. 403)
poliovirus (p. 403)
intercostal
  nerves (p. 403)
cervical (p. 403)
"artificial lung" (p. 403)
medulla (p. 403)
motor nerves (p. 403)
respiratory center (p. 404)
brainstem (p. 404)
chemoreceptors (p. 404)
cerebrospinal fluid (p. 404)

Objectives

1. Explain how breathing occurs by listing the principal structures involved in inhalation and exhalation and describing the behavior of each during each process.

2. List the structures that are involved in detecting carbon dioxide levels in the blood and regulating the rate of breathing. Name the location of each structure.

3. Describe the relationship of the human lungs to the pleural sac and the thoracic cavity.

Self-Quiz Questions

Fill-in-the-Blanks

During inhalation, the (1) _____ moves downward and flattens, and the (2) _____ _____ moves outward and upward; when these things happen, the *chest cavity volume* (3) ○ increases ○ decreases, and its internal pressure (4) ○ rises ○ drops ○ stays the same. The rate of breathing is governed by a respiratory center in the (5) _____, which monitors signals coming in from arterial walls, blood vessels, and other brain regions. The (6) _____ _____ surrounds each lung. In order, air passes through the nasal cavities, pharynx, and (7) _____, past the epiglottis into the (8) _____, which is the space between the true vocal cords; into the trachea, then to the (9) _____, (10) _____, and alveolar ducts. Exchange of gases occurs across the epithelium of the (11) _____.

Summary

As oxygen moves into the blood plasma and then into red blood cells, four oxygen molecules form weak bonds with hemoglobin, forming oxyhemoglobin, and thus the blood can carry about fifty times as much oxygen as plasma alone could. The waste product carbon dioxide diffuses into red blood cells and is converted into bicarbonate ions (which diffuse back into the plasma) and carbaminohemoglobin; in this form it is transported to the lung capillaries, where it is again converted to CO_2 and then exhaled.

Bronchitis is characterized by degradation of bronchial tissue and the formation of scar tissue. If excess CO_2 becomes trapped by scar tissue in the alveoli, the normally pink, elastic lung sacs become dry and perforated; the outcome is *emphysema*, with a severe reduction in the efficiency of gas exchange. *Lung cancer* is believed to be caused largely by compounds found in coal tar and cigarette smoke, which are chemically modified in the body into carcinogens that interfere with normal gene expression.

Key Terms

basement membrane (p. 404)
interstitial fluid (p. 404)
atmospheric pressure (p. 404)
760 mm Hg (p. 404)
partial pressure (p. 404)
passive diffusion (p. 404)
hemoglobin (p. 404)
oxyhemoglobin (p. 404)
hypoxia (p. 405)
hemoglobin binding sites (p. 405)
saturated (p. 405)
carbaminohemoglobin, $HbCO_2$ (p. 405)
bicarbonate ions, HCO_3^{-1} (p. 405)
carbonic acid, H_2CO_3 (p. 405)
carbonic anhydrase (p. 405)
buffer (p. 405)
dissociates (p. 406)
affinity (p. 406)
systemic highways (p. 406)

ciliated epithelium (p. 406)
mucus-secreting goblet cells (p. 406)
"smoker's cough" (p. 406)
chronic bronchitis (p. 407)
emphysema (p. 407)
lung cancer (p. 407)
methylcholanthrene (p. 407)
coal tar (p. 407)
carcinogens (p. 407)
laryngeal cancer (p. 408)
esophageal cancer (p. 408)
pancreatic cancer (p. 408)
bladder cancer (p. 408)
coronary heart disease (p. 408)
stillbirths (p. 408)
impaired immune system function (p. 408)

Objectives

1. Explain why oxygen diffuses from alveolar air spaces, through interstitial fluid, and across capillary epithelium. Then explain why carbon dioxide diffuses in the reverse direction.

2. Explain why oxygen diffuses from the bloodstream into the tissues far from the lungs. Then explain why carbon dioxide diffuses into the bloodstream from the same tissues.

3. Explain the roles of oxyhemoglobin and carbaminohemoglobin.

4. Describe what happens to carbon dioxide when it dissolves in water under conditions normally present in the human body.

5. State the role played by carbonic anhydrase in carbon dioxide transport.

6. Distinguish bronchitis from emphysema. Then explain how lung cancer differs from emphysema.

Self-Quiz Questions

Fill-in-the-Blanks

Oxygen is said to exert a (1) _____ _____ of 760/21 or 160mm Hg. (2) _____ _____ alone moves oxygen from the alveoli into the bloodstream, and it is enough to move (3) _____ _____ in the reverse direction. (4) _____ is the medical name for oxygen deficiency; it is characterized by faster breathing, faster heart rate, and anxiety at altitudes of 8,000 feet above sea level. Eighty-one percent of the carbon dioxide in the blood is transported as (5)_____. Without (6) _____, the plasma would be able to carry only about two percent of the oxygen that whole blood carries. When oxygen-rich blood reaches a (7) _____ tissue capillary bed, oxygen diffuses outward, and carbon dioxide moves from tissues into the capillaries. With the assistance of (8) _____ _____, carbonic acid dissociates to form water and carbon dioxide. (9) _____ is the distension of lungs and the loss of gas exchange efficiency such that running, walking, and even exhaling are painful experiences. At least ninety percent of all (10) _____ _____ deaths are the result of cigarette smoking; only about ten percent of afflicted individuals will survive.

INTEGRATING AND APPLYING KEY CONCEPTS

Consider the amphibians: animals that generally have aquatic larval forms (tadpoles) and terrestrial adults. Outline the changes that you think must occur as an aquatic tadpole metamorphoses into a land-going juvenile.

26

DIGESTION AND ORGANIC METABOLISM

26–I FEEDING STRATEGIES   409-410

Summary

Nutrition is a function that includes ingesting, digesting, absorbing, and assimilating
nutrients. Nutrition depends on digestion, circulation, and respiration as well as on
obtaining nutrients in the first place. An *incomplete digestive system* has only one
opening, which serves both as an entrance for newly obtained food and an exit for undigested residues. Some food is necessarily wasted in the two-way (incoming and outgoing)
traffic in this unspecialized system. The more highly evolved animals have a *complete
digestive system*, with one-way traffic from mouth to anus; the food passes through
different regions specialized for particular digestive functions along the way. *Ingestion* involves simply taking the food in. *Digestion* mechanically and chemically reduces
foods into substances small enough to diffuse into the circulatory system. *Absorption*
passes digested nutrients from the gut lumen into the blood or lymph, which distribute them throughout the body. *Assimilation* moves the nutrients into the cytoplasm of
cells. *Elimination* expels indigestible residues from the body.

nutrition (p. 409)
gut (p. 409)
systems integration (p. 409)
aerobic metabolism (p. 409)
digestive system (p. 409)
digestion (p. 409)
absorption (p. 409)

incomplete digestive system (p. 409)
elimination (p. 410)
pharynx (p. 410)
complete digestive system (p. 410)
crop (p. 410)
gizzard (p. 410)
discontinuous feeding
  patterns (p. 410)

intestine (p. 410)
anus (p. 410)
cloaca (p. 410)
rectum (p. 410)
stomach (p. 410)
ruminants (p. 410)
symbionts (p. 410)

Objectives

1. Define *nutrition* and tell why it is not synonymous with *digestion*.

2. List the items that leave the digestive system and enter the circulatory system during the process of absorption.

3. Explain how the respiratory system is involved in nutrition.

4. Distinguish between incomplete and complete digestive systems and tell which is characterized by (a) specialized regions, (b) two-way traffic, and (c) discontinuous feeding.

5. Define and distinguish *digestion, absorption,* and *assimilation*.

Self-Quiz Questions

Fill-in-the-Blanks

(1) _____ includes taking food into the digestive tract, breaking it down, and transporting the broken-down food into the circulatory system and from there into cells.

(2) _____ is the term for taking food into the digestive tract, but (3) _____ means breaking the food into its simplest components. The passage of food components into the circulatory system is (4) _____, but the components are not (5) _____ until they participate in reactions inside cells.

Summary

The human digestive system is divided into mouth, oral cavity, pharynx, esophagus, stomach, small and large intestines, and anus. Secretory cells in glandular organs such as the salivary glands, the liver, and the pancreas add a variety of digestive substances to the food being broken down. In the oral cavity, incoming food is moistened and mechanically degraded by the teeth and tongue; polysaccharide digestion is initiated by enzymes in the saliva. Peristalsis propels the food ball down the esophagus into the stomach, which initiates protein digestion by means of hydrochloric acid and enzymes. The stomach regulates the rate of food movement into the small intestine, where the digestion and absorption of most nutrients occur. Bile secreted by the liver and stored in the gallbladder emulsifies fats in the small intestine, and the resulting particles are subsequently digested by lipase there. Carbohydrate and protein digestion and absorption continue in the small intestine with the aid of secretions from the pancreas and the intestinal wall. Absorption of water and minerals occurs in both the small and the large intestine. Undigested food residues, together with bacteria, are coated with mucus and move on to the anus.

Key Terms

gastrointestinal
 tract (p. 411)
mucosa (p. 412)
submucosa (p. 412)
muscle layer (p. 412)
serosa (p. 412)
peristalsis (p. 412)
segmentation (p. 412)
sphincters (p. 412)
pyloric
 sphincter (p. 412)
gastrin (p. 413)
secretin (p. 413)
cholecystokinin (p. 413)
gastrin-inhibitory
 peptide, GIP (p. 413)
mouth, oral
 cavity (p. 413)
tooth (p. 413)

molars (p. 413)
incisors (p. 414)
cuspids (p. 414)
saliva (p. 414)
salivary glands (p. 414)
salivary amylase (p. 414)
mucin (p. 414)
bolus (p. 414)
pharynx (p. 414)
esophagus (p. 414)
stomach (p. 414)
hydrochloric
 acid HCl (p. 414)
chyme (p. 414)
pepsinogen (p. 414)
pepsins (p. 414)
trypsin (p. 414)
chymotrypsin (p. 414)
peptidases (p. 414)

peptic ulcer (p. 414)
small intestine (p. 415)
duodenum (p. 415)
jejunum (p. 415)
ileum (p. 415)
villus, villi (p. 415)
microvillus, -villi (p. 415)
pancreas (p. 416)
liver (p. 417)
bile (p. 417)
gallbladder (p. 417)
bile salt (p. 417)
micelle (p. 417)
colon, large,
 intestine (p. 417)
feces (p. 417)
rectum (p. 417)
anus (p. 417)
appendix (p. 418)
appendicitis (p. 418)

Objectives

1. List all the parts (in order) of the human digestive system that food actually passes through. Then list the auxiliary organs that contribute one or more substances to the digestive process.

2. Explain how, during digestion, food is mechanically broken down. Then explain how it is chemically broken down.

3. Describe the mechanism that prevents food from entering the lungs.

4. Explain how the rate at which food is passed from the stomach to the small intestine is regulated.

5. Tell which foods undergo digestion in each of the following parts of the digestive system, and state what the food is broken into: oral cavity, stomach, small intestine, large intestine.

6. List the enzyme(s) that act in (a) the oral cavity, (b) the stomach, and (c) the small intestine. Then tell where each enzyme was originally made.

7. Describe the cross-sectional structure of the small intestine and explain how its structure is related to its function.

8. Describe how the digestion and absorption of fats differs from the digestion and absorption of carbohydrates and proteins.

9. Explain what processes would be disrupted if you took antibiotics that killed all the bacteria in your colon.

10. State which processes occur in the colon (large intestine).

Self-Quiz Questions

Fill-in-the-Blanks

Saliva contains an enzyme ((1) _____) that hydrolyzes starch. Contractions force the larynx against a cartilaginous flap called the (2) _____, which closes off the trachea. The (3) _____ is a muscular tube that propels food to the stomach. Any alternating progression of contracting and relaxing muscle movements along the length of a tube is known as (4) _____. The (5) _____'s most important function is to regulate the rate at which food reaches the intestine. Most digestion and absorption of nutrients occurs in the (6) _____ _____. (7) _____ is an enzyme that works in the stomach. (8) _____ is made by the liver, is stored in the gallbladder, and works in the (9) _____ _____. (10) _____ is an example of an enzyme that is made by the pancreas but works in the small intestine.

Label each of the numbered structures of the
human digestive system.

(11) _____

(12) _____

(13) _____

(14) _____

(15) _____

(16) _____

(17) _____

(18) _____

(19) _____

(20) _____

(21) _____

True-False If false, explain why.

____ (22) Amylase digests starch, lipase digests lipids, and proteases break peptide
bonds.

____ (23) ATP is the end product of digestion.

____ (24) Vitamin K is synthesized in the small intestine and is important in the formation of a blood clot.

____ (25) Water and minerals are absorbed into the bloodstream from the lumen of the
large intestine.

____ (26) Gallstones are formed in the kidney when something changes the concentrations
of bile salts, lecithin, and cholesterol in the bile.

26–III HUMAN NUTRITIONAL REQUIREMENTS

Summary

Carbohydrates, fats, and proteins are the fundamental sources of energy and raw materials for growth and bodily maintenance. Vitamins are accessory substances that are required in small amounts for the normal function of key enzymes in metabolism and for the synthesis of substances important in the structure of specific cells. The continual intake of at least thirteen different vitamins is essential in maintaining human health. Minerals are inorganic materials, such as calcium, magnesium, phosphorus, and iron, that are also important in normal enzyme function, in phosphorylation reactions, in maintaining normal osmotic balances, and as components of important substances used in the structure (e.g., bones and teeth) as well as the function (cytochromes and hemoglobin) of healthy organisms. On the average, about 32-42 grams of protein, 250-500 grams of carbohydrates, and 66-83 grams of fat per day in a well-balanced assortment of foods supply the essential vitamins and minerals. Residents of the United States generally eat too much refined sugar, cholesterol, red meat, and salt and too little fiber. As a result, high blood pressure, heart disorders, diabetes, cancer of the colon, and bad teeth have replaced diseases caused by microorganisms as the main afflictions that hasten death and disability today.

Key Terms

glycogen (p. 419)
essential amino
  acids (p. 419)
phenylalanine (p. 419)
tyrosine (p. 419)
isoleucine (p. 419)
leucine (p. 419)
lysine (p. 419)
threonine (p. 419)
tryptophan (p. 419)
cysteine (p. 419)
methionine (p. 419)
valine (p. 419)
net protein
  utilization, NPU (p. 419)
bulk (p. 420)
transit time (p. 420)
colon cancer (p. 420)
irritable colon
  syndrome, colitis (p. 420)
saturated fat (p. 420)

vitamins (p. 4)
minerals (p. 421)
trace elements (p. 421)
phosphorylation (p. 421)
osmotic balances (p. 421)
cytochromes (p. 421)
vitamin B_1, thiamine (p. 422)
vitamin B_2,
  riboflavin (p. 422)
niacin (p. 422)
vitamin B_6,
  pyridoxine (p. 422)
pantothenic acid (p. 422)
folacin, folic acid (p. 422)
vitamin B_{12} (p. 422)
biotin (p. 422)
choline (p. 422)
vitamin C,
  ascorbic acid (p. 422)
RDA (p. 422)
vitamin A, retinol (p. 422)

vitamin D (p. 422)
vitamin E
  tocopherol (p. 422)
vitamin K,
  phylloquinone (p. 422)
calcium (p. 423)
phosphorus (p. 423)
sulfur (p. 423)
potassium (p. 423)
chlorine (p. 423)
sodium (p. 423)
magnesium (p. 423)
iron (p. 423)
fluorine (p. 423)
zinc (p. 423)
copper (p. 423)
iodine (p. 423)
cobalt (p. 423)

Objectives

1. Compare the contributions of carbohydrates, proteins, and fats to human nutrition with the contributions of vitamins and minerals.

2. Distinguish vitamins from minerals and state what is meant by net protein utilization.

3. Name four minerals important in human nutrition and state the specific role of each.

4. Explain how vitamins A, D, E, and K differ from other vitamins.

5. State the contribution of vitamin C to the human diet and tell what happens if there is not enough vitamin C in the diet. Then name two of the B vitamins, state their contributions, and describe symptoms of deficiency.

6. State the major functions in the body of the fat-soluble vitamins and the possible outcomes of (a) deficiency and (b) excess.

7. Summarize the 1979 U.S. Surgeon General's report that presented ideas for promoting health by "eating properly."

Self-Quiz Questions

Fill-in-the-Blanks

A deficiency of vitamin (1) _____ produces rickets in children and osteomalcia in adults; a deficiency of the mineral (2) _____ produces similar outcomes. It is possible to accumulate too much vitamin (3) _____, which is a constituent of rhodopsin (visual pigment), and vitamin (4) _____, which promotes growth and mineralization of bones. Vitamin (5) _____ functions in forming a blood clot. A deficiency of (6) _____ promotes pellagra, a disease characterized by sores on the skin and gut wall and nervous and mental disorders. A deficiency of vitamin (7) _____ gives rise to scurvy. The mineral (8) _____ is a necessary participant in phosphorylation reactions. (9) _____ is needed to make thyroid hormones. (10) _____ is needed to make hemoglobin; a deficiency of it leads to a form of anemia.

Summary

The liver is the largest glandular organ in vertebrates; it detoxifies blood, is active in the metabolism of carbohydrates, fats, and proteins, and regulates the organic components of blood. Lipids reach the liver (where they are assembled, stored, or broken down into compounds such as acetyl-CoA) by way of the general systemic circulation. Excess glucose is stored as glycogen by the liver. Excess amino acids are converted to forms that can be fed into the Krebs cycle as an alternative source of energy. Most hormones are inactivated by the liver, then sent to the kidneys to be excreted. The liver converts toxic ammonia released from amino acid breakdown into urea that can be excreted by the kidneys. Bile, which emulsifies fats in the small intestine, is made in the liver. Specific scarce substances needed for red blood cell maturation are absorbed and stored by the liver. The liver helps to form antibodies for the blood and helps to remove some foreign particles from the blood. The liver, hypothalamus, pituitary, pancreas, adrenal glands, and muscles are all involved in regulating the level of glucose in the blood. *Insulin*, a pancreatic hormone, enhances the uptake of glucose into body cells; *glucagon*, another pancreatic hormone, stimulates the breakdown of glycogen into its glucose subunits.

Key Terms

liver (p. 421)
lymphatics (p. 421)
urea (p. 421)
hepatic portal
  vein (p. 421)
acetyl-CoA (p. 421)
absorptive state (p. 424)
post-absorptive
  state (p. 424)

islets of
  Langerhans (p. 424)
beta cells (p. 424)
insulin (p. 424)
alpha cells (p. 424)
glucagon (p. 424)
inactivation (p. 425)
detoxification (p. 425)
degradation (p. 425)

lactate (p. 425)
epinephrine (p. 425)
norepinephrine (p. 425)
adrenocorticotropic
  hormone, ACTH (p. 426)
adrenal cortex (p. 426)
glucocorticoid hormones
  (p. 426)

Objectives

1. List seven functions of the human liver.

2. Describe how your body protects itself from excessive glucose in the bloodstream.

3. List the sequence of events that occurs as your blood sugar level drops as the result of sustained exertion coupled with no intake of foods.

Fill-in-the-Blanks

Lipids reach the liver by way of the (1) _____ _____ circulation; other nutrients are absorbed across the intestinal wall and are transported directly by capillaries to the (2) _____ _____ _____, which leads to the liver. Amino acid conversions in the liver form (3) _____, which is potentially toxic to cells; the liver immediately converts this substance to (4) _____, a much less toxic waste product that is expelled by the urinary system from the body. Beta cells of the pancreas secrete (5) _____, and alpha cells secrete (6) _____, a hormone that commands liver cells to convert (7) _____ (a storage starch) into glucose subunits. Under hypothalamic commands, the adrenal medulla begins secreting (8) _____ and (9) _____, which stop (10) _____ synthesis in the liver, stop (11) _____ uptake in muscles, and promote the shift from "burning" (12) _____ during cellular respiration to "burning" fats.

INTEGRATING AND APPLYING KEY CONCEPTS

Suppose that you could not eat solid food for two weeks and that you had only water to drink. List in correct sequential order the measures that your body would take to try to preserve your life. Mention the command signals that are given as one after another critical point is reached and tell which parts of the body are the first and the last to make up the deficit.

27
REGULATION OF BODY TEMPERATURE AND BODY FLUIDS

Summary

Almost all animals must maintain their body temperature within the narrow limits of
0°–40°C. The body temperatures of ectotherms rise and fall with environmental changes,
although they can influence them to a minor extent through metabolic and behavioral
responses to temperature change. Although some birds and mammals continuously maintain
a relatively constant body temperature, most are *heterothermic* and do not maintain
precisely the same body temperature at all times. In mammals the hypothalmus is the
seat of temperature control. Skin thermoreceptors monitor shifts in environmental temperature, and core thermoreceptors within the body signal internal temperatures to the
regulatory centers. Two principal types of responses to cold temperatures occur: heat
production and heat retention. In response to excessive heat, blood is shunted to the
skin surface and heat is radiated, sweat gland secretion is enhanced, epinephrine secretion is inhibited, and exothermic reactions in the body are generally suppressed.
Hypothermia— a drop in body temperature below tolerance levels—is a common cause of
death and disability among humans in boating accidents and on cold weather expeditions.

Key Terms

thermophilic
  bacteria (p. 428)
specific heat
  capacity (p. 429)
conduction (p. 429)
convection (p. 429)
radiation (p. 429)
evaporation (p. 429)
heat balance (p. 429)

ectothermic (p. 429)
endothermic (p. 429)
heterothermic (p. 429)
acclimation (p. 430)
thermal
  regulation (p. 430)
core temperature (p. 430)
hypothalamus (p. 430)
skin thermoreceptors (p. 430)

core thermoreceptors (p. 430)
epinephrine (p. 430)
norepinephrine (p. 430)
thyroxin (p. 430)
shivering response (p. 430)
hypothermia (p. 432)
ventricular
  fibrillation (p. 432)

Objectives

1. Distinguish between ectotherms and endotherms; give two examples of each group.

2. Explain how endotherms maintain their body temperatures if environmental temperatures fall.

3. Explain how endotherms maintain their body temperatures if environmental temperatures rise three to four Fahrenheit degrees above standard body temperature.

4. List the ways in which ectotherms are disadvantaged by not being able to maintain a particular body temperature. Describe the things that ectotherms can do to lessen their vulnerability.

5. Define *hypothermia* and state the situations in which a human might experience the disorder.

Self-Quiz Questions

Fill-in-the-Blanks

In mammals, the (1) _____ is the seat of temperature control. Thermoreceptors located deep in the body are called (2) _____ thermoreceptors. The adrenal medulla secretes (3) _____ when heat production is called for. A drop in body temperature below tolerance levels is referred to as (4) _____.

True-False If false, explain why.

____ (5) Some modern reptiles are ectotherms.

____ (6) Tuna fish are endotherms.

Summary

Animals must have mechanisms for maintaining within some fairly narrow range whatever body fluid concentration and composition are necessary for cell functioning. All animals must regulate individual ion concentrations in the body, and all animals must eliminate all waste products that are potentially toxic.

The mammalian kidney, an intricately structured organ that regulates water and ion content, is part of a tubular network called the urinary system. The discharged fluid—urine—contains water, salts, nitrogen-containing wastes such as urea, and substances that the body cannot metabolize. The composition and volume of urine are adjusted as internal conditions change. Nephrons are tubular components of the kidney that filter and purify a large volume of blood each day. *Filtration* is an entirely passive process in which essentially protein-free plasma is driven by a relatively high pressure difference from the knot of capillaries that make up each *glomerulus* across two layers of cells into the cavity of the *Bowman's capsule*. With the plasma go substances both useful and not useful to the human body. As the plasma passes along the length of the nephron tubule, the useful substances (water and most solutes) are *reabsorbed* by both active and passive means into the surrounding *peritubular capillaries*. Substances not useful (excess water, solutes) are *secreted* from peritubular capillaries into the nephron by both active and passive transport as well. Filtration, reabsorption of water and solutes, and secretion of excess and toxic substances by the nephrons that are surrounded by capillaries in the kidney all influence the ultimate composition and volume of urine; thus, they influence how much water and solute the body conserves. The longer the loop of Henle, the greater is the animal's capacity to conserve water and to concentrate solutes for excretion in urine.

Key Terms

extracellular
   fluid (p. 431)
urea (p. 431)
thirst mechanism (p. 432)
kidneys (p. 432)
mammalian urinary
   system (p. 432)
cortex (p. 433)
medulla (p. 433)
renal pelvis (p. 433)
ureters (p. 433)
urinary bladder (p. 433)

urethra (p. 433)
urine (p. 433)
urination (p. 433)
nephron (p. 433)
glomerulus (p. 433)
Bowman's capsule (p. 433)
proximal tubule (p. 433)
loop of Henle (p. 433)
distal tubule (p. 433)
peritubular
   capillaries (p. 433)
filtrate (p. 434)
solute (p. 434)

glomerular
   filtration (p. 434)
hydrostatic
   pressure (p. 434)
tubular reabsorption (p. 434)
tubular secretion (p. 434)
uric acid (p. 434)
creatinine (p. 434)
renal capsule (p. 435)
collecting duct (p. 435)
renal artery (p. 435)
renal vein (p. 435)

Objectives

1. List the parts of the human urinary system in their order in the path of urine formation and excretion.
2. Locate the processes of filtration, reabsorption, and tubular secretion along a nephron and tell what makes each process happen.
3. List four soluble by-products of animal metabolism that are potentially toxic.

Self-Quiz Questions

True-False If false, explain why.

____ (1) All animals must regulate individual ion concentrations in the body.

____ (2) Water reabsorption into capillaries is achieved by active transport.

____ (3) When the body rids itself of excess water, urine is highly dilute.

____ (4) The subunits in the human kidney that process urine are called *nephridia*.

____ (5) Countercurrent multiplication establishes a sodium concentration gradient in the loop of Henle.

Fill-in-the-Blanks

(6) _____ is an entirely passive event that begins in the Bowman's capsule.

Membrane (7) _____ and the (8) _____ difference across the membrane are the reasons kidneys can (9) _____ an astonishing volume of fluid each day. During (10) _____, all but about one percent of the water and most of the (11) _____ that entered the nephron are returned to the bloodstream. Some substances that appear in urine in greater amounts than were carried into the nephron to begin with are actively transported from the second capillary set into the nephron tubules, an event known as (12) _____ _____. (13) _____ carry urine away from the kidney to the (14) _____ _____, where it is stored until it is released via a tube called the (15) _____, which carries urine to the outside.

Label each of the structures indicated.

(16) _____

(17) _____

(18) _____

(19) _____

(20) _____

(21) _____

(22) _____

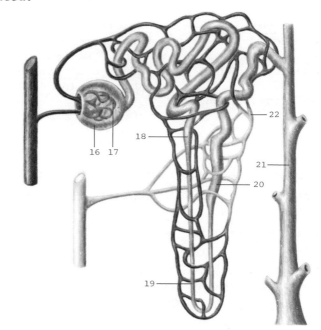

27–III   Controls Over Fluid Volume and Composition   435–436

COMMENTARY: Kidney Failure, Bypass Measures, and Transplants   437–439

Summary

Changes in diet, muscular activity, and metabolic activity alter body fluid volume and
composition. Urine composition is regulated by two homeostatic controls, both governed
by the hypothalamus: (1) a hormonal action, which affects the amount of water and solutes excreted in the urine, and (2) a thirst mechanism, which affects water intake.
Although *antidiuretic hormone* (ADH) is the principal hormone that regulates fluid movement through both vascular and urinary tubes, other hormones such as *aldosterone* also
influence solute concentrations. A sodium concentration gradient is established along
the loops of Henle through a process called *countercurrent multiplication*. If sodium
concentrations and extracellular fluid volume drop, an auxiliary *renin-angiotensin system* will come into play and cause *aldosterone* to be secreted from the *adrenal cortex*.
Aldosterone enhances the reabsorption of sodium by the distal tubules and collecting
ducts which, in turn, causes the blood pressure in the glomeruli to rise.
    When kidneys malfunction, solute levels get out of balance. Substances may accumu-
late in the bloodstream until they reach toxic levels that can lead to nausea, fatigue,
loss of memory, and, in advanced cases death. A kidney dialysis machine can be used to
purify blood and restore the proper solute balance through a process called hemodialysis. An alternative approach is a kidney transplant.

antidiuretic hormone   angiotensin (p. 436)   hemodialysis (p. 437)
  ADH (p. 436)   baroreceptors (p. 436)   cellophane (p. 437)
aldosterone (p. 436)   kidney "stones" (p. 437)   countercurrent
thirst center (p. 436)   kidney dialysis   multiplication (p. 439)
renin (p. 436)   machine (p. 437)
plasma protein (p. 436)

Objectives

1. List some of the factors that can change the composition and volume of body fluids.

2. Describe the process of countercurrent multiplication and explain its specific role in excretion.

3. Name and describe two homeostatic mechanisms that regulate the composition of urine.

4. State explicitly how the hypothalamus, posterior pituitary, and distal tubules of the nephrons are interrelated in regulating water and solute levels in body fluids.

5. List two kidney disorders and explain what can be done if kidneys become too diseased to work properly.

Self-Quiz Questions

    Fill-in-the-Blanks

(1) _____ _____ is the principal hormone that regulates fluid movement

through vascular and urinary tubes. (2) _____ is a Greek word meaning urination.

The (3) _____ contains a nerve cell cluster that is sensitive to concentrations

of sodium and some other solutes in blood; it also contains a (4) _____ center

that detects a rise in salt levels (or a drop in water volume). A kidney (5) _____

machine can be used to restore the proper solute balance when kidneys are too diseased

to function properly.

INTEGRATING AND APPLYING KEY CONCEPTS

The hemodialysis machine used in hospitals is expensive and time-consuming. So far, artificial kidneys that would enable people whose kidneys have failed, to purify their blood by themselves, without having to go to a hospital or clinic, have not been developed. Which aspects of the hemodialysis procedure do you think have presented the most problems in developing a method of home self-care? If *you* had an unlimited budget and were appointed head of a team to develop such a procedure and its instrumentation, which strategy would you pursue?

28
PRINCIPLES OF REPRODUCTION AND DEVELOPMENT

Summary

Animal reproductive strategies range from asexual processes, through development of
unfertilized eggs into adults, through self-fertilization, to sexual reproduction between separate male and female forms. Asexual processes include regeneration and budding; neither case promotes genetic variability because the offspring cells are
identical copies of the parents. Fission and mitosis are considered asexual forms of
reproduction because reproductive cells are not produced. Asexual reproduction is advantageous to organisms that are highly adapted to their surroundings as long as the
surroundings remain stable. Sexual processes include parthenogenesis, hermaphrodism
with self- or cross-fertilization, and separate male forms releasing sperms either
outside or inside separate female forms. Sexual strategies are considered advantageous
to populations of organisms that inhabit unstable surroundings.

Key Terms

zygote (p. 440)
fertilization (p. 440)
cleavage (p. 440)
embryo (p. 440)
larva (p. 440)
sexual reproduction (p. 441)
gametes (p. 441)
asexual
  reproduction (p. 441)
fission (p. 441)

budding (p. 441)
gemmules (p. 441)
clones (p. 441)
parthenogenesis (p. 441)
salinity (p. 441)
hermaphrodite,
  -dism (p. 441)
cross-
  fertilization (p. 441)

synchronous (p. 442)
external fertilization (p. 442)
internal fertilization (p. 442)
testes (p. 442)
penis (p. 442)
vagina (p. 442)
ovary (p. 442)
yolk (p. 442)

Objectives

1. Distinguish between regeneration and budding. State which organisms can accomplish each process.

2. Define *parthenogenesis* and state the organisms in which it occurs. List one advantage and one disadvantage of parthenogenesis.

3. Describe the conditions in which hermaphrodism with cross-fertilization is a more successful strategy than separate sexes using internal fertilization. Identify an organism that uses hermaphrodism with cross-fertilization.

4. Describe the conditions in which hermaphrodism with self-fertilization would be a more successful strategy than separate sexes using internal fertilization. Identify an organism that uses hermaphrodism with self-fertilization.

5. Describe the conditions in which separate sexes using external fertilization would be more successful than separate sexes using internal fertilization.

6. Explain why evolutionary trends in many groups of organisms tend toward developing more complex sexual strategies rather than retaining simpler, asexual strategies.

Self-Quiz Questions

Fill-in-the-Blanks

At the end of fertilization a (1) _____ is formed. (2) _____ is a form of cell division in which the offspring cells do not increase in size. In many animals the embryo develops into a motile, independent (3) _____, which extends the food supply and range of the population. Asexual processes of reproduction include (4) _____, which is common in prey animals, and (5) _____, which is common in animals such as *Hydra* and other cnidarians. (6) _____ is the cleavage and subsequent differentiation of an unfertilized egg into an adult. Earthworms and parasitic tapeworms are (7) _____; each organism has both male and female reproductive organs and produces both eggs and sperms. Eggs are produced in (8) _____, and sperms are produced in (9) _____.

Summary

Gametogenesis follows meiosis and enlarges and polarizes the eggs of most complex eukaryotes; it also includes in the egg many maternal messages, which not only control development until gastrulation begins and the embryonic genotype is activated but also lay the groundwork for all development that follows. *Polarization* is the unequal distribution of structures and substances in the oocyte (and mature egg), including RNA transcripts of the maternal DNA. This unequal distribution is the initial determinant in the structural patterning of the embryo. Sperm penetration into the egg activates a series of events within the egg cytoplasm, and the mingling of DNA from both parents establishes the diploid genotype; all of these activities constitute *fertilization*.

Gene expression also varies through time in animals. During gametogenesis, sperms acquire different structures and function differently than eggs. Fertilization is completed when sperm and egg nuclei fuse, restoring the diploid number of chromosomes. *Cleavage* compartmentalizes the zygote into tiny cells, but no growth or expression of the embryonic genotype happens yet. *Gastrulation* forms two or more layers of relatively undifferentiated embryonic tissues, the embryonic genotype now begins to be transcribed and translated, and groups of cells move about. *Organogenesis* follows and rudimentary organs result.

Finally, by further growth and tissue specialization, the various body parts acquire the structural and physiological properties necessary for performing specialized tasks.

Key Terms

gametogenesis (p. 443)
fertilization (p. 443)
cleavage (p. 443)
gastrulation (p. 443)
germ layers (p. 443)
body axis (p. 443)
organogenesis (p. 443)
growth (p. 443)
tissue specialization (p. 443)

oviparous (p. 444)
viviparous (p. 444)
ovoviviparous (p. 444)
oocyte (p. 444)
Golgi complex (p. 445)
maternal DNA (p. 445)
RNA transcripts (p. 445)
histones (p. 445)

maternal messages (p. 445)
polarity (p. 445)
animal pole (p. 448)
vegetal pole (p. 448)
pigment granules (p. 448)
egg cortex (p. 448)
cytoplasmic lattice (p. 448)
gray crescent (p. 448)

Objectives

1. Characterize gametogenesis, fertilization, and cleavage according to the principal events that occur in each.

2. Explain how gastrulation differs from cleavage.

3. List sequentially all the major phases in animal development and define the activities that make each phase unique.

4. Define *oviparous, viviparous,* and *ovoviviparous.* Cite an example of an animal that goes through each of the three developmental strategies.

5. Explain what causes polarity to occur during oocyte maturation in the mother and state how polarity influences later development.

True-False If false, explain why.

____ (1) In complex eukaryotes, development until gastrulation is governed by DNA in the nucleus.

____ (2) Maternal messages consist principally of extra amounts of DNA from the mother that are included in the cytoplasm of the newly formed egg.

____ (3) Sperm penetration into the cytoplasm of the egg brings about specific structural changes and chemical reactions.

Fill-in-the-Blanks

In animals, gene expression also varies through time. (4) _____, the formation and maturation of eggs and sperm, is considered the first stage of animal development. Rich stores of substances become assembled in localized regions of the (5) _____ cytoplasm. When sperm and egg unite and their DNA mingles and is reorganized, the stage is referred to as (6) _____. (7) _____ includes the repeated mitotic divisions of a zygote that segregate the egg cytoplasm into a cluster of cells; the entire cluster is known as a (8) _____. (9) _____ is the process that regroups cells and arranges them into embryonic tissue layers. Alligators show (10) _____: fertilization is internal; the fertilized eggs develop inside the mother's body, without additional nourishment; and the young are born live. Birds show (11) _____: eggs with large yolk reserves are released from and develop outside the mother's body. (12) _____ _____ activated at fertilization direct the initial stages of development until gastrulation occurs. In amphibian eggs, sperm penetration on one side of an egg causes pigment granules on the opposite side of the egg to flow toward the (13) _____ _____. A lightly pigmented area called the (14) _____ _____ results. It is a visible marker of the site where the (15) _____ _____ will be established and where gastrulation will begin.

Summary

Following fertilization, animal development proceeds through stages of *cleavage* (the subdivision of the zygote into cellular compartments), *gastrulation* (the formation of embryonic tissue layers), and *organ formation* and growth. Initially, development is guided by regionally positioned messages in the egg cytoplasm. The destiny of cell lineages is in part established according to which sector of the egg cytoplasm is inherited by the first embryonic cells formed during cleavage. Cleavage is the compartmentalization of the zygote into many cells that differ because of their differing assemblages of cytoplasmic constituents. Even at this time, there are structural and functional differences between cells that result from the translation of RNA that was transcribed from maternal DNA, and from the variable amounts of yolk. During cleavage, the zygote is divided into many tiny compartments by mitosis to form a *blastula*. Gastrulation involves the formation of two or more layers of embryonic tissues: generally *ectoderm* which forms the body covering and nervous system; endoderm, which forms the inner lining of the gut and gut derivatives; and (in all but sponges) mesoderm, from which are formed the muscles, the skeleton, and the circulatory system. With gastrulation, the embryonic genotype is transcribed and translated into its own unique assemblage of proteins, and the influences of the maternal system wane. Gastrulation also is characterized by the movement of masses of cells; endoderm ends up on the interior of the embryo and ectoderm on the outside. When the embryonic tissues have been established, organogenesis occurs by growth, tissue movements, and specialization, forming the rudimentary organs.

Key Terms

dorsal lip of	morula (p. 449)	ectoderm (p. 449)
blastopore (p. 446)	blastula (p. 449)	archenteron (p. 449)
yolk plug (p. 446)	blastocoel (p. 449)	somites (p. 451)
neural plate (p. 447)	blastodisk (p. 449)	primitive streak (p. 451)
neural fold (p. 447)	inner cell mass (p. 449)	amniotic fold (p. 451)
notochord (p. 447)	blastocyst (p. 449)	eye vesicle (p. 451)
gut cavity (p. 447)	gastrulation (p. 449)	heart vesicle (p. 451)
neural canal (p. 447)	gastrula (p. 449)	neural tube (p. 451)
cleavage (p. 448)	endoderm (p. 449)	wing bud (p. 451)
microvilli (p. 448)	accessory glands (p. 449)	leg bud (p. 451)
incomplete cleavage (p. 449)	mesoderm (p. 449)	

Objectives

1. List and describe the events that occur during cleavage.

2. Explain how the amount of yolk in an ovum can influence an animal's cleavage pattern.

3. Define *gastrulation* and state what process begins at this stage that did not happen during cleavage.

4. Name each of the three embryonic tissue layers and the organs formed from each.

5. Compare the early stages of frog and chick development (Figs. 28.6 and 28.8) with respect to (a) egg sizes and (b) type of cleavage pattern (incomplete or complete).

Self-Quiz Questions

    <u>*True-False*</u> If false, explain why.

____ (1) During gastrulation, maternal controls over gene activity are activated and begin the process of differentiation in each cell's nucleus.

    <u>*Fill-in-the-Blanks*</u>

The third stage of animal development, (2) _____, is characterized by the sub-dividing and compartmentalization of the zygote; no growth occurs at this stage, and usually a hollow ball of cells, the (3) _____, is formed. The fourth stage, (4) _____, is concerned with the formation of ectoderm, mesoderm, and endoderm, the (5) _____ layers of the embryo; at the end of this stage, the (6) _____ is formed. Ectoderm eventually will give rise to skin epidermis and the (7) _____ system; endoderm forms the inner lining of the (8) _____ and associated digestive glands. Mesoderm forms the circulatory system, the (9) _____, and the muscles. Cleavage of a frog's zygote is total because there is little enough yolk that cleavage membranes can subdivide the entire cytoplasmic mass; in the chick, however, there is so much yolk that cleavage membranes cannot. Cleavage is therefore said to be (10) _____, and the chick grows from a primitive streak on the surface of the (11) _____ mass into a chick embryo complete with wing and leg buds and beating heart during the first (12) _____ days.

28–IV   MORPHOGENESIS AND GROWTH

Summary

Morphogenesis (the growth, shaping, and spatial coordination of tissues and organs according to predefined patterns) depends on controls over the expression of genes, which affect the extent, direction, and rate of growth. Morphogenesis also depends on chemical communication links, physical attachments, and physical restraints imposed by neighboring cells in the developing embryo.

During morphogenesis, three types of *morphogenetic* movements cause embryonic parts to move from one site to another. Individual cells may follow specific paths in response to chemical gradients or may respond to adhesive cues. Epithelial sheets fold inward or outward, creating tissue layers, body cavities, and evaginations. In some cases, whole organs move.

During the formation of organs, one body part differentiates according to signals it receives from an adjacent body part; this process is known as *embryonic induction*. In all animals, the coordinated development of body parts in specific regions depends on some combination of (1) the influence of cytoplasmic substances within the unfertilized egg, and (2) inductive interactions between cells at later stages of development.

In morphogenesis, *controlled cell death* causes the folding of parts, the development of slits, the hollowing-out of tubes, the shaping of bones, and the opening of mouth, nostrils, and ears. Sometimes a mutation blocks the capacity of specific cells to respond to the signal from elsewhere to die, so those cells delay dying until they are no longer supported by other parts of the body. During development, *embryonic regulation* compensates for missing or extra parts that result from damage or cell division that occurs too quickly or too slowly.

Key Terms

morphogenesis (p. 450)
differential
  growth (p. 450)
morphogenetic
  movements (p. 450)
pseudopods (p. 450)
chemotaxis (p. 450)
extracellular
  matrix (p. 450)
adhesive cues (p. 450)
epithelial sheets (p. 450)
postnatal (p. 451)

pattern formation (p. 451)
ooplasmic
  localization (p. 451)
nonuniform
  shrinkage (p. 452)
cell elongation (p. 452)
embryonic
  induction (p. 452)
retina (p. 452)
forebrain (p. 452)
Spemann (p. 452)
optic cup (p. 452)

lens (p. 452)
induced (p. 453)
Mangold (p. 453)
dorsal lip of
  the blastopore (p. 453)
blastocoel (p. 453)
Siamese twins (p. 453)
inductive interactions (p. 453)
controlled cell death (p. 453)

Objectives

1. Define *morphogenesis*. If it differs from organogenesis, indicate how.

2. List three of the developmental factors that control and guide morphogenesis.

3. Explain how substances released from cells can induce changes in nearby cells.

4. Describe the relationship that develops between the retina and the lens as a result of embryonic induction.

5. State what Spemann's transplant of the optic cup demonstrated and indicate how Mangold's dorsal lip transplants advanced our understanding of the ways that parts are fit together during development.

6. Give three examples of structures in you that were formed by controlled cell death. Describe how those structures would have appeared if controlled cell death had not occurred.

7. Define *embryonic regulation* and distinguish it from *embryonic induction*.

8. Indicate how the capacity for embryonic regulation can be demonstrated to exist in certain organisms.

True-False If false, explain why.

____ (1) Body parts become folded; tubes become hollowed-out; and eyelids, lips, noses, and ears all become slit or perforated by controlled cell death.

____ (2) A process of predictable cellular deterioration is built into the life cycle of all organisms that consist of differentiated cells that show considerable specialization.

Fill-in-the-Blanks

The capacity to compensate for missing or extra parts is known as (3) _____ _____. If a zygote is split, either naturally or experimentally, into two equal parts, the result is (4) _____ _____. Differentiation requires (5) _____ gene expression in certain cells at certain times. (6) _____ involves the growth, shaping, and spatial coordination necessary to form functional body units. Embryonic cells move in response to adhesive cues and (7) _____ _____. Sometimes entire organs (such as testes in human males) change position in the developing organism but the inward or outward folding of (8) _____ _____ is seen more often. Spemann demonstrated that the process known as (9) _____ _____ occurs in salamander embryos, where one body part differentiates because of signals it receives from an adjacent body part. Mangold further demonstrated by her transplants of dorsal lip tissue from one salamander to a host salamander that the dorsal lip organizer, when transplanted to a different site, can induce some of the (10) _____ _____ to develop in extraordinary ways. The influence of cytoplasmic substances in particular locations of the unfertilized egg on development after fertilization is called (11) _____ _____. A kitten's eyes opening after birth is an example of (12) _____ _____ _____ in action.

28–V CELL DIFFERENTIATION   454-456

Summary

The intricate life cycles of multicelled eukaryotes are typified by changes in form and function that depend not only on cell growth and division but also on *differentiation*. In this process, cells having identical sets of DNA become structurally and functionally different from one another over time. Different portions of the total set of instructions are used in different cells at different times. Controls over this gene expression lead to variations in the way that materials are patterned and arranged as the life cycle unfolds.

Which genes are expressed in a given cell depends on the nature of cytoplasmic changes induced by the environment and by interactions with other cells.

In *regulative* patterns of development, genes that were shut down in differentiated cells can become activated again when missing body parts must be replaced.

In *mosaic* patterns of development, more genes have been shut down by the time the egg matures than in regulative development, and those genes cannot be activated again.

Key Terms

cell differentiation (p. 454)
"equipotential" (p. 454)
regulative pattern of
  development (p. 454)

mosaic pattern of
  development (p. 454)
stem cells (p. 454)
regeneration (p. 454)

regulation (p. 454)
identical twins (p. 456)
sea squirts (p. 456)
chordate (p. 456)

Objectives

1. Define *differentiation* and give two examples of cells in a multicellular organism that have undergone differentiation.

2. Explain why the differentiation of cells in a multicellular organism goes hand-in-hand with division of labor and the integration of life processes.

3. Contrast the regulative and mosaic patterns of development.

Self-Quiz Questions

True-False If false, explain why.

____ (1) Muscle cells that contract and nerve cells that relay messages are examples of cells that have undergone differentiation from the initial ball of embryonic cells.

____ (2) In the regulative pattern of development most of the genes have been inactivated in the mature egg and those genes cannot be activated again.

28–VI METAMORPHOSIS   454

AGING AND DEATH   456, 458

COMMENTARY: Death in the Open   457–458

Summary

Developmental strategies in multicellular organisms vary from *indirect development*, which includes a larval stage to *direct development*, in which the embryo is nourished either externally or internally by the mother.

Normal cells of complex eukaryotes have limited lifespans characteristic of their species. Following growth and division, cells begin to deteriorate in a predictable manner, a process called *aging*. Aging is built into the life cycle of all organisms in which differentiated cells show considerable specialization. Aging is not simply the result of an accumulation of environmental damage. Aging and death may be coded largely in DNA; perhaps signals from the cytoplasm of cells nearby or some distance away in the differentiated body activate those messages, informing the cell that it is time to die. Every cell that lives must eventually die or at least be incorporated into its offspring cells. Although billions of living beings die each day, we are generally unaware of these deaths. Death should be accepted as part of the necessity to renew life in the continual quest for beings perfectly adapted to their changing environments.

Key Terms

direct development (p. 454)   metamorphosis (p. 454)   nymph (p. 454)
indirect development (p. 454)   molts (p. 454)   aging (p. 456)
larva (p. 454)   pupa (p. 454)   Moorhead and Hayflick (p. 456)

Objectives

1. Distinguish direct from indirect development.

2. Define what is meant by *larva*. Distinguish metamorphosis from morphogenesis.

3. State any evidence that suggests that either DNA or the reception of external signals may be involved in aging and death.

4. Define *aging* and state why the death of cells is believed not to be caused solely by an accumulation of environmental insults.

5. Contrast the degree of finality in the death of prokaryotes with that of complex multicellular eukaryotes.

6. Describe the behavior of elephants and humans with regard to the disposal of the dead bodies of their own kind. Then state which populations, human or elephant, do a more efficient job of recycling their atoms through the ecosystem of which they are part.

Self-Quiz Questions

    True-False If false, explain why.

____ (1) The aging and death of a cell may be coded in large part in its DNA; external signals activate those DNA messages and tell the cell that it is time to die.

    Fill-in-the-Blanks

A larva necessarily must undergo (2) _____ in order to become a juvenile. Any

development program that includes an adaptive larval stage is called (3) _____

development.

INTEGRATING AND APPLYING KEY CONCEPTS

If embryonic induction did *not* occur in a human embryo, how would the eye region appear? What would happen to the forebrain and epidermis?

    If controlled cell death did not happen in a human embryo, how would its hands appear? And its face?

29

HUMAN REPRODUCTION AND DEVELOPMENT

Summary

In human males and females, as in other separately sexed animals, the primary reproductive organs are testes and ovaries; the remaining reproductive structures are accessory reproductive organs. Ovaries and testes not only produce gametes by meiosis, they also secrete important sex hormones that maintain gamete production, normal sexuality, and secondary sex characteristics.

Sperm that have matured in the *seminiferous tubules* of the testes during gametogenesis move into the *epididymis* (plural: epididymides), finish maturation, and become motile. Sexual activity propels them along the *vas deferens* (plural: vasa deferentia) upward and posteriorly to the junction behind the urinary bladder where the two vasa deferentia merge with the *ejaculatory duct*. From here, the sperm are propelled along the urethra, receiving secretions from the *seminal vesicles, prostate,* and *bulbourethral glands*. Collectively, the sperm and secretions (mucus, nutrients, water, and prostaglandins) are called *semen*.

Testosterone, produced by the *interstitial cells* of the testes, stimulates *spermatogenesis*, governs the form and function of the male reproductive tract, develops and maintains male sexual behavior, and maintains male *secondary sexual characteristics* (deep voice and growth of beard and pubic hair).

Key Terms

primary reproductive
  organs (p. 459)
gonads (p. 459)
accessory reproductive
  organs (p. 459)
sex hormones (p. 459)
secondary sexual
  trait (p. 459)
puberty (p. 459)
testis, -es (p. 459)
scrotum (p. 459)
seminiferous
  tubules (p. 459)
interstitial cells (p. 459)

testosterone (p. 459)
spermatogenesis (p. 460)
spermatogonia (p. 460)
basement membrane (p. 460)
lumen (p. 460)
primary spermato-
  cytes (p. 460)
secondary spermato-
  cytes (p. 460)
spermatids (p. 461)
mature sperm (p. 461)
sperm (p. 461)
acrosome (p. 461)
midpiece (p. 461)

epididymis (p. 461)
vas deferens (p. 461)
peristaltic (p. 461)
ejaculatory duct (p. 461)
urethra (p. 461)
seminal vesicles (p. 461)
prostate gland (p. 461)
bulbourethral glands (p. 461)
semen (p. 461)
prostaglandins (p. 461)
follicle-stimulating hormone,
  FSH (p. 461)
luteinizing hormone (p. 461)
gonadotropins (p. 461)

Objectives

1. Distinguish between primary and secondary sexual traits, and between gonads and accessory reproductive organs.

2. List in order the stages that compose spermatogenesis.

3. Follow the path of a mature sperm from the seminiferous tubules to the urethral exit. List every structure encountered along the path and state its contribution to the nurture of sperm.

4. Diagram the structure of a sperm, label its components, and state the function of each.

5. Name the three hormones that directly or indirectly control male reproductive function. Diagram the negative feedback mechanisms that link the hypothalamus, anterior pituitary, and testes in controlling gonadal function.

Self-Quiz Questions

Fill-in-the-Blanks

Spermatogenesis occurs in the (1) _____ _____, but sperms become motile

in the (2) _____. The interstitial cells of the testis produce (3) _____.

Accessory reproductive organs of the human male include the (4) _____ _____,

which either stores most of the sperm or, during sexual activity, moves them along by

peristaltic action. The seminal vesicles and (5) _____ contribute secretions

that make up most of the seminal fluid. During ejaculation, a (6) _____ closes off the bladder so that urine cannot mix with semen. A cap over most of the head of each sperm contains (7) _____ _____ that function in egg penetration. The (8) _____ of each sperm contains mitochondria, which provide the energy necessary for motility. (9) _____ secreted by the prostate gland into semen stimulate uterine contractions in the female. In the female reproductive tract, sperms can live for (10) _____ to _____ hours.

29–II FEMALE REPRODUCTIVE SYSTEM

Summary

The *ovarian cycle* consists of oocyte maturation in a fluid-filled follicle within the ovary, the transformation of that cavity into a corpus luteum (endocrine tissue) when the follicle ruptures and the mature ovum escapes from it, and a concurrent production of ovarian hormones. The *menstrual cycle* consists of profound changes in the endometrium (epithelial lining) and muscles of the uterus, as well as glandular secretions, all of which prepare the uterus for implantation. In the absence of implantation, the ovarian and menstrual cycles are repeated monthly.

Some evidence suggests that the *midcycle surge* of *luteinizing hormone* stimulates production of enzymes that can dissolve the thin membranes between the follicle and the surroundings; thus, the follicle ruptures and ovulation occurs. If fertilization does not occur, the corpus luteum degenerates during the last days of the menstrual cycle and the levels of estrogen and progesterone fall rapidly. Deprived of its hormonal support, the highly developed endometrium disintegrates and the menstrual period begins. Coitus (sexual intercourse) generally involves stimulation of the glans penis and vulva by friction, collection of blood in the spongy vascular tissue of both sexes, and rhythmic muscular contractions that force seminal fluid from the vesicles and prostate into the male urethra and thence into the vagina during ejaculation. Orgasm, with its involuntary muscular contractions and associated sensations of warmth and release of tension may or may not be reached by either sex during coitus, but if the male ejaculates, the female can become pregnant whether or not she experiences orgasm.

Sometimes in vitro fertilization is recommended for women who are unable to become pregnant due to blockages of their oviducts. Preovulatory eggs are detected by laparoscopy and removed by suction. Freshly obtained diluted sperm are combined with the ova, and a few hours later fertilization occurs. The cleavage stages are stored for two to four days in a saline solution that supports development; then one is drawn up in a fine tube and transferred to the uterus, where implantation may occur and development proceeds normally.

ovaries (p. 462)
oocytes (p. 462)
ovum, ova (p. 462)
estrogen (p. 462)
progesterone (p. 462)
oviducts (p. 462)
fallopian tubes (p. 462)
lactation (p. 462)
mammary glands (p. 462)
estrous cycle, estrus (p. 462)
menstrual cycle (p. 462)
uterus (p. 462)
cervix (p. 462)

vagina (p. 462)
external genitalia (p. 462)
vulva (p. 462)
clitoris (p. 462)
labium major (p. 463)
labium minor (p. 463)
menopause (p. 464)
endometrium (p. 464)
FSH (p. 464)
LH (p. 464)
primoridal-follicle (p. 464)
chorionic
    gonadotropin (p. 464)

midcycle surge of
    LH (p. 465)
ovulation (p. 465)
corpus luteum (p. 465)
implantation (p. 466)
penis (p. 467)
coitus (p. 467)
glans penis (p. 467)
sphincter (p. 467)
orgasm (p. 467)
pregnancy (467)
laparoscope (p. 472)

Objectives

1. Indicate the ways that the human menstrual cycle differs from estrus as it occurs in most mammals.

2. Distinguish the ovarian cycle from the menstrual cycle and explain how the two cycles are synchronized by hormones from the anterior pituitary, hypothalamus, and ovaries.

3. State which hormonal event brings about ovulation and which other hormonal events bring about the onset and finish of menstruation.

4. Trace the path of a sperm from the seminiferous tubules to where fertilization normally occurs. Mention in correct sequence all major structures of the male and female reproductive tracts that are passed along the way and state the principal function of each structure.

5. Compare the function of the interstitial cells of the testis with the function of the ovarian follicle and corpus luteum.

6. List the physiological factors that bring about erection of the penis during sexual stimulation and bring about ejaculation.

7. List the similar events that occur in both male and female orgasm.

8. Describe how in vitro fertilization occurs.

Self-Quiz Questions

Fill-in-the-Blanks

Ovaries produce the important sex hormones (1) _____ and (2) _____. In

the female, (3) _____ are passageways that channel ova from the ovary into the

(4) _____, which houses the embryo during pregnancy. In most mammals, (5)

_____ is a predictably recurring time when the female becomes sexually receptive

to the male. Each oocyte is contained in a spherical chamber called a (6) _____.

At (7) _____, the follicle ruptures, and the ovum escapes from the ovary and is swept by ciliary action into the (8) _____, the road to the uterus. The ruptured follicle now changes into secretory tissue called the (9) _____

_____. Estrogen causes the (10) _____ (epithelial uterine lining) to thicken. (11) _____ stimulates glands in the thickened tissues to secrete various substances. The midcycle peak of (12) _____, the level of which overshoots (13) _____, triggers ovulation. The decrease of all of the hormones that regulate the monthly cycle brings about (14) _____.

29–III   FROM FERTILIZATION TO BIRTH

Early Embryonic Development

Case Study: Mother as Protector, Provider, Potential Threat

Summary

During the first month of human development, cleavage, gastrulation, and organ formation occur. By the end of the first month, the embryo has begun taking on recognizably human characteristics. Two months of relatively slow development for the main organs follows. At the end of the first three-month period (trimester) the term *embryo* is replaced by *fetus* because the developing individual now resembles a miniature adult. By the end of the seventh month, fetal development appears to be relatively complete, but fewer than ten percent of infants born at this stage survive; by the ninth month, survival chances increase about ninety-five percent. The embryo is particularly vulnerable to damaging substances from the maternal bloodstream during the first six weeks after fertilization because that is the critical period of organ formation. Poor nutrition during the last trimester affects all fetal organs but is most damaging to the brain; in the weeks just before and just after delivery, the human brain undergoes its greatest growth. Mothers who smoke cigarettes every day throughout pregnancy tend to produce smaller infants who risk heart disease and decreased reading abilities.

Key Terms

morula (p. 468)	chorionic gonadotropin (p. 468)	second trimester (p. 469)
blastocyst (p. 468)	amnion (p. 468)	third trimester (p. 469)
inner cell mass (p. 468)	yolk sac (p. 468)	embryo (p. 469)
trophoblast (p. 468)	allantois (p. 469)	fetus (p. 469)
chorionic villi (p. 468)	placenta (p. 469)	German measles (p. 471)
amniotic cavity (p. 468)	umbilical cord (p. 469)	thalidomide (p. 471)
chorion (p. 468)	embryonic disk (p. 469)	"reading age" (p. 473)
	first trimester (p. 469)	

Objectives

1. Describe the events that occur during the first month of human development. State how long cleavage and gastrulation require, when organogenesis begins, and what is involved in implantation and placenta formation.

2. State when the embryo begins to be referred to as a fetus and at what point at least ten percent of births result in survival.

3. Explain why the mother must be particularly careful of her diet, health habits, and life-style during the first trimester after fertilization (especially the first six weeks) and during the last trimester.

Self-Quiz Questions

Fill-in-the-Blanks

Fertilization generally takes place in the (1) _____; five or six days after

conception, (2) _____ begins as the blastocyst burrows inside the endometrium.

Extensions from the chorion fuse with the endometrium of the uterus to form a (3)

_____, the organ of interchange between mother and fetus. By the beginning of

the (4) _____ trimester, all major organs have formed; the offspring is now

referred to as a(n) (5) _____.

29–IV   CONTROL OF HUMAN FERTILITY   473–476

Some Ethical Considerations   473–474
Possible Means of Birth Control   474–476

Summary

Each week the earth's human population grows by about 1,330,560 people and outstrips the resources of many countries. A variety of birth control methods are used to reduce this astounding birth rate. Extremely effective birth control methods include abstinence, abortion, and sterilization. Highly effective methods include the Pill, IUDs, diaphragms with spermicide, and condoms plus spermicide. Effective methods include vaginal foams, good brands of condoms, and diaphragms alone. Douches are unreliable. Cheap brands of condoms, withdrawal, and the rhythm method not based on temperature typically are only seventy to below sixty percent effective. At present, sterilization is generally irreversible.

Key Terms

abstention (p. 474)
rhythm method (p. 474)
fertile period (p. 474)
withdrawal (p. 474)
viable (p. 474)
douching (p. 474)
spermicidal foam   (p. 474)

spermicidal jelly (p. 474)
diaphragm (p. 475)
condoms (p. 475)
venereal disease (p. 475)
the Pill (p. 475)
vasectomy (p. 475)
tubal ligation (p. 475)

cauterized (p. 475)
intrauterine device, IUD
  (p. 476)
copper T (p. 476)
abortion (p. 476)
miscarriage (p. 476)

Objectives

1. Identify the factors that encourage and discourage methods of human birth control.
2. Identify the three most effective birth control methods used in the United States.
3. Identify the four least effective birth control methods.
4. State which birth control methods help to prevent venereal disease.
5. Describe two different types of sterilization.
6. State the physiological circumstances that would prompt a couple to try *in vitro* fertilization.

Self-Quiz Questions

Fill-in-the-Blanks

On earth each week, (1) _____ more babies are born than people die. Each year in the United States, we still have about (2) _____ unwed teenage mothers and (3) _____ abortions. The most effective method of preventing conception is complete (4) _____. (5) _____ are about eighty-five to ninety-three percent reliable and help prevent venereal disease. A (6) _____ is a flexible, dome-shaped disk, used with a spermicidal foam or jelly, that is placed over the cervix. The most widely used contraceptive is the Pill—an oral contraceptive of synthetic (7) _____ and (8) _____ that suppress the release of (9) _____ from the pituitary and thereby prevent the cyclic maturation and release of eggs. Two forms of surgical sterilization are vasectomy and (10) _____ _____.

INTEGRATION AND APPLYING KEY CONCEPTS

What rewards do you think a society should give a woman who has at most two children during her lifetime? In the absence of rewards or punishments, how else can a society encourage women not to have abortions and yet ensure that the human birth rate does not continue to increase?

CROSSWORD NUMBER FOUR

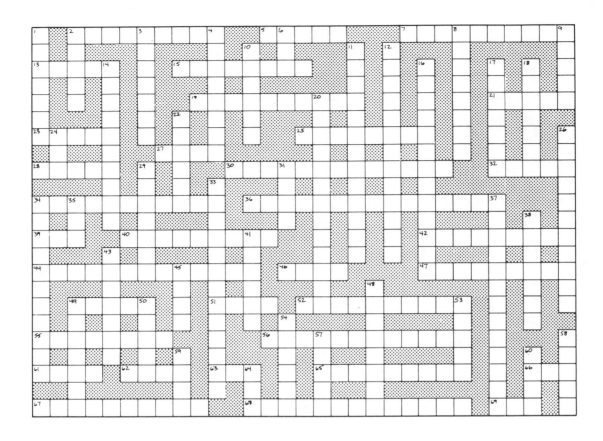

Note: This puzzle covers material from Unit 3, as well as from Unit 5. You may want to refer back to those chapters for some of the answers.

Across

2. one-half of a replicated chromosome
5. material surgically attached to or inserted into a bodily part
7. transmission of disease from an original site to one or more sites elsewhere in the body, as in tuberculosis or cancer
13. a branch modified in the form of a sharp woody spine
15. a pattern used as a guide in making something accurately
19. when DNA is in the form of chromatin
21. a speech form that is peculiar to itself within the usage of a particular language
23. a haploid cell that develops directly into a new haploid body
25. gives rise to digestive glands and the gut lining
27. an individual entity regarded as an elementary constituent of a whole
28. specific DNA regions that tell how to build different kinds of RNA

30. groups of similar cells produce structures of predefined shapes
32. physical characteristic
34. characteristics of two phenotypes appear
36. cells undergo ___ as they become more specialized
39. a kind of nucleic acid
40. an arrangement and grouping of chromosomes on the basis of size, number, and shape
42. embryonic ___: one body part differentiates according to signals it receives from an adjacent body part
44. many-celled
46. ___ cross, in which a dominant-appearing individual of unknown genotype is crossed with a known homozygous recessive individual
47. organelle that makes the precursors to ribosomes
49. process of predictable cellular deterioration
51. ___ Grey, 1875-1939; American author
52. independent ___: the random distribution of chromosomes
55. any heritable change in the genes or chromosomes of an organism
56. at the end of this process, three layers of embryonic tissues will exist
61. an herb; also a wise person
62. a long-lived population of cancerous human cells kept in culture
63. eggs
65. designating an abnormal growth that tends to metastasize
66. male sheep
67. the structure that holds two chromatids together
68. dictated in advance of some other event
69. Middle English "no"

Down

1. the separation of duplicated DNA into two equivalent parcels
2. an organism that is genetically identical to its parent
3. a process that reduces the parental diploid number of chromosomes to the haploid state
4. a ___ allele masks expression of the other allele
6. a successful rodent
8. produced in response to an antigen
9. male gamete
10. one of several alternative forms of a gene
11. the production of sex cells
12. said of partners that, together, make up a whole
14. a bead on a nucleoprotein fiber
16. happens as a result of fertilization
17. a sequence of three nucleotide bases that specifies a particular amino acid
18. a region of apparent contact between two homologous chromosomes
20. a method of sampling uterine fluid and fetal cells
22. not malignant
24. American short-story writer, poet, and journalist, 1809-1849
26. cytoplasmic division that generally follows nuclear division
29. all genes located on a given chromosome compose a ___ group
31. a region of a polytene chromosome with open loops from homologous bands extending outward
33. the fusion and mingling of the hereditary material of two gametes
34. distinct threadlike structures that bear hereditary information in eukaryotes
35. a type of nucleic acid
37. a substance made of amino acids and nucleotides
38. having two complete sets of chromosomes
41. the larger of the two kinds of nucleotide bases

43. a protistan with hairlike organelles
45. discovered that, in mammalian females, only the genes of one X chromosome are expressed
48. more ornamental
49. stimulates antibody production
50. ___, composer of the Grand Canyon Suite
53. belief
54. haploid sex cell
57. a circumscribed, noninflammatory growth that serves no useful physiological function
58. a clear watery fluid that contains white blood cells and acts to remove bacteria and certain proteins from the tissues
59. an entire X chromosome that remains compact during interphase is a ___ body
60. that which a predator consumes
64. an energy storage molecule

30

INDIVIDUALS, POPULATIONS, AND EVOLUTION

Summary

When Carl von Linné studied the diversity of organisms, he believed that *species* didn't change. In general, a species encompasses all of those actually or potentially inter-breeding populations that are reproductively isolated from other such groups. Linné's approach was *typological*: an individual was selected as being the perfect type for the species, on the basis of a somewhat arbitrary choice of what "perfect" physical features were. Now the *population concept*, which holds that variation in populations not only encourages evolutionary change but also is the product of evolutionary change, is followed. In a sexually reproducing population, all of the genotypes taken together compose a *pool of alleles*. For any one locus on the chromosome, some alleles occur more often than others; thus variation can be thought of in terms of *allele frequencies*.

When allele frequencies for a given gene locus remain constant through succeeding generations, the population is said to be at *Hardy-Weinberg equilibrium*, which is maintained only when all of a set of specific conditions occur. Rarely are all these conditions met, but many populations are close enough to the ideal population that equilibrium equations are applicable. If all individuals in a population have an equal chance of surviving and reproducing, then the frequency of each allele in the population should remain constant from generation to generation (that is, be in *genetic equilibrium*). In nature, allele frequencies may change all the time. Some disturbances that cause changes are (1) mutation, (2) genetic drift, (3) gene flow, and (4) the selection pressures that bring about differential survival and reproduction of genotypes within a population. Selection pressures generally favor one allele such that other alleles for the same locus are suppressed. However, two or more alleles for the same locus may persist simultaneously in a population at a frequency greater than can be accounted for by newly arising mutations alone; such a situation is called *polymorphism*.

Key Terms

Carl von Linné (p. 478)
typological (p. 478)
specimen (p. 479)
gene mutation (p. 479)
chromosome mutation (p. 479)
genetic recombination (p. 479)
individual (p. 479)
population (p. 479)
discontinuous variation (p. 479)
continuous variation (p. 479)
bell-shaped curve (p. 479)
$p + q = 1$ (p. 480)
p (p. 480)
q (p. 480)
genotypic frequencies (p. 480)
p^2 (p. 480)
q^2 (p. 480)
$2pq$ (p. 480)
Hardy-Weinberg equilibrium (p. 480)
allele (p. 481)
locus (p. 481)
allele frequencies (p. 481)
baseline (p. 481)

isolation (p. 481)
random mating (p. 481)
viability (p. 481)
fertility (p. 481)
crossing-over (p. 481)
independent assortment (p. 481)
genetic equilibrium (p. 481)
gene flow (p. 481)
genetic drift (p. 481)
masked (p. 482)
allelic partner (p. 482)
fixed (p. 482)
founder effect (p. 483)
bottlenecks (p. 483)
baboon troop (p. 483)
allele pool (p. 483)
natural selection (p. 484)
artificial selection (p. 484)
reproductive potential (p. 484)
competition (p. 484)
adaptive (p. 484)
differential reproduction (p. 484)

Objectives

1. Distinguish the typological approach to diversity from the currently used population concept.

2. Distinguish the terms *population* and *species*.

3. Explain how frequency distributions of traits can be useful in studies of evolution.

4. Show how the Hardy-Weinberg equations allow geneticists to measure changes in allele frequencies in population. In Figure 30.4, determine what would happen to *AA, Aa,* and *aa* if the value of *p* changed to 0.8 and the value of *q* changed to 0.2.

5. Give an example illustrating that genotype frequencies can change even though allele frequencies have remained the same. Then relate the stability of allele ratios to genetic equilibrium.

6. List four conditions that must be met before Hardy-Weinberg equilibrium can occur. Then explain why, if these conditions are seldom met in natural populations, geneticists use the Hardy-Weinberg equations.

7. Define *mutation* and explain how mutations cause allele frequencies to change in natural populations.

8. Define *genetic drift* and explain how it can cause allele frequencies to change in natural populations.

9. List two of the causes of gene flow and explain how gene flow can cause allele frequencies to change in natural populations.

Self-Quiz Questions

Fill-in-the-Blanks

In a(n) (1) _____ approach to studying diversity in populations, an individual is selected as the perfect representative type for the species, on the basis of a somewhat arbitrary choice of what "perfect" representative features were. A(n) (2) _____ is a group of individuals of the same species that occupy a given area at a specific time. The (3) _____ _____ _____ allows researchers to establish a theoretical reference point (baseline) against which changes in allele frequency can be measured. Variation can be expressed in terms of (4) _____ _____: the relative abundance of different alleles carried by the individuals in that population. The stability of allele ratios that would occur if all individuals had equal probability of surviving and reproducing is called (5) _____ _____. Over time, allele frequencies tend to change through infrequent, but inevitable (6) _____, which are the original source of genetic variation. Random fluctuation in allele frequencies over time due to chance occurrence alone is called (7) _____ _____; it is more pronounced in small populations than in large ones. (8) _____ _____ associated with immigration and/or emigration also changes allele frequencies. Populations in which two or more forms of a trait persist are said to show (9) _____ at that gene locus. When individuals of different phenotypes in a population differ in their ability to survive and reproduce, their alleles are subject to (10) _____ _____.

(11) In a population, eighty-one percent of the organisms are homozygous dominant, and one percent are homozygous recessive. Find (a) the percentage of heterozygotes, (b) the frequency of the dominant allele, and (c) the frequency of the recessive allele.

(12) In a population of 200 individuals, determine how many individuals are (a) homozygous dominant, (b) homozygous recessive, and (c) heterozygous for a particular locus if $p = 0.8$.

30–II   EXAMPLES OF NATURAL SELECTION

Summary

Natural selection is no longer in the realm of pure theory; examples of natural selection have been demonstrated in different populations. The transient polymorphism observed among peppered moths appears to be a case of *directional selection*: because of a specific change in the environment, a heritable trait occurs with increasing frequency, and the whole population tends to shift in a parallel direction. The development of pesticide-resistant pests is another example of directional selection.

Disruptive selection, in which two or more distinct polymorphic varieties are favored and become increasingly represented in a population, tends to split up a population. Batesian mimicry among females of the African swallowtail butterfly and their foul-tasting models provides an example of disruptive selection that is probably caused by predation.

A process in which a form already well adapted to a given environment is selected for and maintained, even as extreme variants are selected against, is called *stabilizing selection*. For more than 400 million years, the form of horseshoe crabs has remained essentially the same. Horseshoe crabs probably represent a balanced system of the best working combination of traits in an environment that has remained relatively stable. Presumably, extremely variant individuals were continually eliminated, so little structural change has occurred through evolutionary time.

The survival value of variant alleles must be weighed in the context of the environment in which they are being expressed; they are not advantageous or disadvantageous in themselves. The HbA (normal hemoglobin) and HbS (sickle-cell hemoglobin) form a *balanced polymorphism* that has been maintained in populations throughout West and Central Africa for as long as malaria has been prevalent.

Key Terms

stabilizing selection (p. 485)   horseshoe crab (p. 486)   Kettlewell (p. 487)
directional selection (p. 485)   peppered moth (p. 486)   African swallowtail
disruptive selection (p. 485)   lichens (p. 487)   butterfly (p. 488)
dimorphism (p. 485)   mark-release-   mimicry (p. 488)
polymorphism (p. 485)   recapture (p. 487)   mimic (p. 488)

model (p. 488)
selective advantage (p. 488)
morph (p. 488)
sexual dimorphism (p. 488)
sexual selection (p. 488)

leks (p. 489)
balanced polymorphism (p. 490)
sickle-cell
  polymorphism (p. 490)
Plasmodium falciparum (p. 490)

sporozoites (p. 490)
differential
  mortality (p. 490)
differential
  fertility (p. 490)

Objectives

1. Define *directional selection* and give two examples.

2. Define *disruptive selection* and give an example.

3. Explain how horseshoe crabs have managed to remain essentially the same for 400 million years.

4. Define *balanced polymorphism*. Explain why inheriting a gene for HbS (sickle-cell hemoglobin) might be advantageous for some humans living under certain situations.

5. Contrast transient polymorphism and balanced polymorphism.

Self-Quiz Questions

Fill-in-the-Blanks

(1) _____ _____ occurs when a specific change in the environment causes a heritable trait to occur with increasing frequency and the whole population tends to shift in a parallel direction. (2) _____ _____ favors the development of two or more distinct polymorphic varieties such that they become increasingly represented in a population and the population is split up into different phenotypic variations. (3) _____ _____ provide an excellent example of stabilizing selection because they have existed essentially unchanged for 400 million years. Because deaths of HbA homozygotes due to malaria were balanced by deaths of HbS homozygotes brought about by sickle-cell anemia, (4) _____ _____ at the sickle-cell locus was maintained in populations in regions where malaria was prevalent.

30–III EVOLUTION OF SPECIES

Summary

Evolution may be thought of as successive changes in allele frequencies brought about by such occurrences as mutation, genetic drift, gene flow, and selection pressure.

Usually studies of evolution focus on some local breeding unit: a population of localized extent within a larger population system. The usual barriers to allele exchange between local breeding units of a larger population are those that create geographic isolation: severe storms, major floods, earthquakes, the uplift of mountain ranges, and the separation of continents. Following geographic isolation, mutation, genetic drift, and selection pressures may operate on the different local breeding units. Over time, these forces may lead to *divergence*: a buildup of differences in allele frequencies between isolated populations of the same species.

Speciation has occurred when some restriction on interbreeding between populations is followed by enough genetic differentiation that, even if individuals from those populations do make contact, they cannot or will not interbreed. *Isolating mechanisms* are aspects of structure or function that prevent interbreeding between populations that are undergoing or have undergone speciation. Differences in reproductive structure or physiology, in timing of reproduction, in behavior, and in ecological factors may serve as isolating mechanisms. Local populations that are geographically separated are called allopatric. *Allopatric speciation*, in which geographic separation and gradual divergence lead to the reproductive isolation of two populations, is the most common pattern in nature; it is generally caused by a disturbance in the environment. When distinct local breeding units coexist in the same geographic range, they are called sympatric. In *sympatric speciation*, two or more populations occupying the same distribution range are thought to undergo reproductive isolation before genetic differentiation transforms them into separate species. Especially among plants, polyploidy and/or hybridization are two other speciation routes; about forty percent of all flowering plant species are polyploid. Wheat is a successful polyploid, Kentucky bluegrass hybridizes with a number of related species.

Key Terms

species (p. 490)
reproductive isolation (p. 490)
demes (p. 490)
divergence (p. 491)
speciation (p. 491)
reproductive
  isolating mechanism (p. 491)
hybrid zygotes (p. 491)
mechanical isolation (p. 491)
incompatibility (p. 492)

gamete isolation (p. 492)
temporal isolation (p. 492)
behavioral
  isolation (p. 492)
hybrid inviability (p. 492)
hybrid infertility (p. 492)
hybrid breakdown (p. 492)
allopatric
  speciation (p. 492)

timing of
  reproduction (p. 493)
parapatric
  speciation (p. 493)
sympatric
  speciation (p. 493)
polyploidy (p. 493)
vegetative
  propagation (p. 493)
hybridization (p. 493)

Objectives

1. Describe the usual way that divergence occurs in animal populations and species evolve.

2. Define *local breeding unit* and explain its place in studies of evolution.

3. List three potential causes of geographic isolation and indicate how each might promote divergence.

4. Contrast partial reproductive isolation with complete reproductive isolation.

5. List four examples of isolating mechanisms.

6. Contrast allopatric and sympatric speciation.

7. Define *polyploidy* and indicate how human technology has utilized polyploidy.

8. Explain how adaptive radiation is related to evolution.

Fill-in-the-Blanks

The usual barriers to allele exchange between local breeding units of a larger population are those that create (1) _____ _____, which occurs during severe storms, earthquakes, floods, geologic uplift and subsidence, and the long-term separation of continents. Subsequently, mutation, (2) _____ _____, and selection pressures may operate on the different local breeding units. Over time, these forces may lead to (3) _____: a buildup of differences in allele frequencies (or genetic differentiation) between isolated populations of the same species.

(4) _____ is said to have occurred when some restriction on interbreeding between populations is followed by enough genetic differentiation that, even if individuals from those populations do make contact, they cannot or will not interbreed. When related plant species come into flower at nonoverlapping times, their (5) _____ isolation is complete. (6) _____ speciation, the most common pattern in nature, occurs when geographic separation of two populations, accompanied by gradual divergent evolution between them, leads to reproductive isolation. Especially among plants, (7) _____ and/or hybridization are two other speciation routes; about forty percent of all flowering plants are (8) _____.

30–IV MACROEVOLUTION

A Time Scale for Macroevolution
Rates and Patterns of Change

Summary

The geologic time scale initially was defined as a progression of four broad eras: the *Proterozoic* ("very first life"), the *Paleozoic* ("ancient life"), the *Mesozoic* ("between-ancient-and-modern life"), and the *Cenozoic* ("modern life"). Within the past three decades, fairly firm time boundaries have been assigned to these geologic intervals by using radioactive dating methods on an enormous number of rock samples taken from all over the earth. All of these methods are based on comparing the amounts of radioactive isotopes (such as uranium, thorium, potassium, and strontium) with the amounts of their stable decay products in different kinds of rocks. Each radioactive element has its own characteristic decay rate that cannot be modified by physical or chemical changes in the environment.

As we move from the most ancient fossilized cells that resemble bacteria (from South Africa, about 3.2 billion years old) and blue-green algae (from Rhodesia, about 3.1 billion years old), the fossil organisms preserved within younger and younger rocks become progressively more abundant, more diverse, and more like modern forms. As environments change, so selection pressures change—and so the population changes in character. If one population somehow expands into two different settings, different traits will be selected for in each setting. With time, the two populations will accumulate more and more structural and behavioral differences until members of the two populations can no longer interbreed when brought back together; two different species have evolved from one ancestral species.

When populations of the same species become adapted for exploiting different resources in the environment in specialized ways, they are said to have undergone *adaptive radiation*; the dispersal of Darwin's finches is an example of adaptive radiation.

Key Terms

macroevolution (p. 493)	Mesozoic (p. 495)	selection pressures (p. 496)
radioactive isotopes (p. 494)	Cenozoic (p. 495)	convergent
characteristic	lineage (p. 495)	evolution (p. 496)
decay rate (p. 494)	divergence (p. 495)	parallel evolution (p. 496)
half-life (p. 494)	adaptive radiation (p. 496)	partitioning (p. 496)
geologic time scale (p. 495)	adaptive zone (p. 496)	homologous (p. 497)
Proterozoic (p. 495)	adaptive traits (p. 496)	analogous (p. 497)
Paleozoic (p. 495)		

Objectives

1. Identify the four major eras of the geologic time scale, their time spans, and the principal organisms that lived during each era.

2. Explain how radioactive dating is used to estimate the age of rocks. Use terms such as *isotope, decay product,* and *half-life* in your explanation.

3. List the fossil evidence suggesting that living cells existed on earth 3.2 billion years ago.

4. Explain how speciation can occur.

Self-Quiz Questions

Fill-in-the-Blanks

The (1) _____ era extended from 3.2 billion years ago to about 570 million years ago; initially heterotrophic and autotrophic (2) _____ existed, but between 900 and 750 million years ago (3) _____ had appeared. The (4) _____ era was characterized by "ancient life," with invertebrates, fishes, and amphibians coexisting principally with aquatic plants. As environments change, (5) _____ pressures also change, and so populations change in character. If subgroups of one original population that have been living separated for many generations can no longer interbreed when brought back together, (6) _____ is said to have occurred.

When populations of the same species become adapted for exploiting different aspects

of the environment in specialized ways, they are said to have undergone (7) _____

_____.

30–V EVOLUTIONARY TRENDS

Summary

Change within an established lineage (as, for example, in titanotheres) is called
phyletic evolution. Gradual changes in traits such as horn size occur as early genera
evolve into advanced ones. According to the *gradualistic model*, phyletic evolution is
the main trend in the history of life. Not all transitions in the fossil record are
gradual; some major groups of organisms appeared relatively abruptly and were already
highly developed and diverse when they did appear. Several explanations have been
proposed to account for such abrupt appearances. *Quantum speciation* is the rapid crossing of adaptive thresholds, which, the *punctualistic model* states, occurs in times of
intense selection pressure in small, isolated populations. This model also holds that
quantum speciation has been the principal mode of speciation. Sometimes important gaps
occur in the fossil record.
    Not all organisms become fossilized when they die. The probability that represent-
atives of any group will be preserved depends partly on the number of individuals in
the group. Abrupt climatic or geographic changes encourage some populations to expand
and diversify rapidly.
    Extinction, the disappearance of groups of organisms, is one of the major patterns
of evolution. Some factors as the absence of genetic diversity or having a narrow distribution range or small population size may place a group on the brink of extinction.
In recent times, human-induced disturbances are pushing an alarming number of species
over the edge at an accelerated rate that far exceeds anything that has occurred in
the past.

Key Terms

australopithecines (p. 497)	branch points (p. 497)	key character (p. 497)
phyletic evolution (p. 497)	quantum evolution (p. 497)	extinction (p. 497)
titanotheres (p. 497)	adaptive thresholds (p. 497)	taxon (p. 498)
Eocene period (p. 497)	punctuational model (p. 497)	epochs (p. 499)
gradualistic model (p. 497)	mosaic evolution (p. 497)	Archaeopteryx (p. 499)

Objectives

1. Contrast the gradualistic model with the punctuational model.

2. Explain why you think there are "gaps" in the fossil record.

3. Explain why transitions in the fossil record don't always occur gradually.

4. Define *selection pressure*, give two examples of selection pressures, and indicate how selection pressures can promote differential survival and reproduction of genotypes within a population.

Self-Quiz Questions

Fill-in-the-Blanks

Evolution can be viewed either by the (1) _____ model, which envisions speciation as accounting for only a small amount of large scale change in the manner of (2) _____ evolution, or by the (3) _____ model, which sees higher taxa originating by the rapid crossing of adaptive thresholds in the manner of (4) _____ _____ . (5) _____ was a fossil link between reptiles and birds; deBeer coined the term (6) _____ evolution to describe a situation in which such a fossil link has developed one or more (7) _____ _____ that enable a new evolutionary path to be pursued. In this case, the feature that enabled the evolution of birds was the appearance of (8) _____ . (9) _____ is the disappearance of an entire taxon.

INTEGRATING AND APPLYING KEY CONCEPTS

Can you imagine any way in which directional selection may have occurred or may be occurring in humans? Which factors(s) do you suppose are the driving force(s) that sustain(s) the trend? Do you think the trend could be reversed? If so, by what factor(s)?

31
ORIGINS AND THE EVOLUTION OF LIFE

Summary

A cohesive theory linking the flow of earth's history and the history of life is being developed. Billions of years in the past, our solar system was born from an exploding star. Our planet took form through the gravitational compression of dust and debris that swirled around the primordial sun. Five billion years ago Earth was a cold, homogeneous mass, but through contraction and radioactive heating it developed a core that was dense and hot. By 3.7 billion years ago, a crust that periodically erupted volcanic outpourings from the inferno below had formed, and a dense atmosphere of hydrogen, nitrogen, methane, ammonia, hydrogen sulfide, and water vapor was being retained by earth's gravitational field as our planet settled into an orbit around the sun. Temperatures on the planet favored the development of life, and somewhere between 3.7 and 3.5 billion years ago, living organisms appeared in the mineral-rich primeval seas. A variety of experiments by Miller, Fox, and others have demonstrated that all the building blocks required for life can form under abiotic conditions given the primitive earth atmosphere and an abundant supply of energy. The mechanism by which independently formed organic molecules combined into a permanent reproducible living system is not yet known.

Key Terms

crust (p. 500)
plastic mantle (p. 500)
continents (p. 500)
stellar
  explosions (p. 501)
accretion (p. 501)
gravitational
  compression (p. 501)
oxidation (p. 501)
free oxygen (p. 501)
gravitation
  mass (p. 501)
hydrogen (p. 502)
methane (p. 502)
ammonia (p. 502)

Stanley Miller (p. 502)
organic "soup" (p. 502)
templates (p. 503)
Cairns-Smith (p. 503)
clay crystals (p. 503)
amino acids (p. 502)
nucleotides (p. 502)
adsorbed (p. 503)
replication
  enzymes (p. 503)
activators of
  replication (p. 503)
membrane-bound
  protocells (p. 503)

information storage (p. 503)
stereoisomers (p. 503)
left-handed forms (p. 503)
microspheres (p. 504)
differentially
  permeable (p. 504)
lipid-protein (p. 504)
liposomes (p. 504)
self-assembly (p. 504)
aqueous
  compartments (p. 504)
fluidity (p. 504)
elasticity (p. 504)
primordial (p. 503)

Objectives

1. List in sequence the events and processes that cooperated to make earth a planet suitable for supporting life. Indicate how long ago these processes were occurring.

2. List the principal elements needed to support life and tell the form they took in the atmosphere and/or in the oceans.

3. Describe the basic ideas of the theory of the possible chemical evolution of the first primitive organism. Present evidence that supports and undermines the theory.

4. State any alternatives to the proposed theory of chemical evolution, and the evidence that supports or undermines those alternatives.

Self-Quiz Questions

Fill-in-the-Blanks

The earth's crust has been divided into vast (1) _____, which have been moving

about uneasily on top of a plastic (2) _____ and carrying the continents with

them. About (3) ◯ 10 ◯ 5 ◯ 3.7 ◯ 3.2 billion years ago, our planet took form as a cold

homogeneous mass. By (4) ◯ 5 ◯ 3.7 ◯ 3.2 ◯ 2 billion years ago, the earth had a hot,

dense core and was hurtling through space as a thin-crusted inferno. Rain fell;

minerals were stripped from rocks, and the oceans were formed. According to

Cairns-Smith, (5) _____ _____ probably were the adsorbing agents that

served to assemble (6) _____ _____ into proteins. (7) _____ are differentially permeable assemblages of polypeptides that accumulated greater

concentrations of certain substances inside than were found on the outside.

(8) _____ are microscopic spherical and tubular structures that are assembled

from simple lipids that have single-chain tails and hydrocarbon heads.

Summary

Until about 3.7 billion years ago, the earth's crust may have been too unstable or too
thin--and perhaps the heat flow from the interior was too great--for permanently stable
land masses to form. Nevertheless, there were probably many volcanic islands rising
above the primeval seas. It may have been in the shallow, nearshore waters of these
islands that life appeared. Reeflike formations, perhaps formed by calcium-secreting
algae, date from about 3.5 billion years ago. For about 2 billion years or so,
organisms resembling modern bacteria and blue-green algae had the world much to
themselves. Little or no oxygen was in the atmosphere, so anaerobic pathways must have
been the style of the day. Between 2 billion and 1 billion years ago, both heterotrophic
and autotrophic prokaryotes grew steadily more abundant, if not more diverse.

     Between 1.4 billion and 750 million years ago, the first eukaryotes arose.
Predatory forms were among them. From that time onward, fossils of both predators and
prey are increasingly larger. With the rise of predatory eukaryotes, the blue-green
algae of the seas began to decline. By 570 million years ago, astonishingly diverse
eukaryotes dominated the scene. The combined activity of billions of photosynthetic
microorganisms generated significant levels of oxygen--oxygen that could be used to
advantage by predatory heterotrophic mobile eukaryotes as they chased after their prey,
and oxygen that, through aerobic pathways, could be used to harness energy in ATP to
build mineralized protective hard parts in both predator and prey organisms alike.

Key Terms

sedimentary (p. 504)
anaerobic
   prokaryotes (p. 505)
heterotrophic (p. 505)
stromatolite (p. 505)
Desulfovibrio (p. 506)
sulfate (p. 506)
sulfide (p. 506)
autotrophic (p. 506)
transport chains (p. 506)
spontaneously (p. 506)
cyanobacteria (p. 506)

Eubacteria (p. 506)
Archaebacteria (p. 506)
methane-producing
   bacteria (p. 506)
urkaryotes (p. 506)
Age of Prokaryotes (p. 506)
Margulis (p. 506)
symbiotic origin
   of eukaryotes (p. 507)
symbiosis (p. 507)
host daugher cells (p. 507)
mesosomes (p. 508)

plate tectonic
   processes (p. 508)
metazoans (p. 508)
mineralized
   skeletons (p. 508)
seafloor
   spreading (p. 509)
continental
   drift (p. 509)
oceanic ridges (p. 509)
thermal convection
   in mantle (p. 509)

Fill-in-the-Blanks

For about (1) ○ 1/2 ○ 1 ○ 2 ○ 5 billion years, it seems, organisms resembling modern bacteria and blue-green algae had the world much to themselves. Little or no free (2) _____ was in the atmosphere, so (3) _____ pathways must have been the usual metabolic pattern. Between (4) ○ 3.2 ○ 1.4 ○ 1 billion and 750 million years ago, the first eukaryotes appeared. The primitive eukaryotes were mobile, and some chased after their prey; (5) _____ respiration would clearly have been advantageous in providing energy sources for their movements. Mats of cells that resemble modern cyanobacteria and that have been found in rock formations more than two billion years old are called (6) _____. *Oscillatoria limnetica* can carry out (7) _____ _____ as well as anaerobic respiration. The (8) _____ theory of eukaryote origins suggests that anaerobic amoeboid cells ingested aerobic bacteria that weren't later digested.

31–III   LIFE DURING THE PALEOZOIC
508–514

Summary

The *Paleozoic* was characterized by several massive inundations of continents and retreats by shallow seas. By the late Cambrian, there apparently were three major land masses mostly covered by shallow seas. Two would eventually become North America and Europe, and the third (Gondwana) would be fragmented into Africa, South America, Australia, Antarctica, and parts of Asia and the eastern North American coast. All but one *invertebrate* phylum had arisen by the close of the Cambrian, and the shallow seas of the Ordovician encouraged evolutionary experimentation in the invertebrate groups. Toward the end of the Ordovician, volcanic outpourings in the mobile zone along the eastern edge of the North American land mass created the immense ancestral Appalachian Mountains as the European plate moved in. During late Silurian times, some life forms developed traits that would permit them to live on land: tough or waxy surface layers to keep from drying out, the potential for exploiting new water supplies, systems for taking in oxygen or carbon dioxide from the surrounding air, varying frameworks needed to support the body, new methods to maintain the balance of salts (ions) in body fluids, and new modes of reproduction and offspring dispersal.

  As the Silurian gave way to the Devonian some 400 million years ago, the continents collided, producing a dramatic increase in dry land area; plants and then animals began their tentative adaptive forays into this vacant environment. Certain fish, left behind in drying tide pools or in evaporating freshwater ponds, were able to use their swim bladders as lunglike organs and their fins as simple limbs; with such adaptations, they could crawl from pond to pond. Thus began an evolutionary trend that gave rise to the amphibians: organisms that could spend part of their lives on land but had to return to water to lay eggs, for their eggs had no protective shell to keep from drying out. The Carboniferous, with land masses being submerged and drained no less than fifty times, was the heyday of the amphibians. Adaptations appeared that allowed gymnosperms and reptiles to move onto higher and drier land: gymnosperms developed seeds, and reptiles developed shelled eggs and internal fertilization; now both groups could complete their reproductive cycles without free-standing water.

As the Carboniferous gave way to the Permian, collisions of crustal plates brought all land masses together into one vast supercontinent, called Pangea. Lowlands and humid uplands emerged in the north, glaciation occurred in the south, and the shallow seas drained from the massive continent. Everywhere in the shrinking seas, huge numbers of marine forms became extinct.

Key Terms

Paleozoic Era (p. 508)	Silurian (p. 512)	labyrinthodonts (p. 513)
Cambrian (p. 508)	Tethys Sea (p. 512)	amphibian (p. 513)
Gondwana (p. 508)	armor-plated	Carboniferous (p. 513)
Laurentia (p. 508)	fishes (p. 512)	gymnosperms (p. 514)
trilobites (p. 508)	gill arches (p. 512)	seed-bearing (p. 514)
Burgess shale (p. 512)	jaw (p. 512)	reptiles (p. 514)
Ordovician (p. 512)	Devonian (p. 512)	shelled eggs (p. 514)
inundations (p. 512)	Laurasia (p. 513)	insects (p. 514)
brachiopods (p. 512)	Rhyniophyta (p. 513)	Permian (p. 514)
ectoprocts (p. 512)	lycopods (p. 513)	Pangea (p. 514)
cephalopods (p. 512)	lungs (p. 513)	glaciers (p. 514)
vertebrates (p. 512)	lobe-finned	
jawless fishes (p. 512)	fishes (p. 513)	

Objectives

1. Describe the principal events that occurred in each of these periods of the Paleozoic era: Cambrian, Ordovician, Silurian, Devonian, Carboniferous, Permian. State whether aquatic or terrestrial conditions prevailed and which organisms were most abundant and conspicuous during each period.

2. List the principal ideas of the plate tectonic theory and state which earthly phenomena are explained by the theory.

3. Name the changes that permitted plants and animals to shift from an aquatic environment to moist land, and from moist land to dry land.

4. Name the animals and plants that were the pioneers in the transition to land.

Self-Quiz Questions

Fill-in-the-Blanks

(1) _____ was a supercontinent that became fragmented into Africa, South

America, Australia, Antarctica, and parts of Asia and the eastern North American coast.

By the late Cambrian, (2) _____, mud-burrowing, mud-crawling scavengers, were

abundant in the shallow marine environment. All but one (3) _____ phylum had

appeared by the close of the Cambrian. (4) _____ fishes appeared by the late

Ordovician, and the Appalachian Mountains were being built. The Devonian period is known for giving rise to the (5) _____: organisms adapted to life both in water and on land. The (6) _____ period is characterized by flat, low coastlines where immense swamp forests became established. Collisions of crustal plates during the Permian brought all land masses together into one vast supercontinent: (7) _____; the shallow seas were drained and massive extinctions occurred.

Summary

During the ensuing *Mesozoic* era, gymnosperms and reptiles (especially the dinosaurs) had the competitive edge as Pangea began to break up and the three major land masses began moving their separate ways creating the basin of the Atlantic Ocean in the process. Mammals arose in the Mesozoic, but they did not become well established until the predatory dinosaurs disappeared suddenly from the earth at the end of the Cretaceous period. The reasons for this sudden demise are not known; there were no apparent profound shifts in climate and no sudden, massive upheavals in the crust. By *Cenozoic* times major reorganization was occurring among all the crustal plates that remained from the breakup of the supercontinent. North America, Europe, and Africa were moving their separate ways; brittle fragmentation of coastlines, severe vulcanism, and massive uplifting of mountain ranges promoted a variety of cooler, drier environments. Sea levels changed, as did the earth's overall temperature. Grasslands were formed; plant-eating animals and their predators evolved and occupied them. With the land essentially cleared of major reptilian predators, the *mammals* and *birds* eventually diversified, along with the *flowering plants*. The tropical forests were fragmented into a patchwork of new environments, and many forest inhabitants were forced into new life-styles in mountain highlands, deserts, and plains. One such evicted form gave rise to the human species.

 Diversity means adaptations to different kinds of predators, different prey, different competitors or cooperative groups after the same resource, and different behaviors and patterns of fur and feather for attracting mates and assuring reproductive success. Adaptive success is ensured only as long as there is responsiveness to the environment, and only as long as there is dynamic stability between the requirements and the demands of organisms making their homes together. If the record of earth history tells us anything at all, it is that life in one form or another has survived disruptions of the most cataclysmic sort.

Key Terms

Mesozoic Era (p. 514)	pterosaurs (p. 515)	phytoplankton (p. 516)
Jurassic (p. 514)	plesiosaurs (p. 515)	Cenozoic Era (p. 516)
therapsids (p. 514)	Cretaceous (p. 516)	Eocene (p. 517)
thecodont reptiles (p. 515)	Alvarez (p. 515)	Pliocene (p. 517)
dinosaurs (p. 515)	asteroid (p. 515)	

Objectives

1. Describe the tectonic events that were shaping land environments during Mesozoic and Cenozoic times and state the types of organisms that evolved to exploit the different types of environments.

2. Explain how a knowledge of earth's geologic history helps explain why there is such a great diversity of organisms on earth today.

3. State whether or not you think humans are successfully adapted to their environments and provide some evidence to support your statement.

Self-Quiz Questions

Fill-in-the-Blanks

In the Mesozoic, widespread (1) _____ activity marked the birth of the Atlantic basin, as North America, Europe, and Africa began moving their separate ways. (2) _____ occurred along the coasts, and (3) _____ gave rise to mountains along the margins of massive (4) _____ and along zones where some plates were thrust under others. Never before had there been such worldwide (5) _____ _____, as crustal plates collided with and jockeyed for position relative to one another. Glaciation caused changes in (6) _____ _____ as well as in the earth's overall (7) _____. In much of the world, the concurrent shifts in climate led to the emergence of extensive, semiarid, cooler (8) _____, into which herbivores and their predators radiated. During the Cenozoic era, (9) _____ and (10) _____ plants evolved and dominated most environments. (11) _____ and (12) _____ had the competitive edge during Jurassic times; the climate was warm and humid with plains, mountains, and vast lagoons. Their ultimate challengers were to be the little ratlike (13) _____ scurrying through the shrubbery of the Jurassic and Cretaceous periods, whose populations exploded into the biotic vacuum established by the abrupt extinction of the dinosaurs at the end of the Cretaceous.

INTEGRATING AND APPLYING KEY CONCEPTS

As the earth becomes increasingly loaded with carbon dioxide and various industrial waste products, how do you think living forms on earth will evolve to cope with these changes?

32
VIRUSES, BACTERIA, AND PROTISTANS

Summary

Some bacteria (chemosynthetic autotrophs) extract energy from inorganic molecules, some are photosynthetic, and others are heterotrophs (decomposers and parasites). Two groups of bacteria have distinctively different cell walls and cell membranes as well as different transfer RNAs and RNA polymerases. Archaebacteria such as methanogens and halophiles tolerate low-oxygen environments. Eubacterial cells all have *peptidoglycan* in their cell walls; in addition, cyanobacteria have the same kinds of chlorophyll molecules as do plants. Oxygen is lethal for obligate anaerobes, but *facultative anaerobes* use oxygen when it is present and switch to anaerobic pathways when it isn't. Even though all bacteria have a rigid cell wall surrounding a plasma membrane, they differ somewhat in shape; *cocci* are spherical, *bacilli* are rodlike, and *spirilla* are spiral. Although each bacterium is a functionally independent unit, bacteria often remain linked together following division. Growth, reproduction, and dormancy depend on resource availability. Bacteria generally divide by asexual binary fission, although many bacteria form endospores, which resist drying out, irradiation, disinfectants, and acids.

    Cholera, diptheria, bubonic plague, and tuberculosis are diseases caused in humans by heterotrophic bacteria. Although many bacteria are agents of human diseases, the benefits derived from their recycling activities far outweigh their harmful effects.

Cyanobacteria generally exist as filaments or single cells and carry out photosynthesis and nitrogen fixation in freshwater, saltwater, and land environments. They need very few raw materials and little energy. *Mycoplasmas* are disease agents. They are the smallest independent organisms known that are capable of metabolism and of transforming energy into usable forms; they also have a complete genetic system for maintaining and reproducing themselves. *Rickettsiae* are obligate parasitic bacteria that grow inside host cells. Rocky Mountain spotted fever and typhus are two rickettsial diseases that are spread by arthropod bites.

Key Terms

mucilaginous
  sheath (p. 521)
coccus, cocci (p. 520)
bacillus, bacilli (p. 523)
rods (p. 520)
spirillum, spirilla
  spirals (p. 520)
nucleoid (p. 520)
bacterial
  flangella (p. 520)
peptidoglycan (p. 521)
Gram-positive
  bacteria (p. 521)
Gram-negative
  bacteria (p. 521)
Rhodospirillum (p. 522)
obligate
  anaerobes (p. 521)
facultative
  anaerobes (p. 521)
heterotrophs (p. 521)
decomposers (p. 521)
parasites (p. 521)
pathogenic (p. 521)
Rhizobium (p. 521)
autotrophs (p. 521)
chemosynthetic (p. 521)

methanogenic
  bacteria (p. 521)
cyanobacteria (p. 521)
green bacteria (p. 521)
purple bacteria (p. 521)
heterocysts (p. 521)
Anabaena (p. 521)
phototaxis (p. 522)
chemotaxis (p. 522)
Escherichia coli (p. 524)
chemoreceptors (p. 522)
runs and tumbles (p. 522)
magnetite (Fe_2O_3) (p. 522)
Archaebacteria (p. 522)
thermoacidophiles (p. 522)
Sulfobolus (p. 522)
denaturation (p. 523)
eubacteria (p. 523)
nitrogen-fixing
  cells (p. 523)
bacterial fission (p. 523)
clostridia (p. 523)
Bacillus
  thuringiensis (p. 523)
Clostridium
  tetani (p. 523)
mycoplasmas (p. 524)

endospores (p. 523)
autoclaves (p. 523)
neurotoxin (p. 524)
tetanus (p. 524)
botulism (p. 524)
Rickettsiae (p. 524)
obligate
  parasite (p. 524)
Rocky Mountain
  spotted fever (p. 524)
typhus (p. 524)
vector (p. 524
bubonic plague (p. 524)
cholera (p. 524)
diphtheria (p. 524)
tuberculosis (p. 524)
bacillary
  dysentery (p. 524)
fire blight (p. 524)
soft rot (p. 524)
bacterial wilt (p. 524)
Lactobacillus (p. 525)
Streptococcus (p. 525)
ruminants (p. 525)
rumen (p. 525)
Staphylococcus (p. 525)
Candida albicans (p. 525)

Objectives

1. Distinguish chemosynthesis from photosynthesis and obligate anaerobes from facultative anaerobes.

2. Describe the diversity of body plans that bacteria of different groups have.

3. List three essential ways that Archaebacteria differ from Eubacteria and give three examples of each group.

4. Explain how endospore formation by bacteria can concern humans.

5. List some of the ways that bacteria are involved in the cycling of carbon and nitrogen in natural communities.

6. State the principal characteristics of mycoplasmas and distinguish them from bacteria.

Fill-in-the-Blanks

(1) _____ autotrophs extract energy from inorganic molecules. Two types of bacterial heterotrophs are (2) _____, which break down dead organisms, and (3) _____, which obtain their nutrients from a host organism. For (4) _____ _____, exposure to oxygen is lethal. Eubacterial cell walls are composed of (5) _____, a substance that never occurs in eukaryotes. (6) _____ include (7) _____ such as *Halobacterium* and thermoacidophiles such as *Sulfobolus*. (8) _____ have the same kind of chlorophyll pigments found in plants. Heterocysts are cells in *Anabaena* that carry out (9) _____ _____. When environmental conditions become adverse, many bacteria form (10) _____, which resist moisture loss, irradiation, disinfectants, and even acids. (11) _____ are the smallest living organisms known that are capable of metabolism and of transforming energy into usable forms. (12) _____ are the only obligate parasitic bacteria that grow *inside* host cells; they are generally transmitted by arthropod (ticks, lice, fleas) bites. Unlike other photosynthetic bacteria, (13) _____ have the same kind of chlorophyll pigments found in all true algae and plants; like bacteria, they reproduce by (14) _____ _____, and, when environmental conditions become unfavorable, many can form (15) _____.

Summary

Viruses have a set of nucleic acids sheathed in a protective protein coat, but they are not capable of metabolism and they cannot reproduce themselves unless they are inside a host cell. Viruses are an expression of parasitic simplification. They are the causative agents of human diseases such as smallpox, influenza, polio, and possibly cancer, and they also damage livestock and crops. However, they play an important role in keeping certain bacterial populations within bounds.

Viroids are infectious pieces of single-stranded RNA that lack protein coats and are much smaller than the smallest virus. They are suspected to cause *slow-virus* diseases that have incubation periods that may last for years. They afflict the chrysanthemum and coconut crops and may be implicated in the development of cancer.

Several sexually transmitted diseases (gonorrhea, syphilis, and chlamydial infections) are caused by bacteria; others such as genital herpes are caused by viruses.

Key Terms

viruses (p. 526)
viral capsid (p. 526)
host cell (p. 526)
tobacco mosaic
  virus, TMV (p. 526)
bacteriophages (p. 526)
influenzaviruses (p. 526)
pandemic (p. 526)
herpesviruses (p. 528)
latent infections (p. 528)
virion (p. 526)
adenoviruses (p. 526)
parvoviruses (p. 526)
poxviruses (p. 526)
rhinoviruses (p. 526)

pink-eye (p. 526)
warts (p. 526)
chickenpox (p. 526)
shingles (p. 526)
yellow fever (p. 526)
German measles,
  rubella (p. 526)
rabies (p. 526)
meningitis (p. 526)
leukemia (p. 526)
gonorrhea (p. 527)
pelvic inflammatory
  disease (p. 527)
syphilis (p. 527)
spirochaete (p. 527)

chancre (p. 527)
treponeme (p. 527)
chlamydial
  infection (p. 528)
genital
  herpes (p. 528)
acquired immune
  deficiency syndrome,
  AIDS (p. 528)
Kaposi's
  sarcoma (p. 528)
viroids (p. 529)

Objectives

1. State the principal characteristics of viruses, indicate how they might have evolved, and list three diseases caused by viral agents.

2. Describe viroids and discuss how their existence might affect humans.

3. Contrast the characteristics of syphilis with those of gonorrhea. Compare each of these diseases with genital herpes and AIDS in terms of causative agents, methods of prevention, and treatment.

Self-Quiz Questions

Fill-in-the-Blanks

Individual viral particles are called (1) _____; each consists of a central

(2) _____ _____ surrounded by a (3) _____ _____.

(4) _____ contain the blueprints for making more of themselves but cannot carry

on metabolic activities. Influenza viruses cause (5) _____ diseases that occur

as worldwide epidemics. Chickenpox and shingles are two infections caused by a DNA

virus from the (6) _____ category. Nucleic acids that lack a

protein coat are called (7) _____. Syphilis is caused by a motile

(8) _____ that produces a localized ulceration called a (9) _____.

Euglenids 530
Chrysophytes and Diatoms 530–531
Dinoflagellates 531

Summary

Dinoflagellates are primitive photosynthetic eukaryotes that are motile members of the phytoplankton; many populations "bloom" suddenly in response to suddenly available nutrients. *Golden algae* have armor made of cellulose, pectin, and sometimes silicon dioxide. Diatoms, despite their glassy shells, are the major producers of marine communities. Photosynthetic flagellates are related to both plants and animals; although they can photosynthesize many of their own organic molecules, they need certain nutrients such as vitamins, which they absorb in a funguslike manner. The rarely encountered photoheterotrophs need both light for photosynthesis and food molecules to survive. The resemblances between photosynthetic and heterotrophic flagellates, together with the existence of photoheterotrophs, could signify divergence of many modern flagellates from a common ancestor.

Key Terms

protistans (p. 529)
protozoans (p. 529)
Euglena (p. 529)
euglenids (p. 530)
pellicle (p. 530)
eyespot (p. 530)
longitudinal
  fission (p. 530)

chrysophytes (p. 530)
pectin (p. 530)
silicon (p. 530)
fucoxanthin (p. 530)
Cretaceous chalk
  deposits (p. 530)
golden algae (p. 530)
phytoplankton (p. 531)

diatomaceous
  earth (p. 531)
dinoflagellates (p. 531)
cellulose plates (p. 531)
bioluminescence (p. 531)
Gonyaulax (p. 531)
red tides (p. 531)
neurotoxin (p. 531)

Objectives

1. Explain how heterotrophic protistans could have acquired the capacity for photosynthesis and state the evidence to support your explanation.

2. State the principal characteristics of dinoflagellates and indicate their ecological roles.

3. Explain what causes red tides and bioluminescence.

4. Tell how golden algae resemble diatoms, how the two groups differ from each other, and the ecological importance of each group.

5. Discuss how photosynthetic flagellates, heterotrophic flagellates, and photoheterotrophs are related. Then state how the existence of these three groups and their relationship may have promoted the establishment of the kingdom Protista.

Self-Quiz Questions

   Fill-in-the-Blanks

(1) _____ is mutually rewarding dependence between two organisms.

(2) _____ could be called primitive photosynthetic eukaryotes because

their chromosomes resemble those of prokaryotes; many organisms of this group exhibit

(3) _____, the production of light caused by the metabolic reaction of ATP

with specific compounds. The (4) _____ _____ have armor made of

cellulose, pectin, and sometimes silicon; their close relatives, the (5) _____,

are the principal prey on which many marine communities are built. Except for their

chloroplasts, *Euglena* and its relatives are virtually identical with several

(6) _____ flagellates.

32–IV Protozoa   532–537

Summary

There are four main types of animal-like protistans: the *heterotrophic flagellates*
discussed above, the *amoebas* and their relatives, the *sporozoans*, and the *ciliates*. All
are unwalled single cells that live by pinocytosis and phagocytosis. Some species are
nearly as small as bacteria; others can be seen with the naked eye. Amoebas move by
pseudopodia and phagocytize their food by surrounding it and engulfing it; they are
thought to have evolved from heterotrophic flagellates that capitalized on the ability
to distort the cell surface, extending and withdrawing it. In amoebas, actin and myosin
proteins are attached in an apparently random fashion to the cell membrane; when
contraction occurs in one region, a pseudopodium is slowly extended. *Foraminiferans* are
amoeboid relatives that secrete a hardened case of limestone that is peppered with tiny
holes through which sticky pseudopodia extend and trap bits of phytoplankton that float
by. *Sporozoans* are all parasitic, living within the bodies of animals; the host obtains
the food. *Plasmodium* is the causative agent of the disease malaria; in order to
complete its life cycle, its various developmental stages must spend time in human or
mosquito hosts.
　　Ciliates are covered with thousands of cilia and have voracious appetites. They
can move rapidly through their liquid environments, crawl, or simply stay put. Most
ciliates have a cavity lined with cilia, which set up movements that sweep food
particles into the cavity, where food vacuoles are formed. Digestion occurs in the food
vacuoles, and nutrients are absorbed by the cytoplasm across the vacuole membrane.
Paramecium, Didinium, and *Vorticella* are examples of ciliates.

Key Terms

protozoa (p. 532)
Mastigophora (p. 532)
Rhizopoda (p. 532)
Sporozoa (p. 532)
Ciliophora (p. 532)
flagellate
　protozoans (p. 532)
trypanosomes (p. 533)
Trypanosoma
　rhodesiens (p. 533)
amoebas (p. 533)

foraminiferans (p. 533)
pseudopods (p. 533)
Trichonympha (p. 533)
Amoeba proteus (p. 533)
contractile
　vacuoles (p. 533)
Entamoeba
　histolytica (p. 533)
sporozoans (p. 534)
heliozoans (p. 534)
radiolarians (p. 534)

Plasmodium (p. 534)
vector (p. 534)
host (p. 534)
Trypanosphaera
　transformata (p. 534)
malaria (p. 534)
merozoites (p. 535)
gametocytes (p. 535)
ciliate protozoans (p. 534)
cilia (p. 534)

African sleeping
  sickness (p. 533)
tsetse fly (p. 533)
trichosomes (p. 533)
Trichomonas
  vaginalis (p. 533)
amoeboid
  protozoans (p. 533)

amoebic
  dysentery (p. 533)
cyst (p. 533)
microtubular
  rods (p. 534)
silicon (p. 534)
parasites (p. 534)

Paramecium (p. 535)
gullet (p. 535)
anal pore (p. 535)
Didinium (p. 536)
trichocysts (p. 537)
Vorticella (p. 536)

Objectives

1. State the principal characteristics of the amoebas, radiolarians, and foraminiferans. Indicate how they generally move from one place to another and how they obtain food.

2. Characterize the sporozoan group, identify the group's most prominent representative, and describe the life cycle of that organism.

3. List the features common to most ciliates. Then describe the activities of two ciliates that have different ways of locomotion or obtaining food.

Self-Quiz Questions

Fill-in-the-Blanks

Amoebas move by sending out (1) _____, which surround food and engulf it.

Amoebas move by using (2) _____ and myosin, which form a jellylike clot when

mixed together and contract if (3) _____ is added. (4) _____

secrete a hard exterior covering of limestome that is peppered with tiny holes through

which food-trapping sticky pseudopodia extend. (5) _____ is a famous

(6) _____ that causes malaria. When a particular mosquito draws blood from an

infected individual, (7) _____ of the parasite fuse to form zygotes, which

eventually develop into spindle-shaped (8) _____ within the mosquito. *Paramecium*

is a ciliate that lives in (9) _____ environments and depends on (10) _____

_____ for eliminating the excess water constantly flowing into the cell.

Paramecium has a (11) _____, a cavity that opens to the external watery world.

Once inside the cavity, food particles become enclosed in (12) _____

_____, where digestion takes place. (13) _____ are long contractile

filaments anchored within the cell that are used to hold onto and paralyze its dinner.

Summary

A photosynthetic bacterium (*Prochloron*) that contains chlorophyll *a*, chlorophyll *b*, and carotenoids (the same pigments as are in all plants) may have given rise to the chloroplasts of green algae after symbiotic merger with a heterotrophic flagellate. Multicellular organisms are believed to have appeared when single-celled organisms— perhaps flagellates—divided and the newly formed daughter cells failed to separate. To the extent that staying together increased size, deterred smaller predators, improved motility, or offered resistance to strong currents, the factors causing adherence were selected for and perpetuated. Colonial forms have persisted to the present. In a colony, each cell benefits from the loose association, but each acts independently and is incapable of changing its behavior according to what is happening to its neighbors. All cells feed, reproduce, and respond the same way. When there is division of labor and interdependence, one type of cell cannot exist without the other. *Trichoplax*, a tiny blastulalike marine animal, is one of the simplest multicelled animals.

Key Terms

matrix (p. 537)	Pandorina (p. 537)	multicellularity (p. 537)
Prochloron (p. 537)	Gonium (p. 537)	Trichoplax
Chlamydomonas (p. 537)	Volvox (p. 537)	adhaerens (p. 537)
volvocines (p. 537)	daughter colonies (p. 537)	blastula (p. 538)

Objectives

1. Trace a sequence of events that may have transformed heterotrophic prokaryotes into photosynthetic eukaryotes.

2. Outline a possible route from unicellularity to a multicelled state with division of labor that could have been traveled by protistan ancestors.

3. Distinguish between colonial and truly multicellular organisms.

Self-Quiz Questions

    Fill-in-the-Blanks

(1) _____ is thought to resemble an ancestral prokaryote that may have given

rise to the chloroplasts of green algae. A very simple flagellated green alga is

(2) _____; cells that resemble it occur among the (3) _____,

colonial organisms that straddle the fence between protistans and (4) _____.

(5) _____ _____ is a tiny blastulalike marine animal that may be

reminscent of simple multicelled animals that made their entrance during the

Proterozoic. In a (6) _____, each cell benefits from a loose association with other cells but each acts independently. In a (7) _____ organism, labor is divided up among the cells, and there is interdependence to the extent that one type of cell cannot exist without the other.

INTEGRATING AND APPLYING KEY CONCEPTS

Suppose that genetic engineers could successfully introduce chloroplasts into human zygotes and that this new combination established a symbiosis. Describe all changes that you can imagine would occur in the appearance and behavior of the resulting individuals.

33

FUNGI AND PLANTS

Summary

Fungi are eukaryotes. Because their vegetative bodies are branched and filamentous, allowing them to come in contact with a large volume of their surroundings, they have access to raw materials that may be dilute or scarce. Like plants, many fungi have cells with large central vacuoles, and their body cells generally have walls. Also like plants, spore formation is an important part of their life cycle. However, fungi lack chloroplasts and differ from plants in other ways. Most fungi rely on enzyme secretion that promotes digestion *outside* the fungal body, followed by nutrient absorption across the plasma membrane of individual cells. Between 80,000 and 200,000 different fungal species are recognized. Some (for example, those that cause most plant diseases and athlete's foot) are parasitic, but most fungal species are saprophytic, feeding on the remains of dead organisms or their by-products. By bringing about the decay of organic material, they help recycle such vital substances through communities of organisms.

    The vegetative body of most *true fungi* is a mycelium, a mesh of branched, tubular filaments called *hyphae*. Most true fungi can reproduce asexually through spore formation, fission, budding, or fragmentation from a parent mycelium. Many also can reproduce sexually. *Egg fungi* (Oomycetes), which have motile reproductive cells, include saprophytic water molds, parasitic downy mildews, and late blight. *Zygote fungi* (Zygomycetes) are generally saprophytic and simply constructed; they include fly fungi and bread molds. *Sac fungi* (Ascomycetes) bear spores in saclike structures called *asci*, which are often concentrated in a complex fruiting body called an *ascocarp*. Yeasts, morels, truffles, powdery mildews, and *Trichophyton* are examples of sac fungi. *Club fungi* (Basidiomycetes) are structurally the most complex of all fungi. They include edible mushrooms as well as extremely toxic ones, shelf fungi, puffballs, many rusts, and smuts. Their complex reproductive structures are called *basidiocarps*.

    Phytophthora infestans is a parasitic fungus that causes late blight in potatoes and tomatoes. An example of its potentially devastating effect was seen in Ireland between 1845 and 1860, as the fungus got out of control in the uninterrupted fields of susceptible host plants—one-third of the population starved to death, died in the outbreak of typhoid fever that followed as a secondary effect, or fled the country. Currently as much as ten percent of the United States wheat crop is lost to the wheat rust fungal disease because very few resistant wheat strains exist and they are planted as monocrops.

Key Terms

fungi (p. 539)
saprobes (p. 539)
parasites (p. 539)
decomposers (p. 539)
mycelium (p. 539)
hypha, hyphae (p. 540)
chitin, (p. 540)
Precambrian (p. 540)
fungal spores (p. 540)
fragmentation (p. 540)
binary fission (p. 540)
divisions (p. 540)
Oomycota (p. 540)
oomycetes (p. 540)
egg fungi (p. 540)
water molds (p. 540)
downy mildews (p. 540)
Phytophthora
  infestans (p. 540)
late blight (pp. 540-541)

monocrop (p. 541)
Zygomycota (p. 541)
zygospores (p. 541)
Rhizopus
  stolonifer (p. 541)
Ascomycota (p. 541)
ascus, asci (p. 541)
ascocarp (p. 541)
sac fungi (p. 541)
Neurospora
  crassa (p. 541)
powdery mildew (p. 542)
Dutch elm
  disease (p. 542)
chestnut blight (p. 542)
truffles (p. 542)
morels (p. 542)
yeasts (p. 542)
Saccharomyces (p. 542)

Trichophyton (p. 542)
athlete's foot (p. 542)
Basidiomycota (p. 542)
club fungi (p. 542)
cup fungi (p. 542)
Amanita (p. 542)
shelf fungi (p. 542)
puff balls (p. 542)
basidiospores (p. 542)
primary mycelium (p. 542)
secondary mycelium (p. 542)
dikaryotic hypha (p. 542)
tertiary mycelium (p. 542)
basidiocarp (p. 542)
mushroom (p. 542)
rusts (p. 542)
smuts (p. 542)
wheat rust (p. 542)
basidia (p. 542)

Objectives

1. Describe the general structure of fungi and its relation to their method of obtaining nutrients.

2. Distinguish between parasitic and saprobic fungi. Mention one way that parasitic fungi harm humans and a way saprobic fungi benefit humans.

3. List the ways that fungi can reproduce.

4. List the four principal groups of fungi and give one example of an organism in each group.

5. Distinguish between the vegetative body and the reproductive part (which is sometimes referred to as the fruiting body).

6. State the fundamental contribution of fungi to ecosystems.

7. Give two examples of parasitic fungi that have played havoc with the production of crop plants.

8. Suggest ways in which American agricultural methods could be changed to prevent massive crop destruction by parasitic fungi.

Fill-in-the-Blanks

Like (1) _____, many fungi have cells with large central vacuoles, and their

body cells generally have cell walls; (2) _____ formation is also an important

part of their life cycle. Some fungal species, such as late blight, are

(3) ○ parasitic ○ saprophytic. The vegetative body of most true fungi is a

(4) _____, which is a mesh of branched, tubular filaments called (5) _____.

Fly fungi and bread molds are examples of (6) _____ fungi. Sac fungi bear spores

in saclike structures called (7) _____ in a complex fruiting body called a(n)

(8) _____. Morels, truffles, and yeasts are examples of (9) _____ fungi.

Rusts, smuts, puffballs, and shelf fungi are examples of (10) _____ fungi.

Currently as much as (11) _____ percent of the United States wheat crop is lost

to the wheat rust fungal disease.

33–II   Mycorrhizal Mats and the Lichens   544
Species of Unknown Affiliations   545

Summary

Almost all complex land plants depend on intimate association with fungi (*mycorrhizae*)
that help them absorb certain vital nutrients from the soil. *Lichens* are "composite"
organisms comprising an alga and a fungus living interdependently: The fungus obtains
photosynthetically derived food and the alga enjoys improved water conservation,
mechanical protection from being blown away, better gas exchange, and less overlap
between individual algal cells (such as there would be in an algal crust). In their
natural environment, *parasitic fungi* normally attack organisms that are damaged or
weakened, as with age.

   Slime molds have diverse shapes; some are masses of cells and others are multi-
nucleate masses of cytoplasm that have grown in size without cell division.

Key Terms

mycorrhizae (p. 544)	Penicillium (p. 545)	Distyostelium
lichen (p. 544)	roquefort (p. 545)	discoideum (p. 545)
cyanobacteria (p. 544)	camembert (p. 545)	cellular slime
green algae (p. 544)	penicillin (p. 545)	mold (p. 545)
Cladonia (p. 544)	predators (p. 545)	pseudoplasmodium (p. 545)
reindeer "moss" (p. 544)	slime molds (p. 545)	acellular slime
old man's beard (p. 544)	amoebalike protistans (p. 545)	molds (p.545)
Fungi Imperfecti (p. 545)	vegetative body (p. 546)	plasmodium (p. 545)

Objectives

1. Give two examples of fungi that participate in symbiotic relationships. Describe the contributions to the relationship of the fungus and of the other participant.

2. Explain why the Fungi Imperfecti are viewed as "imperfect."

3. Explain why taxonomists have so many problems in categorizing the slime molds.

Self-Quiz Questions

   Fill-in-the-Blanks

(1) _____ help almost all complex land plants absorb certain vital nutrients

from the soil. (2) _____ are "composite" organisms that comprise an alga and a

fungus living interdependently. When food is scarce, *Dictyostelium discoideum* amoebas

swarm together to form a (3) _____ that crawls around for a while, then

differentiates to form a (4) _____ _____ that eventually discharges

spores.

Summary

Among the aquatic plants that survive in water bordering lowlands are forms that grow as filaments or sheets only one or two cells thick. For them (the red algae, the brown algae, and the green algae) simple diffusion and passive transport carry materials and wastes across external cell membranes that directly contact the surrounding water. The surrounding water supports much of the body weight of these algal forms. Even in large complex species, support tissues are devoted more to keeping the plant body intact during wave action than to holding it upright. The dominant plant body of many existing aquatic species is composed only of haploid cells, some of which divide mitotically and produce gametes during sexual reproduction. The single-celled zygotes that result from the fusion of two gametes each enter a resting stage that carries them through adverse conditions. When favorable conditions return, the zygotes undergo meiosis, which leads to the formation of haploid spores that give rise to a new, conspicuous haploid body.

   The diverse body forms of the *red algae* (Rhodophyta) are adapted to a wide range of communities, but they are found especially in deep marine tropical waters. They resemble the blue-green algae with respect to their photosynthetic membranes and pigments, and may have been evolutionarily derived from them. The *brown algae* (Phaeophyta), which are almost all marine, include kelps, some of which are anchored by holdfasts, and some of which thrive in the open sea. Most of the complexly organized kelps show alternation of generations, and one species of giant kelp has developed phloemlike tissue. The diverse *green algae* (Chlorophyta) generally dwell in freshwater communities, but there are also many species that live in salt water or in moist soil. Their photosynthetic system and pigments are identical with those in complex land plants. Green algae show alternation of generations and variation in the type of gametes produced. For these reasons, ancestral green algae are believed to have given rise to land plants about 400 million years ago.

Key Terms

haploid
  dominance (p. 546)
diploid
  dominance (p. 546)
isogamy (p. 546)
oogamy (p. 546)
vascular
  tissues (p. 546)
gametophyte (p. 546)
haploid spores (p. 546)
sporophytes (p. 547)
homosporous,
  homospory (p. 547)
heterosporous,
  heterospory (p. 547)
megaspores (p. 547)
microspores (p. 547)

pollen grains (p. 547)
embryo (p. 547)
germination (p. 547)
seed (p. 548)
nonvascular
  plants (p. 548)
algae (p. 548)
red algae,
  Rhodophyta (p. 548)
accessory
  photosynthetic
  pigments (p. 548)
phycobilins (p. 548)
thylakoids (p. 548)
Porphyra (p. 548)
brown algae,
  Phaeophyta (p. 548)

kelps (p. 548)
Fucus (p. 548)
green algae,
  Chlorophyta (p. 548)
Bryophyta,
  bryophytes (p. 548)
xanthophylls (p. 549)
holdfasts (p. 549)
Sargassum (p. 549)
stipe (p. 550)
floats (p. 550)
carotenoids (p. 550)
starch (p. 550)
Chlamydomonas (p. 550)
sea lettuce, Ulva (p.550)
thallus (p. 550

Objectives

1. State how long ago the evolution of dry-land plants began. Then state the types of aquatic ancestors that are believed to have given rise to land plants and determine how long they have existed on earth.

2. Distinguish the red, brown, and green algae from one another.

3. Explain why neither red nor brown algae are thought to have given rise to land plants.

4. Tell how the plant body of a green alga would have to be modified in order to to survive in a dry-land environment.

5. Describe the basic differences between the life cycle of a green alga and the life cycle of a dry-land flowering plant.

6. List five key reproductive trends that have occurred in the evolution of plant life cycles.

Fill-in-the-Blanks

(1) _____ million years ago, the earth's crustal plates collided, and vast

regions were elevated above sea level. In algae, simple diffusion and (2) _____

transport carry materials and wastes across external cell membranes into the

surrounding water. The dominant plant body of many multicellular algae consists only of

haploid cells and produces gametes by mitosis; this is the (3) _____ generation.

Zygotes formed by the fusion of two gametes undergo (4) _____, which leads to the

formation of haploid (5) _____, each of which can then give rise to a new,

conspicuous haploid body. (6) _____ algae resemble the blue-green algae with

respect to their photosynthetic membranes and (7) _____, and may have been

evolutionarily derived from them. (8) _____ algae are almost all marine and

include the giant (9) _____, some of which are anchored by a (10) _____

and some of which thrive in the open sea. The (11) _____ algae are found in the

greatest diversity of environments; they have photosynthetic systems and pigment

systems that are identical with those in complex (12) _____ plants.

33–IV   The Bryophytes   551–552
Ferns and Their Allies: First of the Vascular Plants   552–556

Summary

400 million years ago the earth's crustal plates began colliding, and vast regions
became elevated above sea level. As shallow seas drained away, many aquatic forms must
have been structurally and functionally equipped in ways that allowed them to survive
on dry land.
     Transitional land plants *(bryophytes)* had to develop structures for getting and
conserving water even though *they never developed vascular tissue*, roots, or the
ability to reproduce sexually in dry-land environments. The 24,000 species of
bryophytes include mosses, liverworts, and hornworts, which grow close to the soil
surface of moist environments. Rhizoids anchor the photosynthetic gametophyte, which
supports the dependent, stalked sporophyte. The development of a protected embryo
sporophyte--attached to and nourished by gametophyte tissue--must have been an
important advance for life on land; it was first seen among the bryophytes.
     The primitive vascular plants (horsetails, lycopods, and ferns) generally have
underground stems (rhizomes) that produce aerial branches and leaves; true roots anchor
the rhizome in the soil. Such plants survive largely because they can take in soil
water and nutrients through roots, transport water through vascular tissue, and control
water loss from the stems and leaves. The development of supportive tissue enabled the

primitive vascular plants to grow taller and compete more effectively for available sunlight than the nonvascular transitional plants. Although the spores of primitive vascular plants typically are dispersed by wind, the flagellated sperms require water to reach the female structures. And although the gametophyte is more conspicuous and dominant and bryophyte life cycles, the sporophyte dominates vascular plant life cycles. All bryophytes and most primitive vascular plants produce only one kind of spore, which grows into a gametophyte that produces either male or female sex organs (or both); this strategy is referred to as a *homospory*.

Key Terms

Bryophyta,
  bryophytes (p. 551)
archegonium, -ia (p. 551)
antheridium, -ia (p. 551)
sporangium, -gia (p. 551)
cuticle (p. 551)
stomata (p. 551)
moss (p. 551)
liverworts (p. 551)
hornworts (p. 551)
vascular
  tissues (p. 551)
rhizoids (p. 552)

parenchymal
  cells (p. 552)
filamentous (p. 552)
green algae (p. 552)
lycopods,
  Lycophyta (p. 552)
Lycopodium (p. 552)
club mosses (p. 552)
scalelike leaves (p. 552)
strobilus,
  strobili (p. 552)
horsetails (p. 553)

Sphenophyta (p. 553)
Equisetum (p. 553)
rhizomes (p. 553)
aerial branches (p. 553)
nodes (p. 553)
Selaginella (p. 554)
ferns (p. 553)
Pterophyta (p. 553)
sporophyte (p. 554)
heart-shaped
  gametophyte (p. 556)
sorus, sori (p. 556)

Objectives

1. State how the primitive vascular plants differ from bryophytes.

2. Compare the life cycles of ferns and mosses.

3. Describe lycopods and horsetails.

4. Explain what a strobilus is.

Self-Quiz Questions

Fill-in-the-Blanks

Even though bryophytes never developed complex (1) _____ tissue, roots, or the

ability to reproduce (2) ◯ sexually ◯ asexually in dry-land environments, they did

develop structures for getting water and for conserving it. The 24,000 species of

bryophytes include (3) _____, liverworts, and hornworts. The leafy green plant

that comes to mind when the word *moss* is mentioned is the (4) ◯ sporophyte

◯ gametophyte generation. From it grows the (5) ◯ sporophyte ◯ gametophyte generation.

Byrophytes produce (6) ◯ homospores ◯ heterospores, which germinate and grow into the

(7) ◯ sporophyte ◯ gametophyte generation; this generation produces (8) ◯ spores

◯ gametes. When two gametes fuse, the resulting zygote divides mitotically, producing a

(9) ◯ haploid ◯ diploid (10) ◯ sporophyte ◯ gametophyte. The development of a protected embryo (11) ◯ sporophyte ◯ gametophyte—attached to and nourished by (12) ◯ sporophyte ◯ gametophyte tissue—must have been an important advance for life on land.

The primitive (13) _____ plants include horsetails, lycopods, and (14) _____. Instead of having upright aerial stems, most primitive vascular plants (except for tree ferns) have (15) _____, which run along or just below the surface. The (16) ◯ sporophyte ◯ gametophyte is the conspicuous part of the fern plant life cycle.

Summary

Competition for sunlight set in motion a series of developments that led to strong reinforced stems and vascular tissues, to efficient root systems, and to stomata and waxy leaf coverings. Finding places to live in dry-land environments led to the emergence of a prolonged diploid stage (sporophyte) and the reduction of the haploid stage (gametophyte). Some species of lycopods and ferns and all seed plants are *heterosporous*: two kinds of spores are produced. Megaspores give rise to female gametophytes, and microspores give rise to male gametophytes. In gymnosperms, female gametophytes develop while still attached to the parent plant; water and nutrients required for their development are provided by the sporophyte—the stage that is well adapted for obtaining these resources on dry land. In contrast, immature male gametophytes (pollen grains) are released from the parent plant and travel to the female gametophyte by wind, insects, and the like. Free water is not necessary for this transfer in most species. *Gymnosperms* (cycads, ginkgos, conifers, and gnetophytes) and *angiosperms* (the flowering plants) produce seeds. Each seed contains an *embryo sporophyte*, which generally is surrounded by internal *food reserves* that nourish it during germination and by a *seed coat* that guards against mechanical damage and extreme water loss. Seeds have taken over the dispersal function of spores.

The seeds of gymnosperms (naked seeds) are carried on the surfaces of reproductive structures without being protected by additional tissue layers. *Cycads* (Cycadophyta) resemble squat palm trees that have cones instead of flowers and a slow reproductive rate. *Ginkgos (Ginkgophyta)* are represented by only a single modern species of tree with fan-shaped leaves that grows under protected conditions. *Conifers* (Coniferophyta), which include pine, spruce, fir, hemlock, juniper, cypress, larch, and redwood species, are the most diverse and widely distributed gymnosperms. The sporophyte is a cone-bearing woody tree or shrub with needlelike or scalelike leaves that are generally retained for several years (evergreen). Their heterosporous, seed-bearing life cycles vary in their details, but in general their winged pollen grains are dispersed on the wind. Conifers were dominant land plants during the Mesozoic, and their mechanisms of seed production, protection, and dispersal helped assure their success. Although their numbers are much reduced now, they are still the dominant vegetation in northern regions and high altitudes.

In contrast to the "naked" seeds of gymnosperms, the angiosperm seed is contained within the ovary as it develops. *Angiosperms* (flowering plants) are grouped into two classes: monocots and dicots. The 50,000 or so species of monocots include palms, lilies, and orchids, as well as the principal crop plants (wheat, rice, corn, oats, rye, and barley) that support human populations. There are at least 200,000 dicot species (for instance, roses, beans, oaks, squash), and they are the most diverse group of flowering plants.

The vegetative bodies of plants and fungi are immobile, yet by relying on the wind, the splashing rain, and insects for dispersal, both groups have moved across the plains, to the mountains . . . to land everywhere.

Key Terms

gymnosperm (p. 556)
cycad (p. 556)
ginkgo (p. 556)
gnetophyte (p. 556)
Cycadophyta (p. 556)
cones (p. 556)
Ginkgophyta (p. 556)
conifer (p. 557)
Coniferophyta (p. 557)
evergreens (p. 557)
sporangium, -gia (p. 557)
microspores (p. 557)
megaspores (p. 557)
megasporangium,
  -gia (p. 558)

cone scale (p. 558)
megaspore mother
  cell (p. 558)
tissue-layer
  cover (p. 558)
ovule (p. 558)
pollination (p. 558)
fertilization (p. 558)
microspore mother
  cells (p. 558)
meiosis (p. 558)
pollen grain (p. 557)
pollen tube (p. 558)
male gametophyte (p. 558)
embryo sporophyte (p. 558)

seed coat (p. 558)
female gametophyte
  tissue (p. 558)
dispersal (p. 558)
angiosperms (p. 558)
Anthophyta (p. 558)
flowering plants (p. 558)
monocot (p. 559)
dicot (p. 559)
palms (p. 559)
orchids (p. 559)
corn (p. 559)
barley (p. 559)
vector (p. 559)

Objectives

1. Define *homosporous* and *heterosporous*, and distinguish one type of life cycle from the other. State which types of plants are homosporous.

2. Define *gymnosperm* and name the four divisions that are in the gymnosperm category.

3. Describe a representative cycad, ginkgo, and conifer.

4. Outline the principal steps of the conifer life cycle. Tell where the spores and gametes are formed. Distinguish pollination from fertilization.

5. List the three parts of a seed and explain how each is produced.

Self-Quiz Questions

Fill-in-the-Blanks

Some species of lycopods and ferns are (1) _____: two kinds of spores are produced. In gymnosperms, megaspores give rise to (2) ◯ male ◯ female gametophytes; these develop while still attached to the parent plant, which provides water and nutrients for their development. In contrast, immature male gametophytes (commonly known as (3) _____ _____) are released from the parent plant and are carried to the female plant by (4) _____, insects, and so on. Gymnosperms and

(5) _____ (the flowering plants) produce (6) _____, each of which

contains an embryo sporophyte that generally is surrounded by internal food reserves,

which nourish it during germination, and by a (7) _____ _____, which

guards against extreme water loss and mechanical damage. The seeds of (8) _____

are carried on the surfaces of reproductive structures without being protected by

additional layers. (9) _____ resemble squat palm trees that have cones instead

of flowers and have a slow reproductive rate. (10) _____ are represented by only

a single modern species of tree with fan-shaped leaves. (11) _____ have

needlelike or scalelike leaves that are generally retained for several years; they were

the dominant land plants of the (12) _____ era and are still dominant in

northern regions and high altitudes.

INTEGRATING AND APPLYING KEY CONCEPTS

Explain why totally submerged aquatic plants that live in deep water never developed
heterosporous life cycles.

34
ANIMAL DIVERSITY

Summary

More than ninety-nine percent of all the kinds of animals in the world are backboneless invertebrates. Somewhere between 2 billion and 1 billion years ago, the first animals arose; by the dawn of the Cambrian, they were already well-developed multicellular animals (*metazoans*). One of the early animal forms is thought to have led to the *sponges*, which are relatively simple vaselike animals that inhabit shallow seas. Seawater flows in through pores in the body wall, then out through the opening at the top; the interior is lined with flagellated collar cells, which trap and ingest microscopic organisms carried in on the water flowing through the various pores.

Sponges may have evolved from colonial protistans that resemble collar cells, but because sponges have developed four distinct cell types that cannot exist independently, they are truly multicellular even though communication between cells and integration of activities is poorly developed. Sponges have no nerve cells, no muscles, and no gut.

Cnidarians (jellyfish, corals, *Hydra*, and sea anemones) have simple organs distributed in a radial symmetry. Stinging cells (*nematocysts*) aid in food capture and defense. Most cnidarian life cycles have a *planula* larval stage: a roving, ciliated undifferentiated mass of cells that eventually settles down and becomes a sedentary, attached form called a *polyp*. For most cnidarians, the polyp stage eventually gives rise to free-swimming, bell-shaped forms called *medusae*, which produce gametes in sex organs; the gametes later fuse, forming a zygote that divides mitotically to form the planula larva. Even though the body plan of both polyp and medusa is essentially two connected epithelial layers separated by a jelly layer that contains the few simple organs, cnidarians do have a true gut. The gut is little more than an epithelial cavity; enzymes from glands in the epithelium break down food that has been stuffed into the cavity. No circulatory system transports materials. Undigested residues are simply expelled through the mouth. These animals also show integration of neural and muscular activity, having nerve nets, muscle tissue layers, and some sensory organs.

Key Terms

vertebrates (p. 561)
invertebrates (p. 561)
metazoans (p. 562)
multicellularity (p. 562)
radial
   symmetry (p. 562)
bilateral
   symmetry (p. 562)
blind sac (p. 562)
tube-within-
   a-tube (p. 562)
mesenchyme (p. 562)
coelom (p. 562)
acoelomate (p. 562)

pseudocoelomate (p. 562)
coelomate (p. 562)
peritoneum (p. 563)
segmentation (p. 563)
cephalization (p. 563)
Porifera (p. 563)
sessile (p. 563)
spicules (p. 563)
Cambrian (p. 563)
atrium (p. 566)
osculum (p. 566)
water-canal
   systems (p. 566)

collar cells (p. 566)
gemmule (p. 566)
Cnidaria (p. 567)
Obelia (p. 567)
polyps (p. 568)
medusae (p. 568)
planula (p. 568)
gastrovascular
   cavity (p. 568)
tentacles (p. 568)
mesoglea (p. 568)
cnidocyte (p. 568)
nematocyst (p. 568)

Objectives

1. Distinguish radial from bilateral symmetry and acoelomate from pseudocoelomate.

2. List two characteristics that distinguish sponges from other animal groups.

3. Describe each of the four cell types found in sponges.

4. Describe the two cnidarian body types.

5. Explain how radial symmetry might be more advantagenous to floating or sedentary animals than bilateral symmetry would.

6. State what nematocysts are used for and explain how they operate.

7. Design an illustration that summarizes a generalized cnidarian life cycle. Include these terms: *egg, sperm, zygote, planula, polyp, medusa.*

8. Describe a generalized cnidarian body plan.

9. Tell how cnidarians obtain and digest food and tell what they do with food they cannot digest.

10. Name several cnidarians.

Self-Quiz Questions

Fill-in-the-Blanks

Ninety-nine percent of all animals on earth are (1) _____. Multicellular animals are called (2) _____. (3) _____ form the most primitive major group of multicellular animals. They are nourished by microscopic organisms extracted from the water that flows in through pores in the body wall by sticky (4) _____ _____. (5) _____ between sponge cells and integration of activities are poorly developed. Sponges lack (6) _____ cells, muscles, and a gut. Cnidarians are (7) _____ symmetrical and have stinging cells called (8) _____, which aid in defense and food capture. Most cnidarian life cycles have a (9) _____ larval stage. Of the two body types, the (10) ◯ polyp ◯ medusa is the sexual stage, in which simple sex organs produce eggs or sperms.

Summary

The most primitive *flatworms* resemble the planula stage of cnidarian life cycles. Perhaps, long ago, such a planuloid kept on crawling (rather than settling down and becoming attached); the leading end of such a mutant would have encountered food and danger more than other body areas, hence nerve and sensory cell clustering at the head end would promote faster sensing and response. Predators of such a crawling planuloid would be more likely to attack from above than from below, so dorso-ventral differentiation would have been encouraged. Food and danger would be found as often on one side of a forward crawler as the other--hence the emergence of bilateral symmetry, which could have led to paired organs of the sort seen in many flatworms. Flatworms include free-living turbellarians, parasitic flukes, and parasitic tapeworms. All have flattened bodies shaped like broad leaves or long ribbons, and they range in size from microscopic to eighteen meters long! Adult parasitic flatworms attach to their hosts by suckers and hooks and are sustained by predigested nutrients from the host. Large free-living flatworms have a much-branched gut with only one opening, well-developed muscles, mesoderm, and simple tubes that pick up dissolved wastes from internal cells and expel them from the body much as kidneys do for other animals. Flatworms lack respiratory systems; gases easily enter and leave their thin bodies by diffusion.

Another major group of animals, the *round worms*, includes nematodes. This group encompasses hundreds of thousands of species, including some destructive parasites and many more free-living recyclers of nutrients. The body plan of nematodes consists of a tube (the gut) within another tube (the body wall). The gut has a mouth *and* an anus, and food travels from mouth to anus. Reproductive cells between the gut and body wall give rise to gametes. Nematodes have only longitudinal muscles and no circular muscles; thus they thrash awkwardly about. A tough cuticle permits them to survive high levels of acidity and alkalinity, temperature extremes, and poisonous compounds in their environment. If oxygen is present, they use it; if not, they switch to anaerobic pathways.

Key Terms

Platyhelminthes (p. 569)	fission (p. 569)	scolex (p. 570)
turbellarians (p. 569)	Schistosoma (p. 570)	proglottids (p. 570)
flukes (p. 569)	miracidia (p. 570)	neck (p. 571)
tapeworms (p. 569)	intermediate	planuloid theory (p. 571)
planarian (p. 569)	host (p. 570)	roundworms,
pharynx (p. 569)	trematodes (p. 569)	nematodes (p. 571)
flame cell (p. 569)	cuticle (p. 569)	Trichinella
protonephridium (p. 569)	schistosomiasis (p. 570)	spiralis (p. 572)
monoecious (p. 569)	cestodes (p. 570)	trichinosis (p. 572)

Objectives

1. Outline how a planula could have given rise to a simple flatworm. Indicate the selective pressures that helped to promote bilateral symmetry.

2. List the three main types of flatworms.

3. State the evolutionary advances seen in flatworms compared to cnidarians.

4. Describe the body plan of round worms, comparing its various systems with those of the flatworm body plan.

Self-Quiz Questions

Fill-in-the-Blanks

A mutant (1) _____ larva is believed to be the ancestor of the free-living

flatworms. A middle tissue layer, (2) _____, gives rise to muscular,

reproductive, and water-and-(3) _____ regulating systems. A shift from radial to

bilateral symmetry could have led to (4) _____ _____ of the sort seen in

many flatworms. Flatworms have no (5) _____ systems, and their (6) _____ system is incomplete. If an organism has one-way traffic through the gut, (7) _____

_____ is possible for more efficient food processing. (8) _____ have

complete digestive tracts, no circular muscles, and only a few longitudinal muscles.

A tough (9) _____ offers some protection from acids, bases, poisons, and extreme

temperatures. If oxygen is not present in sufficient quantities, they switch to

(10) _____ pathways.

34–III TWO MAIN LINES OF DIVERGENCE: PROTOSTOMES AND DEUTEROSTOMES 572–573

Summary

Two groups of animals are descended from ancestral flatworms. In *protostomes* (annelids, arthropods, and mollusks), the first opening into the gut that arises during embroyonic development becomes the mouth; the anus forms later. In *deuterostomes* (echinoderms and all vertebrates), the first opening to the gut becomes the anus, and the second becomes the mouth. The distinction between protostomes and deuterostomes seems trivial, but it helps us to identify two major lines of animal evolution that diverged long ago.

Key Terms

protostomes (p. 572)   deuterostomes (p. 572)   radial cleavage (p. 572)
mollusks (p. 572)   echinoderms (p. 572)   spiral cleavage (p. 572)
annelids (p. 572)   chordates (p. 572)   pouches (p. 573)
arthropods (p. 572)   blastopore (p. 572)   archenteron (p. 573)

Objectives

1. Define *protostome* and *deuterostome* and give examples of each group.

2. Explain why zoologists care to make what seems to be such a trivial distinction between the two groups.

3. Reproduce from memory the basic evolutionary relationships presented in Figure 34.4.

Self-Quiz Questions

Fill-in-the-Blanks

(1) _____ include echinoderms and vertebrates; in this group, the first

opening to the gut becomes the (2) _____, and the second one to appear becomes

the (3) _____. The situation is reversed in the (4) _____, which includes annelids (such as (5) _____), arthropods (such as (6) _____ and crabs), and mollusks (such as (7) _____). The distinction between these two major groups seems trivial, but it helps us to explain examples of (8) _____ evolution.

Summary

Mollusks include about 100,000 species of clams, oysters, abalones, limpets, snails, slugs, squids, octopuses, and an array of other diverse species. Unlike that of the arthropods, the body wall in mollusks has stayed flexible, and its musculature has become well developed in some. The mantle secretes substances that form a hard shell. In chitons, limpets, and snails, the shell is a protective shield from all but the most persistent, clever, and forceful predators. In other molluscan groups--notably predatory squids and octopuses--the shell has been reduced to a tiny internal structure and the mantle converted into a highly muscularized, conical, jet-propulsive structure, making these animals the most magnificent swimmers of the invertebrate world. The nervous systems of squids and octopuses represent the peak of invertebrate complexity; in terms of size and complexity, their brains approach those of mammals that have the capacity for storing information and learning new behaviors. Like vertebrates, these animals have acute vision and refined motor control. Because of similar evolutionary pressures, the separately evolved vertebrate and cephalopod lines developed many similar neural and sensory structures with similar functions.

Key Terms

Mollusca (p. 573)
Precambrian (p. 573)
Gastropoda (p. 573)
nudibranch,
  sea slug (p. 573)
Bivalvia (p. 573)
scallops (p. 573)
Cephalopods (p. 573)
radula (p. 573)

rasping (p. 573)
foot (p. 573)
visceral mass (p. 573)
mantle (p. 573)
mantle cavity (p. 573)
nephridia (p. 573)
torsion (p. 573)
filter feeders (p. 574)

siphons (p. 574)
giant geoduck (p. 574)
shipworms (p. 575)
nautilus (p. 575)
water propulsion (p. 575)
squids (p. 575)
parallel evolution (p. 575)
cuttlefish (p. 575)

Objectives

1. List eight examples of mollusks.

2. Define *mantle* and tell what role it plays in the molluscan body.

3. Contrast (a) the selection pressures and adaptations that have helped to make the cephalopods the most complexly evolved invertebrates with (b) the adaptations and selection pressures that have shaped the other mollusks.

4. Define *parallel evolution* and indicate the similar evolutionary pressures that have apparently caused parallel evolution in invertebrates and cephalopods.

Self-Quiz Questions

  Fill-in-the-Blanks

(1) _____ also gave rise to another successful group: the (2) _____, which include abalones, limpets, squids, and chambered nautiluses. The most highly evolved invertebrates are generally considered to be the (3) _____, which include squids and octopuses; in terms of sheer size and complexity, the (4) _____ of these animals approach those of mammals. Like vertebrates, these animals have acute (5) _____ and refined (6) _____ control, which is well integrated with the activities of the nervous system. In the less highly evolved mollusks, a structure known as the (7) _____ secretes one or more pieces of calcareous armor that protects these soft-bodied animals from predation. In the cephalopods, the mantle has become a conical cloak that surrounds the internal organs and secretes a reduced shell; seawater moves in and out of the (8) _____ _____ in a jet-propulsive manner.

34–V   ANNELIDS

Summary

Annelids, the true segmented worms such as earthworms, polychaetes, and leeches, differ from the flatworms in that annelids have (1) a complete digestive system and (2) a *coelom* (a fluid-filled space between the gut and body wall). In contrast, the flatworm gut is joined to the body wall by a solid mass of cells; hence, flatworms are *acoelomate* (without a coelom). Once organs become suspended in a coelom, they are bathed and cushioned in coelomic fluid and become insulated from the stresses of body movement. A coelom permits increased size along with more activity, especially if, as in annelids, a *circulatory system* also is part of the body plan. Through various tubes, a fluid containing nutrients and wastes travels from the body's surface to its inner regions and back again. The circulating fluid (blood) provides a means for transporting materials between internal and external environments. In the annelid system, forceful contractions in muscularized blood vessels ("hearts") keep blood circulating in one direction. In some annelids, hemoglobin, which dramatically increases the blood's oxygen-carrying capacity, is one of the protein components of blood.

    Segmentation, which is a repeating series of bodyparts, is well-developed in annelids. Each annelid segment contains its own allotment of gut, coelom, body wall, muscles, ventral double nerve cord with a segmental ganglion ("brain"), circulatory tubes, nephridia, and bristles (setae) embedded in the body wall. The muscles are arranged in paired, antagonistic sets that move in a coordinated sequence in response to nerve signals relayed through the mini-brains in each segment. Thus earthworms move forward with far more coordination than the thrashing nematodes. Other annelids have fleshy, paddle-shaped lobes called *parapodia*, which project from the body wall. Some annelids with muscle-endowed parapodia can swing them up, down, forward, and backward; thus in some annelids we see paired appendages and a mode of "walking."

Key Terms

Annelida (p. 576)
segmented worms (p. 576)
leeches (p. 576)
polychaetes (p. 576)
oligochaetes (p. 576)
proboscis (p. 576)
septa (p. 577)

hydrostatic
  skeletons (p. 577)
double nerve
  cord (p. 577)
setae (p. 577)
parapodia (p. 577)
"hearts" (p. 577)

coelomic
  chamber (p. 577)
nephrostome (p. 577)
nephridium (p. 577)
nephridiopore (p. 577)
regional
  specialization (p. 577)

Objectives

1. Define *coelom* and list three benefits the development of a coelom brings to an animal.

2. Name the three groups of annelids and give a specific example from each group.

3. Explain how the development of a coelom and circulatory system provided the potential for the development of larger and more massive animal bodies.

4. Define *segmentation* and explain how it is related to the development of muscular and nervous systems.

Self-Quiz Questions

Fill-in-the-Blanks

(1) _____ include truly segmented worms such as earthworms, (2) _____,

and leeches; they differ from flatworms in that they (1) have a complete digestive

system with a mouth and (3) _____, and a (4) _____, which is a fluid-

filled space between the gut and body wall. In a circulatory system, (5) _____

provides a means for transporting materials between internal and external environments.

In some annelids, (6) _____, a protein component of blood, dramatically

increases the blood's oxygen-carrying capacity. (7) _____, which is a repeating

series of body parts, is well developed in annelids. In each segment, there are swollen

regions of the (8) _____ _____ _____ that control local activity

and a pair of (9) _____ that act like kidneys, and bristles embedded in the body
wall. Many polychaetes (marine worms) have fleshy, paddle-shaped lobes called
(10) _____, which project from the body wall.

34–VI ARTHROPODS

Summary

Arthropods (insects, crustaceans, arachnids, and their relatives) have developed a
thickened cuticle and a hardened exoskeleton, all of which are superb barriers to
evaporative water loss, protection for predation, and surfaces to which
antagonistically arranged muscles can attach. Between segments, the cuticle remains
pliable and functions as a hinge. On land, the exoskeleton provides support for an
animal body deprived of water's buoyancy. In the seas and freshwater environments, the
jointed-legged animals known as crustaceans have come to be well-represented. The first
land arthropods were the arachnids; their descendants (spiders, scorpions, ticks, and
mites) still endure. Centipedes and millipedes arose later. Insects are thought to have
evolved from centipedelike ancestors. The larger marine arthropods extract oxygen from
water with gills (thin flaps of tissue richly supplied with blood vessels). Among
insects, tracheal systems—which consist of branching tubes stiffened with chitin—
supply each cell of the insect body with fresh air and afford insects the highest
metabolic rates known. An insect wing is not a modified leg, as is the wing of birds;
rather, insects have a modified flap of cuticle that lacks internal muscles.

Key Terms

Arthropoda (p. 578)   millipedes (p. 578)   tracheae (p. 579)
Trilobita (p. 578)   exoskeleton (p. 578)   lamellae (p. 579)
Chelicerata (p. 578)   thorax (p. 579)   chelicerae (p. 580)
Crustacea (p. 578)   cephalothorax (p. 579)   mandibles,
Uniramia (p. 578)   hemolytic   jaws (p. 580)
trilobites (p. 578)   venom (p. 579)   carapace (p. 580)
horseshoe   brown   compound
  crabs (p. 578)   recluse (p. 579)   eyes (p. 580)
ticks (p. 578)   book lungs (p. 579)   Onychophora (p. 581)
barnacles (p. 578)   spiracle (p. 579)   Peripatus (p. 581)
centipedes (p. 578)

Objectives

1. Explain how the development of a thickened cuticle and hardened exoskeleton
    affected the ways that arthropods lived.

2. Construct a diagram corresponding with data from the geologic time scale that represents the evolutionary pattern of arthropods. State what ancestors are thought to have given rise to the arthropods and whether you think any other major groups of animals descended from the arthropods.

3. List the changes the basic aquatic arthropod plan must undergo in order to adapt to life on land.

4. Contrast tracheal systems with gill systems and explain why land animals cannot utilize gill systems in obtaining oxygen from their environment.

5. Explain how insect wings differ from those of birds.

6. List six different groups of arthropods.

Self-Quiz Questions

Fill-in-the-Blanks

Arthropods developed a thickened (1) _____ and a hardened (2) _____.

In freshwater and saltwater environments, arthropods known as (3) _____ came to

be well represented. The first land arthropods were the (4) _____. Their

descendants ((5) _____, scorpions, ticks, and mites) are still living.

(6) _____ and millipedes arose later. (7) _____ are thought to have

evolved from centipedelike ancestors. The larger aquatic arthropods extract oxygen from

water with (8) _____, whereas most insects utilize (9) _____ systems,

which provide the basis for the highest (10) _____ rates known. Arthropods are

thought to have evolved from (11) _____.

Summary

Of the small number of deuterostomes that have survived to the present, only the echinoderms and chordates are prominent. *Echinoderm* means spiny-skinned, in reference to the bristling spines on some members of the group. In echinoderms, radial symmetry has been overlaid on an earlier bilateral heritage. Some biologists hypothesize that echinoderms are descended from deuterostomes that were evolving at a time when predators were increasing in both size and number. In response to selection by predators, the echinoderm ancestors came to be equipped with heavy, defensive armor of mineralized plates embedded in the body wall. Although heavy armor must have helped them avoid being eaten, it also may have forced them to settle to the sea bottom. Most echinoderms still go through a free-swimming, bilaterally symmetrical larval stage, but adults generally show the radial symmetry typical of bottom-dwelling organisms that cannot move quickly. The echinoderm way of moving about is based on a *water-vascular system* of canals and tube feet. Coordinated extension and retraction of many suckered tube feet enables echinoderms to move body parts in specific directions. Echinoderms include crinoids, sea stars, sea urchins, sand dollars, sea cucumbers, and brittle stars.

Chordates are organisms that, at some time during their life span, develop (1) *gill slits,* (2) a *notochord,* (3) a *dorsal tubular nerve cord,* and some sort of (4) *endoskeleton*. Primitive chordates (such as tunicates and lancelets) tended to be more or less sessile as adults. They obtained food by passing plankton-laden water over sheets of mucus that lined the gill slits; the entrapped plankton was then passed into the gut for digestion. Such a system is known as *filter feeding*. Sessile or sedentary animals generally produce many offspring, and they also produce motile larvae. Chordate larvae are bilaterally symmetrical, with a head end and a tail end. They have a nervous system with a dorsal nerve cord and a rod of stiffened tissue (the notochord) running the length of the body. Muscles attached to the notochord contract rhythmically, which causes the body to bend and move through the water. The notochord was the forerunner to the chordate's internal skeleton (endoskeleton).

A strengthening of the notochord and its attached muscles encouraged the development of vertebrates: animals whose dorsal stiffening rod was segmented into a series of hard bones (vertebrae) arranged in a column that formed a backbone. The first vertebrates were the *jawless fishes*. As the brain region came to be protected by hard structures that eventually developed into a *skull*, various bone and muscle modifications occurred that led to the formation of jaws, which are characteristic of the *true fishes*. As the number of jawed fishes increased, there must have been a selective advantage in being able to detect food from a distance. Trends toward increasingly specialized sensory detectors and complex neural integration were established among the true fishes and were elaborated in the descendent forms that moved onto land. Gills were devices used mainly in filter feeding by primitive chordates, but as predatory activities increased, gills became more important in oxygenating the blood and supplying muscles with oxygen. Because gills can function only if they are kept moist, the first vertebrates that moved onto land during Devonian times relied increasingly on an internal swim bladder to extract oxygen from air.

In fishes that apparently gave rise to amphibians, blood traveled from gills to a two-chambered heart, which pumped it to the rest of the body. Amphibian (frogs, toads, salamanders) hearts are three-chambered. Reptilian (snakes, crocodiles, lizards, turtles) hearts lie partway between three and four-chambered hearts. Both birds and mammals have developed completely separate systemic and pulmonary circulations that work in harmony.

Key Terms

Echinodermata (p. 581)
water-vascular
   system (p. 581)
Precambrian (p. 581)
crinoid (p. 582)
brittle star (p. 582)
sea urchin (p. 582)
sea star (p. 582)

Cephalochordata (p. 582)
lancelets (p. 582)
Vertebrata (p. 582)
lampreys (p. 582)
hagfishes (p. 582)
rays (p. 582)
metamorphosis (p. 583)
gill slits (p. 584)

olfactory
   receptors (p. 585)
Devonian (p. 585)
lungs (p. 585)
swim bladder (p. 585)
dual-circulation
   system (p. 588)
nerve cord (p. 588)

sea cucumber (p. 582)
tube foot (p. 582)
endoskeleton (p. 582)
Chordates (p. 582)
notochord (p. 582)
Urochordata (p. 582)
tunicates, sea
   squirts (p. 582)

placoderms (p. 583)
ostracoderms (p. 583)
cartilaginous
   fishes (p. 583)
atrium (p. 585)
Amphioxus (p. 584)
vertebrae (p. 585)
skulls (p. 585)
bony jaw (p. 585)

countercurrent flow
   mechanism (p. 588)
evolutionary
   accretion (p. 588)
midbrain (p. 589)
R complex, reptilian
   brain (p. 589)
lymbic system (p. 589)
neocortex (p. 589)

Objectives

1. Define *echinoderm* and list five groups of echinoderms.

2. Explain how the radial symmetry seen in adult echinoderms might have arisen
    during the Precambrian.

3. Describe how locomotion occurs in echinoderms.

4. Explain how a sea star might use its tube feet to obtain its food.

5. List three characteristics found only in chordates.

6. Describe the adaptations that sustained the sessile or sedentary life-style
    seen in primitive chordates such as tunicates and lancelets.

7. State what sort of changes occurred in the primitive chordate body plan that
    could have promoted the emergence of vertebrates.

8. Describe the differences between primitive and advanced fishes in terms of
    jaws, muscles, sensory detectors, neural integration, skull, and brain.

9. Describe the changes that enabled aquatic fishes to give rise to land dwellers.

10. State what kind of heart each of the four groups of four-limbed vertebrates has.

Self-Quiz Questions

Fill-in-the-Blanks

Only two prominent groups of (1) _____ have survived to the present—the

(2) _____ and the chordates. In echinoderms, (3) _____ symmetry has been

overlaid on an earlier bilateral heritage; most echinoderms still go through a free-

swimming (4) _____ symmetrical larval stage. Echinoderm locomotion is based on

constant circulation of seawater through a (5) _____-_____ system of

canals and (6) _____ _____. (7) _____ _____ have a rounded

globose body with bristling spines. (8) _____ have many long feather-duster

arms; both mouth and anus lie within the circlet of arms.

(9) _____ are among the most primitive of all living chordates; when they are tiny, they look and swim like (10) _____. A rod of stiffened tissue, the (11) _____ runs the length of the larval body; it was the forerunner of the chordate's (12) _____. Tunicates and (13) _____ practice (14) _____ _____; they draw in plankton-laden water through the mouth and pass it over sheets of mucus, which trap the particulate food before the water exits through the (15) _____ _____. The first vertebrates were the (16) _____ _____. The development of the formidable (17) _____ that are characteristic of the true fishes was surely an important evolutionary force. The first vertebrates that began moving onto land during Devonian times began to rely more and more heavily on using the (18) _____ _____ to acquire oxygen from the air. Fish have (19) ◯ one- ◯ two- ◯ three- ◯ four-chambered hearts. Chambers that receive blood are called (20) _____. Blood flow in birds and mammals is separated into two systems, the systemic and the (21) _____ circulations, which work in synchrony.

INTEGRATING AND APPLYING KEY CONCEPTS

Scan Table 34.1 to verify that most of the most highly-evolved animals have a complete gut, a closed blood-vascular system, both central and peripheral nervous systems, and are dioecious. Why do you suppose that having two sexes in separate individuals is considered to be more highly evolved than the monoecious condition utilized by earthworms? Would it not be more efficient if *all* individuals in a population could produce both kinds of gametes? Cross fertilization would result in both individuals being able to produce offspring.

35
HUMAN ORIGINS AND EVOLUTION

THE PRIMATE FAMILY   Ancient Apelike Forms
                                       Australopithecines
  Primate Origins   The First Humans
  General Characteristics of
    the Primates   COMMENTARY: A Biological Perspective
                                       on Human Origins
EARLY HOMINIDS (PERHAPS) AND
THEIR PREDECESSORS (MAYBE)

Summary

When the world climate began to grow cooler and drier during the Cenozoic, birds and mammals, which have internal and external adaptations to help them maintain a relatively constant body temperature even when environmental temperatures rise and fall, rose to dominance.

    As *primates*, humans share with apes, monkeys, and prosimians a large cerebral cortex, highly integrated systems of muscles and nerves, flexible shoulder joints, eyes with good color, image, and depth perception, fingers and toes adapted for grasping rather than running, and the birth of only one offspring at a time.

Key Terms

metazoans (p. 590)	anthropoids (p. 590)	head rotation (p. 592)
mammalian stem (p. 590)	ceboids (p. 590)	grasping digits (p. 592)
phylogenetic tree (p. 590)	cercopithecoids (p. 590)	opposable thumb (p. 592)
Primates (p. 590)	hominoids (p. 590)	upright vertebral
prosimians (p. 590)	Cenozoic (p. 590)	column (p. 592)
tree shrews (p. 590)	gibbon (p. 591)	depth perception (p. 592)

Objectives

1. State when the earliest mammals evolved and when the shrewlike insectivores underwent adaptive radiation to give rise to the earliest primates.

2. Compare structural and behavioral features of the early primitive primates with those of the cercopithecoids and hominoids. Which anatomical features underwent the greatest changes along the evolutionary line to humans?

3. Name the places where prosimian survivors dwell today.

4. Name the two *key* characters of primate evolution.

Self-Quiz Questions

Fill-in-the-Blanks

The world climate began to grow cooler and drier during the (1) _____ era, when

the birds and mammals rose to dominance. Primates, like all placental mammals,

apparently arose from ancestral forms of the order (2) _____; representatives of

this order (2) living today are (3) _____, which are nighttime omnivores.

Humans, apes, monkeys, and prosimians are all (4) _____; they have excellent

(5) _____ perception due to their forward-directed eyes, and their fingers and

toes are adapted for (6) _____ instead of running.

Summary

The dryopithecine apelike forms of the Miocene are thought to have given rise to the
ancestors of humans. Geologic upheavals in the Great Rift Valley of East Africa during
the late Cenozoic fostered environmental diversity, which in turn fostered distinct
adaptations in the indigenous small, scattered bands of early apes. The first hominids
are believed to have appeared in regions of increasing aridity; they were foragers on
the ground for seeds, fruits, nuts, and the remains of small animals. They also learned
to hunt. The earliest indisputable hominids were the australopithecines, the "southern
ape-men" that apparently emerged during the Pliocene. Approximately 2 million years
ago, there were also members of the genus *Homo*, of which we are the only living
representatives. Since then several distinctly different hominid populations appeared
and disappeared, but by 30,000 years ago only one hominid species (*H. sapiens*)
remained: man, the reasoner.

Key Terms

Fayum Depression (p. 592)	Ramapithecus (p. 592)	Homo sapiens (p. 594)
hominoids (p. 592)	hominid (p. 593)	culture (p. 595)
Aegyptopithecus (p. 592)	Australopithecus (p. 593)	savanna (p. 595)
quadrupedal (p. 592)	australopithecines (p. 593)	enlarged fetal
mosaic (p. 593)	Johanson (p. 593)	cranium (p. 595)
dryopithecines (p. 592)		

Miocene (p. 592)
adaptive
  radiation (p. 592)
gorillas (p. 592)
sivapithecines (p. 592)

Lucy (p. 593)
bipedal (p. 593)
cranial capacity (p. 593)
Homo erectus (p. 593)
Neanderthalers (p. 594)

home base (p. 596)
provisioning (p. 596)
bipedalism (p. 595)

Objectives

1. Describe the environmental changes that occurred in Africa from about 35 million years ago up to about 10 million years ago and tell why these changes would have encouraged adaptive radiation.

2. Consult Figure 35.6 and note the principal points of divergence during the evolution of humans from the primitive primates.

3. Beginning with the primates most closely related to humans, list the main groups of primates in decreasing closeness of relationship to humans.

4. State the selective agents that are believed to have been most important in rendering humans distinct from chimpanzees and gorillas.

5. Explain how you think the human species arose. Make sure your theory incorporates existing paleontological, biochemical, and morphological data.

6. Evaluate Lovejoy's attempts to relate the establishment of maternal home bases and the provisioning of mothers as behaviors that would have promoted bipedalism and culture.

Self-Quiz Questions

  Fill-in-the-Blanks

One of the Fayum hominoids (1) _____ lived 30 million years ago; it was the size of a housecat and had a skull resembling early hominoids, but it was (2) _____ (went about on four legs). About (3) _____ million years ago, the redistribution of land masses led to conspicuous changes in climate; there was a major cooling trend and a decline in (4) _____. Under these conditions, forests began to give way to (5) _____; however, in the interim, the environment was a mosaic of forests and (6) _____. Hominoids of 25 million years ago were collectively known as (7) _____; they underwent (8) _____ into the new ecological niches. One form (9) _____ had teeth and dental arches that were more like those of humans than of apes. The beginning of the hominoid line diverged from the dryopithecine-to-modern ape line sometime between 12 million and (10) _____ million years ago, during the Pliocene period. Johanson's find, Lucy, was one of the earliest (11) _____, a collection of forms that combined ape and human features; they were fully two-legged, (12) _____, with essentially human bodies and ape-shaped heads. The oldest fossils of the genus *Homo* date from approximately (13) _____ million years ago. The (14) _____ were a distinct hominid population that appeared about 100,000 years ago in Europe and Asia; their cranial capacity was indistinguishable from our own and they had a complex culture. By 30,000 years ago, there was only one remaining hominid species: (15) _____.

INTEGRATING AND APPLYING KEY CONCEPTS

Suppose that someone told you that sometime between 12 million and 6 million years ago sivapithecines were forced by predatory larger members of the cat family to flee the forests and take up residence in estuarine, riverine and sea coastal habitats where they could take refuge in the nearby water to evade the tigers. Those that, through mutations, became naked, developed an upright stance, developed subcutaneous fat deposits as insulation, and developed a bridged nose had advantages in watery habitats that other sivapithecines that remained inland never developed. As time went on, predation by the big cats and competition with other animals for available food caused most of the terrestrial sivapithecines to become extinct, but the water-habitat varieties survived as scattered remnant populations, adapting to easily available shellfish and fish, wild rice and oats, various tubers, nuts and fruits. It was in these aquatic habitats that the first food-getting tools (baskets, nets, and pebble tools) were developed, as well as the first words that signified different kinds of food.

How does such a story fit with current speculations and evidence of human origins? How could such a story be demonstrated as being true or false?

CROSSWORD NUMBER FIVE

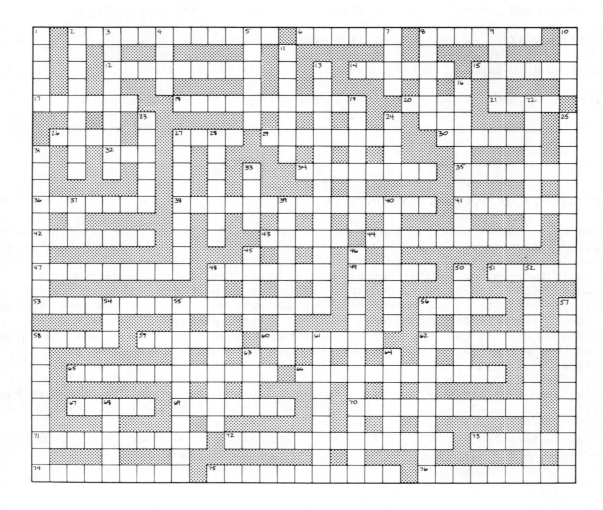

Across

2. an organism in which the second embryonic opening becomes the mouth
6. a bivalve mollusk
8. animallike protistans
12. a snail or slug, for example
14. treeless arctic biome
15. a remnant or trace of an organism of a past geological age
17. a tongue-like rasping organ in various mollusks
18. having only one type of spore
20. ___ worms construct their own "houses"

21. tiny arthropod
26. provides home for parasites
27. an encapsulated form of an organism
29. a severe kind of food poisoning
30. ___ algae, such as diatoms
32. geologic ___, a span of time
34. protistans with cilia
35. type of mammal
36. bivalve mollusks
38. primitive photosynthetic eukaryotes
41. free-swimming marine bivalve
42. club moss
43. a span of geologic time
44. makes all parts work in harmony
47. mutually rewarding dependence
48. same as #55 down
49. fungal sac
51. threadlike filament in fungi
53. earliest hominid type
56. a primitive seed-bearing gymnosperm
58. secreting organ
59. fluid-filled space
60. true segmented worm
62. same as #55 down
65. ___ earth is formed by certain protistans
66. an arrangement of repetitive compartments
67. a type of gastropod
69. ___ stars
70. the process of two different species evolving in response to each other
71. oxygen-bearing blood pigment
72. an asexual reproductive process in which an animal breaks into two or more pieces
73. sweet liquid of flowers or fruits
74. slow addition by deposition
75. disease caused by parasitic nematodes
76. substances that settle to the bottom

Down

1. necessary compound
2. loses leaves
3. ___ have hoofs
4. specific rodents
5. bell-shaped cnidarian
7. part of respiratory system
8. receiving chamber in heart
9. rapid growth of plankton
10. organ that absorbs oxygen from water
11. type of angiosperm
13. what glow-worm exhibits
16. one that breaks down organic matter
19. helical bacteria
22. elongated flexible grasping organs
23. parts of a circulatory system

24. a fungal disease of plants
25. division of brown algae
27. have a notochord sometime during their life histories
28. grassland interspersed with trees
31. smallest complete organism known
33. a fault
37. ____ fungi include *Penicillium* and the cup fungi
39. animals that can live on land or in water
40. sea squirt
45. ____ algae include *Ulva* and *Spirogyra*
46. club fungi
48. disease-causing bacteria
50. rodlike bacteria
52. photosynthetic plankton
54. ____ algae include the coralline algae and Irish "moss"
55. a time span on the geologic time scale
56. squids and octopuses
57. develop into sperm-bearing male gametophytes
58. bacterial venereal disease
61. not deciduous
63. fungal disease of plants
64. type of arthropod
68. aquatic plants

36
POPULATION ECOLOGY

ECOLOGY DEFINED

POPULATION DENSITY AND
DISTRIBUTION

   Ecological Density
   Distribution in Space
   Distribution Over Time

POPULATION DYNAMICS

   Parameters Affecting
     Population Size
   Life Expectancy and
     Survivorship Curves

POPULATION GROWTH

   Biotic Potential
   Exponential Growth
   Environmental Resistance
     to Growth
   Tolerance Limits

Carrying Capacity and
   Logistic Growth

REPRODUCTIVE RESPONSES TO GROWTH
LIMITS

FACTORS THAT REGULATE POPULATION
GROWTH

   Density-Dependent and
     Density-Independent Factors
   Extrinsic Influences
   Intrinsic Influences

HUMAN POPULATION GROWTH

   Doubling Time for the
     Human Population
   Where We Began Sidestepping Controls
   Age Structure and Fertility Rates

PERSPECTIVE

Summary

For humans, social, economic, and political considerations may influence the short-term distribution of resources on which any population depends. Biological principles predict the consequences of competition for scarce natural resources. *Ecology* is the study of the relations between organisms and the totality of the physical and biological factors affecting them or influenced by them. All evolutionary change is influenced by ecological interactions that operate on the *population, community, ecosystem*, and *biosphere* levels.

Crude density is no more than a head count in a total amount of defined space; *ecological density* measures population density in space actually occupied for specific lengths of time. Individuals might be clumped together, spread out rather uniformly, or dispersed randomly in the environment. Distribution may change over time as animals migrate from place to place.

Population density rises or falls with changes in number of births (*natality*), number of deaths (*mortality*), *immigration*, and *emigration*. *Population size (N)* = (births + immigration) − (death + emigration). When *N* remains constant over time, there is *zero population growth*.

Data summarized in life tables can be used to construct *survivorship curves*, which show trends in mortality and survivorship and reveal three basic kinds of population types.

Key Terms

illiterate (p. 599)
compulsory
  sterilization (p. 599)
ecology (p. 599)
ecologists (p. 599)
population (p. 600)
community (p. 600)
ecosystem (p. 600)
abiotic (p. 600)
biotic (p. 600)
producer (p. 600)
consumer (p. 600)
decomposer (p. 600)
autotroph (p. 600)
heterotroph (p. 600)
biosphere (p. 600)
materials
  cycling (p. 600)

selection theory (p. 600)
fitness (p. 600)
crude density (p. 600)
habitat (p. 600)
population
  density (p. 600)
ecological
  density (p. 600)
nonrandom
  distribution (p. 600)
random
  distribution (p. 600)
clumped
  distribution (p. 600)
uniform
  distribution (p. 601)
creosote bushes,
  Larrea (p. 601)

spatial
  distribution (p. 600)
population
  size (p. 601)
natality (p. 601)
mortality (p. 601)
immigration (p. 601)
emigration (p. 601)
zero population
  growth (p. 601)
life expectancy (p. 601)
life tables (p. 601)
survivorship
  curves (p. 601)
age increment (p. 601)
population type I (p. 601)
population type II (p. 601)
population type III (p. 601)

Objectives

1. Understand how the principles of ecology can influence human social, economic, and political considerations.

2. Explain how the kinds of interactions among species can shape the structure of a community.

4. Explain why measurements of *ecological density* are more informative than measurements of *crude density*.

5. List the three ways that individuals can be distributed in space and provide an example of each.

6. Discuss the effects of the four parameters (natality, mortality, immigration, and emigration) on the equation for population size.

7. Explain how the construction of life tables and survivorship curves can be useful to humans in managing the distribution of scarce resources.

Fill-in-the-Blanks

A group of individuals of the same species occupying a given area at a specific time is a(n) (1) _____. The (2) _____ maintains a system of energy use and materials cycling. (3) _____ is the genetically based ability of an organism to perpetuate itself under prevailing environmental conditions. Measurements of (4) _____ _____ are more informative about population density than are measurements of (5) _____ _____. Examples of nonrandom distribution are (a) uniform and (b) (6) _____. *N* = (natality + (7) _____) - (mortality + (8) _____). (9) _____ _____ use information summarized from life tables, which show trends in mortality and life expectancy. Type (10) _____ populations have low survivorship early in life.

Summary

A population is defined in part by population size (its number of individuals) and population density (how many individuals are present in a particular area). Any population that is not restricted in some way will grow in size at an increasingly accelerated rate. Population growth is the difference between the birth rate and the death rate, plus or minus any inward or outward migration. A convenient way to express the growth rate of a given population is known as its *doubling time* (how long it takes to double in size). What starts out as a gradual increase in numbers turns into explosively accelerated increases—a pattern of *exponential growth*. As new individuals are added to a population, the number of reproducing members increases, and the increase enlarges the potential reproductive base. When we plot the course of such exponential growth, we end up with a *J-shaped curve*. As long as the birth rate remains even slightly above the death rate, any population grows exponentially.

Any natural system keeps its populations in check. Initially, a bacterial population goes through an exponential growth phase. Next, population growth tapers off, only to give way to a plateau phase in which population size remains relatively stable. Then the population begins to decline from lack of essential resources and increases in toxic wastes. When any essential resource is in short supply, it becomes a *limiting factor* on population growth. Only if the toxic medium were removed every so often and replaced with a fresh medium would things stabilize. Then we would end up with an *S-shaped curve*: an exponential growth phase that leads into a stable plateau phase.

A population of any species can tolerate only a certain range of environmental conditions. Resource availability, prevailing conditions, and tolerance ranges interact to dictate where population growth must level off. For a given populatin, they define the environment's *carrying capacity*: the maximum density of individuals that can be indefinitely sustained in a given area, under a given set of environmental conditions-- a brake on runaway growth.

Key Terms

biotic potential (p. 602)
exponential growth (p. 602)
maximum rate of
 increase, r_m (p. 602)
doubling time (p. 602)
J-shaped curve (p. 603)
fission (p. 604)
limiting factor (p. 604)
crash (p. 605)
environmental
 resistance (p. 604)
Liebig's "law of the
 minimum" (p. 605)

tolerance limits (p. 605)
Shelford's "law"
 of tolerance" (p. 605)
performance (p. 605)
variation (p. 605)
tolerance curve (p. 605)
optimal
 performance (p. 605)
carrying
 capacity (p. 605)
sigmoid (p. 605)
S-shaped curve (p. 605)

logistic growth (p. 605)
exponential growth
 phase (p. 605)
Pribilof Islands (p. 605)
timing of
 reproduction (p. 606)
r-selection reproductive
 pattern (p. 606)
K-selection reproductive
 pattern (p. 606-607)
clutch size (p. 607)

Objectives

1. List some of the reasons that people give for wanting more than three children per family.

2. Define *zero population growth* and indicate how achieving it would affect the human population of the United States.

3. State how increasing the death rate of a population affects its doubling time.

4. Contrast the conditions that promote J-shaped curves rather than S-shaped curves in populations.

5. Explain why any population grows exponentially if the birth rate exceeds the death rate.

6. Define *limiting factors* and tell how they influence population curves.

7. Distinguish tolerance range from carrying capacity, explaining which is a feature of a population and which is a feature of the population's environment.

Self-Quiz Questions

Fill-in-the-Blanks

Any population that is not restricted in some way will show a pattern of (1) _____ growth because any increase in population size enlarges the (2) _____ base. When the course of such growth is plotted on a graph, a (3) _____ -shaped curve is obtained. If (4) _____ factors (essential resources in short supply) act on a population, population growth tapers off, giving way to a (5) _____ phase. Any population has its (6) _____ _____, which represents its adaptability to environmental conditions. (7) _____ _____ is a feature of the environment that is defined for one or more populations living in that environment and serves as a brake on runaway growth.

The ability to produce the maximum possible number of new individuals is called the population's (8) _____ _____. Sigmoid curves are characteristic of (9) _____ _____. (10) _____-selected populations favor rapid rates of natural increase; weeds that colonize new habitats are consummate examples of these.

36–III   FACTORS THAT REGULATE POPULATION GROWTH

Summary

A natural population is said to be stabilized when its population size typically fluctuates within a predictable range. Feedback mechanisms regulate birth and death rates; the mechanisms are related to population density. There are mechanisms built into the species (intrinsic factors) and ones dictated by the environment (extrinsic factors). *Density-dependent factors* come into play whenever populations change in size; they depress population size when the density of individuals approaches the environment's carrying capacity, and they ease up when population density decreases. *Density-independent factors* work regardless of population size; they are environmental variables (such as periodic rains or temperature extremes) that may shift enough to push a population above or below its tolerance range for a given variable.

Extrinsic limiting factors (such as temperature, rainfall, nutrients, and the presence or absence of other populations) also help control population growth and distribution. One of the most dramatic extrinsic checks is the predation of one species on another.

Intrinsic limiting factors are built-in features of individuals that affect population growth and distribution. They include limitations imposed by the individual's physiology, structure, metabolism, and behavior. Myers's studies of extreme overcrowding among wild rabbits suggest that endocrinological changes accompany increases in population density and that these intrinsic changes in turn influence population distribution and reproductive behavior. In natural settings, intrinsic limiting factors affect how populations come to be distributed in space; *territorial* and *migratory behaviors* are tied to the carrying capacity of the environment, as is the *timing of reproduction.*

Key Terms

density-dependent
  mechanisms (p. 608)
density-independent
  mechanisms (p. 608)
extrinsic factors (p. 608)
abiotic conditions (p. 608)
resource
  availability (p. 608)

intraspecific
  competition (p. 608)
scramble
  competition (p. 608)
contest
  competition (p. 608)
sheep blowflies (p. 609)
territoriality (p. 608)

social dominance (p. 608)
predation (p. 608)
myxomatosis (p. 609)
intrinsic factors (p. 609)
Myers (p. 609)
ovulate (p. 609)
adrenal glands (p. 609)

Objectives

1. Define *density-dependent factors*, give two examples, and indictate how density-dependent factors act on populations.

2. Define *density-independent factors*, give two examples, and indicate how they affect populations.

3. Define *intrinsic limiting factors* and give two examples. Indicate how intrinsic limiting factors can affect population growth and distribution in (a) wild rabbits living under laboratory conditions and (b) territorial or migratory animals living in the wild.

4. Explain how timing of reproduction can affect the degree of intraspecific competition for available resources.

5. Define *extrinsic limiting factors*, and give two examples, and indicate how they can affect population growth and distribution.

Self-Quiz Questions

Fill-in-the-Blanks

Food availability is a density- (1) _____ factor that works to cut back

population size when it approaches the environment's (2) _____ _____.

Environmental disruptions such as forest fires and floods are density- (3) _____

factors that may push a population above or below its tolerance range for a given

variable. Limitations imposed by the physiology, structure, metabolism, and behavior of

individuals also affect population growth and distribution; they are referred to

collectively as (4) _____ _____ factors. External conditions such as

temperature, rainfall, nutrients, and the presence or absence of other species also

control population growth and distribution; they are referred to collectively as

(5) _____ _____ factors.

Matching Choose the most appropriate letter(s) for each. A letter may be used more than once, and a blank may contain more than one letter.

(6) ___ intrinsic limiting factors

(7) ___ density-independent factors

(8) ___ density-dependent factors

(9) ___ extrinsic limiting factors

A. drought, floods, earthquakes

B. reindeer in the Pribilof Islands in 1935

C. food availability

D. adrenal enlargement in wild rabbits in response to crowding

Summary

In 1983, the human population reached 4.7 billion. Between 5 and 20 million people each year now die of starvation or malnutrition-related diseases, and the future is even more grim because the net population growth is increasing ever more rapidly. Net population growth for a given period is the difference between the total number of live births and the total number of deaths occurring in that time span. Values calculated on the basis of the number of births and deaths per 1,000 persons in the population at the mid-point of a given year are used in determining the percent annual growth rate, which can be extended to determine how long it will take the human population to double in size. The world average growth rate in 1983 was 1.8 percent; if this rate continues, world population will double within forty-one years. In 1983, Mexico's growth rate was 2.6 percent; its population will double in twenty years. Food production and supplies or other limiting factors are not likely to double even with the most intensive effort. In the past two centuries, the long-term human growth rate has undergone an astonishing acceleration due to three principal factors: (1) humans developed the capacity to expand steadily into new environments, (2) the carrying capacity was increased, and (3) a series of limiting factors was removed so that more of the available resources could be exploited.

We now have two options: either we make a global effort to limit our numbers so that our population stabilizes according to the environment's carrying capacity, or we passively wait until the environment does it for us. Two important factors influence just how much we can expect to slow down the birth rate. *Age structure*, how individuals are distributed at each age level for a population, can give us an idea of the magnitude of the effort that will be needed to control birth rates. More than a third of the world population is young people about to move into their reproductive years. The *fertility rate* is determined by how many infants are born to each woman during her reproductive years. In industrialized countries, the average number is 2.6; in developing countries, it is 5.7. A world average of 2.5 children per family would stabilize the world population 70 to 100 years from now at about 6.3 billion people if the average were implemented now. A simple way to slow things down would be to encourage *delayed reproduction*—childbearing in the early thirties as opposed to the mid-teens or early twenties. As long as birth rates even slightly exceed death rates, populations experience exponential growth.

radiation (p. 610)

cultural
  communication (p. 610)

metallurgy (p. 611)

bubonic plague (p. 610)

carrying capacity (p. 611)

limiting factors (p. 611)

social
  stratification (p. 611)

labor base (p. 611)

domestication
  of animals (p. 611)

vaccine (p. 611)

antitoxin (p. 611)

antibiotic (p. 611)

diptheria (p. 611)

cholera (p. 611)

age structure (p. 611)

pre-reproductive
  base (p. 611)

fertility rate (p. 611)

replacement
  level (p. 612)

delayed
  reproduction (p. 612)

Objectives

1. Compare the doubling time of Mexico's human population with that of Japan. Then compare both with that of the world.

2. List three possible reasons the human long-term growth rate accelerated.

3. Define age structure and explain why this is the principal reason it would be 70 to 100 years before the world population would stabilize even if the world average became 2.5 children per family.

Self-Quiz Questions

  Fill-in-the-Blanks

The (1) _____ _____ of a population shows how individuals are distributed in each age group. The number of infants born each year per 1,000 women between ages 15 and 44 is known as the (2) _____ _____. A world average of (3) _____ children per family is the estimated rate that would bring us to zero population growth; if that were achieved, it would still require at least (4) _____-_____ years before the human population stops growing because (5) _____ _____. A simple way to slow things down would be to encourage (6) _____ _____. At the current rate of increase, Mexico's population will double in (7) _____ years, whereas Japan's will double in (8) _____ years.

INTEGRATING AND APPLYING KEY CONCEPTS

Assume that the world has reached zero population growth. The year is 2080, and there are 20 billion individuals of *Homo pollutans* on the planet Earth. You have seen stories on the community television screen about how people used to live 100 years ago. List the ways that life has changed, and comment on the events that no longer happen because of the enormous human population.

37
COMMUNITY INTERACTIONS

Summary

To understand the character of populations, we must examine their interactions with
other organisms in the same environment as well as their interactions with
conspecifics. A *community* may be defined as all those populations of different species
that occupy and are adapted to a given area; it differs from an *ecosystem*, in that an
ecosystem is a community plus its physical environment. Different populations generally
interact in three basic ways: (1) *interspecific competion* for limited resources (2)
predation of one species on another, and (3) *coevolution* of two species through time.
A *habitat* is defined as the sort of place where an organism usually resides; it
includes its geographic range and the array of organisms with which it is generally
associated. A *niche* is defined by all of the environmental or ecological requirements
and interactions that influence a particular population and define its
role in the community. The full range of variable conditions under which a species can
function (assuming no interference from other species) is its *fundamental niche*. The
actual range of conditions under which a species exists is its *realized niche*.

    In *symbiotic relationships*, one or both members of two species comes to rely on
the continuing presence of the other for survival. Such relationships vary in the
degree of collaboration, the exclusiveness of the attachments, and the extent to which
one species is helped or harmed by the presence of the other. In two-species interactions there can be a one-way or two-way flow of positive, negative, or neutral
effects. Neutral interactions have no direct effect on either species. Both *protocooperation* and *mutualism* are positive interactions that benefit both species; the
latter is an obligatory relationship, whereas the former is not.

Commensalism is a weak interspecific symbiotic attachment in which a guest species simply lives better in the presence of its host, and the host suffers no harm. Predation, parasitism, and interspecific competition are examples of symbiotic relationships that have a negative effect on one or both species.

Key Terms

sympatric (p. 614)
community (p. 614)
interspecific
  competition (p. 614)
predation (p. 614)
coevolution (p. 614)
habitat (p. 614)
fundamental
  niche (p. 615)

resource
  categories (p. 615)
resource gradient (p. 615)
resource
  utilization (p. 616)
realized niche (p. 616)
symbiosis (p. 617)
neutral
  interactions (p. 617)

protocooperation (p. 617)
obligatory (p. 617)
mutualism (p. 617)
commensalism (p. 617)
parasitism (p. 618)
intraspecific
  competition (p. 618)

Objectives

1. Define and distinguish *community* and *ecosystem*.

2. List the three basic ways that populations interact.

3. Define and distinguish *habitat* and *niche*.

4. Distinguish between *fundamental niche* and *realized niche*. Then show how a realized niche could be visualized and studied.

5. Define and give one example of commensalism, mutualism, and protocooperation.

Self-Quiz Questions

Fill-in-the-Blanks

A(n) (1) _____ is defined as a community and its nonliving physical environment; a(n) (2) _____ is limited to all the different populations that occupy and are adapted to a defined area. The (3) _____ of a population is the sort of place it is typically located, whereas the (4) _____ of a population is defined by its role in a community, including all the ecological requirements and interactions that influence that population in its community. In the real world, a population is prevented from making optimum responses to all the variables of its fundamental niche; the actual range of conditions under which the population exists is its (5) _____ _____. The way that a population (for example, gnatcatchers) responds to a resource category can be plotted in a graph with resource utilization (such as capture rate) on the *Y*-axis versus (6) _____ _____ (such as prey size) on the

X-axis. Robins and fruit flies are (7) _____ with human populations. The

flowering plants and their pollinators form (8) _____ relationships from which

both participating populations benefit.

Summary

When different species have some requirements or activities in common, their niches
overlap. The greater the overlap, the greater the potential for competition between
them.

In *competition*, one population or individual exploits the same limited resources
as another or interferes with another enough to keep it from gaining access to the
resource. When two populations are competing for the same resource, one tends to
exclude the other from the area of overlap. According to the theory of *competitive
exclusion*, no two species can be exactly the same in their activities or in their
ability to get and use resources. To a greater or lesser extent, one would have the
advantage and the other would be forced to modify its niche. In this view, no two
species in a community can simultaneously occupy the same niche for very long. The
abundance of resources and population density influence whether or not exclusion
occurs. Balancing the tendency to minimize niche overlap is a tendency for each species
to expand its niche. Coexistence of two or more populations that compete in the same
resource category may occur when each specializes within a narrow range of the resource
gradient. Coexistence can also occur through resource partitioning, whereby populations
share the same resource in different ways, in different areas, or at different times.
Although requirements for different species might be similar they cannot be identical,
so interspecific competition is usually less intense than intraspecific competition.

A *predator* is an organism that actively seeks out and/or captures another living
organism, then ingests either the whole thing or parts of it. Plants are the *prey* of
herbivores and animals the prey of carnivores; any organism can be the prey of
omnivores. Predators generally focus on prey that are large enough or numerous enough
to make the hunt and the capture worth the energy outlay.

Key Terms

competitive
   exclusion (p. 618)
giant tortoise (p. 618)
exploitative
   competition (p. 618)
interference
   competition (p. 618)

extrinsic control
   factor (p. 620)
intrinsic control
   factor (p. 620)
mallow (p. 620)
smartwood (p. 620)
territories (p. 620)

cyclic
   oscillations (p. 621)
Canadian lynx (p. 621)
showshoe hare (p. 621)
early successional
   species (p. 622)
toxins (p. 622)

resource
  partitioning (p. 619)
bristly foxtails (p. 620)

scent markings (p. 620)
Lotka and
  Volterra (p. 620)

adventitious
  shoots (p. 622)

Objectives

1. Explain how the concept of competitive exclusion is related to the concept of a niche.

2. State the two tendencies that act to define the extent to which a population influences a community.

3. Describe a study that demonstrates laboratory evidence in support of the competitive exclusion concept.

4. Describe a study that demonstrates interference competition.

5. Present two general ways that two or more different populations can coexist in the same resource categories.

6. Define *predator* and tell how predators benefit both themselves and prey populations.

7. Distinguish herbivores, carnivores, and omnivores from one another.

8. Describe how prey populations avoid predation. Explain how the environment in which prey populations live helps to shape the population's responses to predation.

9. Contrast Batesian and Müllerian mimicry.

Self-Quiz Questions

Fill-in-the-Blanks

In (1) _____, one population or individual exploits the same limited resources as another or intervenes with another sufficiently to keep it from gaining access to the resource. According to the concept of (2) _____ _____, when two species are competing for the same resource, one tends to exclude the other from the area of a niche overlap. To a greater or lesser extent, one would have the advantage; the other would be forced to modify its (3) _____. When one population denies another species access to a limited resource, usually by aggressive behavior, the tactic is called (4) _____ _____.

When two or more populations share resources in different ways, in different areas, or at different times, coexistence is also possible through (5) _____ _____. Populations of Canadian lynx and snowshoe hare undergo (6) _____

_____, thus providing support for the Lotka-Volterra model of predator-prey interactions. Wolves and dogs establish (7) _____ to avoid direct competition.

Herbivores eat (8) _____, carnivores eat (9) _____, and omnivores eat almost (10) _____.

37–III   COEVOLUTION

Summary

Ecologically interacting populations can coevolve through reciprocal selection pressures. For example, predator and prey populations exert continual selection pressures on each other. *Mimicry* and *warning coloration* are two of the coevolutionary results of predator-prey systems. When two or more species of dangerous or noxious organisms resemble each other, the resemblance is Mullerian mimicry; no deception is involved, and there is no division into model and mimic as occurs in Batesian mimicry. Camouflage involves adaptations in form, patterning, color, or behavior that enable an organism to blend with its background and escape detection.

Chemical defenses such as warning odors, repellants, alarm substances and poisons have frequently been developed among a variety of prey species. Prey species avoid predation by adaptations for flight, hiding, fighting, or disguise.

Flowering plants and their pollinators provide examples of *mutualism*, in which positive benefits are exchanged between participating organisms. As time goes on, mutualism may become so mutually beneficial that the two participant species may coadapt structurally, functionally, and/or behaviorally because each acts as a powerful selective agent on the other; this is *coevolution*. Whenever selective advantages lead to specialized structures or behaviors that make each partner coadapted to and totally dependent on the other, the two species have entered an *obligate relationship*.

Parasitism is generally a one-way relationship in which a guest species benefits at the expense of a living host. Parasites spend less energy on survival by forcing their hosts to perform specialized tasks for them; as a result, their structure tends to be simple. Parasitoids, on the other hand, kill their hosts by completely consuming their soft tissues by the time they metamorphose into adults.

Key Terms

coevolution (p. 622)   katydid (p. 625)   krill (p. 626)
aposematic   American Bittern (p. 625)   mycorrhiza (p. 626)
  coloration (p. 622)   warning   mycelia (p. 626)
Mullerian mimic (p. 622)   coloration (p. 624)   yucca moth (p. 626)

Batesian mimic (p. 622)
aggressive
  mimicry (p. 622)
mimicry (p. 624)
camouflage (p. 624)
moment-of-truth
  defenses (p. 624)
Lithiops (p. 624)

bombardier
  beetle (p. 624)
grasshopper mice (p. 624)
tannins (p. 624)
terpenes (p. 624)
pheromone (p. 624)
dispersal agent (p. 625)
opportunistic (p. 626)

yucca plant (p. 626)
true parasite (p. 627)
social parasitism (p. 627)
botflies (p. 628)
pupate (p. 628)
parasitoids (p. 628)
jackpine sawfly
  larvae (p. 628)

Objectives

1. Define *coevolution* and suggest why it might serve the interests of the populations concerned to coevolve.

2. Suggest an outstanding vulnerability that coevolved populations share.

3. Distinguish mimicry from camouflage.

4. Name an animal that utilizes a startle display behavior when cornered.

5. Explain what energy outlay has to do with predation strategy. Contrast the blue whale's strategy with that of the killer whale.

6. Explain how obligate relationships are established through coevolution.

Self-Quiz Questions

  Fill-in-the-Blanks

(1) _____ often occurs through reciprocal selection pressures operating on two ecologically interacting populations. Prey populations attempt to avoid predation by adaptations for flight, hiding, fighting, and/or (2) _____. (3) _____ refers to adaptations in form, patterning, color, or behavior that enable an organism to blend with its background and escape detection. In (4) _____ mimicry, harmful species resemble each other; there is no model or mimic.

  Matching The same letter may be used more than once. Use only one letter per blank.

(5) ____ American bittern

(6) ____ blood flukes (schistosomes) and humans

(7) ____ blue whale and krill

(8) ____ bombardier beetle and grasshopper mouse

(9) ____ brown-headed cowbird, Kirtland's warbler, and botflies

(10) ____ Canadian lynx and snowshoe hare

A. parasitism

B. mutualism

C. predator-prey
      relationship

D. camouflage

E. startle display

(11) ___ fungal mycelia and plant root hairs

(12) ___ jackpine sawfly larvae and other
parasitoid larvae

(13) ___ katydid

(14) ___ killer whales and seals and dolphins

(15) ___ *Lithiops*

(16) ___ short-eared owl

(17) ___ tigers and tall-stalked golden grasses

(18) ___ yucca moth and yucca plant

INTEGRATING AND APPLYING KEY CONCEPTS

Is there a *fundamental niche* that is occupied by humans? If you believe so, describe the *minimal* abiotic and biotic conditions that populations of humans require to live and reproduce. [Note that 'thrive' and 'be happy' aren't being mentioned.] If you do not think so, state why.

These minimal niche conditions can be viewed as resource categories that must be protected by populations if they are to survive. Do you believe that the 'cold war' between the United States and the Soviet Union primarily concerns protection of minimal niche conditions, or do you believe that the 'cold war' is concerned with other more important factors?

(a) If the former, how do you think *minimal* niche conditions might be guaranteed for all humans willing and able to accept certain responsibilities as their contribution toward enabling this guarantee to be met?

(b) If the latter, what do you think those factors are?

38
ECOSYSTEMS

Summary

Feeding relationships among species determine the direction and extent of materials used and energy flow through a community. In its simplest form, the trophic structure of a community includes three levels of participants: *producers*, which most often harness energy from the sun; *consumers*, which include herbivores, carnivores, scavengers, parasites, and omnivores; and *decomposers*, which generally are fungi and bacteria that break down organic debris and thereby help cycle nutrients and ions back to producers.

    The pattern of feeding relationships in a community is best expressed as a *complex food web* rather than a simple, linear food chain. A food web is a network of crossing, interlinked food chains, encompassing primary producers, and an array of consumers and decomposers. Too often we find that our disruptions of natural systems have disastrous consequences not only for our crops but for our economy and even our lives. The massive application of DDT in an attempt to curb malaria in Borneo brought rats, sylvatic plague, and caved-in roofs to the human community. DDT is an exceptionally stable compound that undergoes biological concentration as it moves up through trophic levels.

Key Terms

energy fixation
  base (p. 629)
nutrient concentration
  base (p. 629)
ecosystem (p. 630)
savanna (p. 629)
trophic level (p. 630)
photosynthesizer (p. 630)
chemosynthesizer (p. 630)

herbivore (p. 630)
primary consumer (p. 630)
primary carnivore (p. 630)
secondary
  carnivore (p. 630)
omnivore (p. 630)
decomposer (p. 630)
food chain (p. 630)
food web (p. 630)

krill (p. 631)
petrel (p. 631)
chlorinated
  hydrocarbon (p. 631)
biological
  concentration (p. 631)
sylvatic
  plague (p. 632)

Objectives

1. Define the term *ecosystem* and state how autotrophic organisms are related to ecosystems.

2. List the principal trophic levels in an ecosystem of your choice, state the source of energy for each trophic level, and give one or two examples of organisms associated with each trophic level.

3. Describe how DDT damages ecosystems in general and how it damaged Borneo's rural habitats in particular.

Self-Quiz Questions

Fill-in-the-Blanks

Photosynthetic plants and (1) _____ fix (2) _____ and concentrate

(3) _____ for ecosystems. All of the organisms in a community that obtain energy

primarily from a common source constitute a (4) _____ _____. All

(5) _____ are primary consumers. (6) _____ obtain their energy from the

organic remains and products of other organisms. DDT was sprayed in Borneo to control

(7) _____ which transmitted the organisms that cause (8) _____. Fill

in the blank for this food web:

     mosquitoes, flies, cockroaches ⇥ lizards

                                       ⇥ (9) _____ _____.

               fleas ⇥ rats

Because of its stability, DDT is a prime candidate for (10) _____ _____:

the increasing concentration of a nondegradable substance as it moves up through

trophic levels.

Summary

Primary productivity is the *rate* at which energy becomes stored in organic compounds through photosynthesis. *Gross primary production* is the total amount of solar energy converted to organic compounds during photosynthesis. *Net primary production* is the energy remaining after aerobic respiration by autotrophs, that is stored as organic matter (biomass). This net production is the variable energy base for consumers of the ecosystem. A growing animal must take in about ten kilocalories of photosynthetically derived energy to produce every one kilocalorie of stored energy (or potential food energy for predators). Energy storage becomes more inefficient over time because energy must be expended on growth and maintenance. The faster energy can be transferred from one trophic level to the next, the higher the efficiency of transfer within the community.

Energy flows through ecosystems by way of *grazing food webs* (based on the living tissues of photosynthesizers) and *detrital food webs* (based on organic waste products and remains of photosynthesizers and consumers). Energy generally leaves ecosystems as a result of metabolic activities (for example, respiration) occurring in both producers and consumers *and is not recycled*.

Feeding relationships may change cyclically or permanently over time. The sizes of organisms being counted in each trophic level may vary and must be accurately weighed. With such data, pyramids of either numbers or biomass can be constructed as crude forms of ecosystem analysis. An energy budget can be constructed for a community to determine how much energy flows through in a given time period; H. T. Odum's 1957 study of an aquatic ecosystem in Florida provides a classic example of a "simple" energy budget. More complex ecosystems are more difficult to analyze. If energy budgets are worked out for each entire community, a general principle can be derived: a community cannot survive indefinitely if its members expend more energy than they take in.

Key Terms

primary
   productivity (p. 632)
gross primary
   production (p. 632)
net primary
   production (p. 632)
plankton (p. 632)

detritus (p. 632)
grazing food web (p. 632)
detrital food
   web (p. 633)
detritus feeders (p. 633)
pyramid of
   numbers (p. 635)

pyramid of biomass (p. 635)
phytoplankton (p. 635)
ecosystem analysis (p. 635)
meteorological
   input (p. 635)
leaching of
   nutrients (p. 635)

Objectives

1. Distinguish between net and gross primary production.

2. Contrast the ways in which energy and nutrients pass through an ecosystem. Explain why nutrients can be completely recycled but energy cannot.

3. Compare grazing food webs with detrital food webs. Present an example of each.

4. Explain why determining a pyramid of numbers or a pyramid of biomass for a particular ecosystem is, at best, an inadequate representation of energy flow through an ecosystem. Explain how one goes about preparing an ecosystem analysis, such as the one prepared by H. T. Odum.

Self-Quiz Questions

Fill-in-the-Blanks

The energy remaining after respiration that is contained in an organism's biomass is its (1) _____ _____ _____. (2) _____ typically is expressed as grams (dry weight) of organic matter per unit of area. In a (3) _____ _____ _____, decomposers feed on organic wastes, dead tissues, and decomposed organic matter. Earthworms, millipedes, and fly and beetle larvae are examples of (4) _____ feeders. On an average, a growing animal must take in about (5) _____ kilocalories of energy to produce one kilocalorie stored in its own biomass.

38–III   NUTRIENT CYCLING

Summary

In constructing a model of nutrient flow through an ecosystem, studies of the tundra were used to test the model's ability to predict the path of nutrients. The tundra is a treeless plain—flat, windswept, and cold. Rainfall is minimal, yet for the three summer months, continuous sunlight encourages profuse plant growth. Underlying permafrost inhibits drainage and decomposition. Most of the plant biomass resides in only ten species. And the brown lemming is the single main herbivore in the tundra.

    Studies of this ecosystem formed the basis of the nutrient recovery hypothesis, which states that interlocking feedback loops exist between ecosystem variables such as population size for predators and for prey, the total plant biomass, and the decomposition rate for organic matter. Each variable is seen to amplify or counteract events and conditions along interconnected pathways.

    All organisms must synthesize proteins, for which they need nitrogen in the form of ammonia (NH_3), ammonium ions (NH_4^+), nitrite (NO_2^-), or nitrate (NO_3^-); these forms nitrogen are supplied by aspects of the nitrogen cycle known as *nitrogen fixation, nitrification,* and *ammonification. Denitrification* converts nitrate to gaseous nitrogen (N_2), thus closing the *nitrogen cycle.* Bacteria and blue-green algae are the microorganisms that can carry out nitrogen fixation. Enormous amounts of petrochemical energy are needed to synthesize fertilizer industrially; in many cases, modern agriculture has been pouring more energy into the soil in the form of fertilizer than we are getting out of it in the form of food or gasohol.

Key Terms

tundra (p. 636)
permafrost (p. 636)
peat (p. 636)
brown lemming (p. 637)
sedges (p. 637)
Pitelka and
  Schultz (p. 637)
nutrient recovery
  hypothesis (p. 637)

leaching (p. 639)
nitrogen fixation (p. 638)
triple covalent
  bond (p. 638)
ammonia, NH_3 (p. 638)
ammonium ion, NH_4^+ (p. 638)
nitrification (p. 638)

nitrite, NO_2^- (p. 638)
nitrate, NO_3^- (p. 638)
ammonification (p. 638)
legumes (p. 638)
nitrogen cycle (p. 638)
denitrification (p. 638)
crop rotation (p. 639)

Objectives

1. Describe the tundra as an example of a simple ecosystem.

2. Explain how the nutrient recovery hypothesis helps ecologists analyze the critical limiting factors operative in the ecosystem and the tolerance ranges of the few different populations that interact there.

3. Define the chemical events that occur during nitrogen fixation, nitrification, ammonification, and denitrification.

4. Explain why United States agricultural methods tend to put more energy into the soil in the form of fertilizers, pesticides, food processing, storage, and transport than we get out of the soil in the form of energy stored in foods.

Self-Quiz Questions

Fill-in-the-Blanks

The principal herbivore of the tundra is the (1) _____ _____.

(2) _____ can assimilate nitrogen from the air in the process known as

(3) _____ _____. In (4) _____, either ammonia or ammonium ions are

stripped of electrons, and (5) _____ (NO_2^-) is released as a product of the

reactions. Under some conditions, nitrate is converted to (6) _____ _____

by denitrifying bacteria. According to the (7) _____ _____ _____,

cyclic variations in population size result from interactions between the vegetation

cover and the herbivores grazing on it; these variations depend on nutrient recovery

and (8) _____ _____ in the soil.

Summary

Of the 4.7 billion humans on earth, only 700 million have adequate diets; 1 billion have enough energy but not enough protein, and for the remaining 3 billion, the situation is desperate. Humans can subsist most efficiently on plant proteins. When domestic cattle eat grains that humans could consume, there is a 10-to-1 conversion loss in both calories and proteins. Although fish harvesting has neared the maximum possible for long-term yields, fish farming may become sufficiently developed to provide a significant supply of protein. Genetic engineering of plants may soon improve the protein content of crop plants. High-yield crops of the green revolution require pesticides, fertilizers, and abundant water for irrigation; these added costs increase the prices of the harvests. Suggestions for opening up new areas for agriculture, equalizing food distribution, and stabilizing the world population generally are met with resistance.

   The integrated structure and stability of existing communities must be disturbed as little as possible during human efforts to exploit resources. Humans should try to live in harmony with the material cycling and energy flow in stable ecosystems and try to encourage sustained long-term yields of resources rather than destroying ecosystems for short-term gain.

Key Terms

"low-grade"
  protein (p. 640)
"high-grade"
  protein (p. 640)
	fish harvesting (p. 641)
	fish farming (p. 641)
	lysine (p. 641)
		genetic engineering (p. 641)
		green revolution (p. 641)
		high-yield crops (p. 641)

Objectives

1. Present your arguments in favor of eating "low on the food chain."

2. Explain the problems associated with increasing fish harvesting from the oceans, using high-yield varieties of crops, and opening up new areas for agriculture.

3. Explain why the world food crisis would be lessened if more people stopped eating beef and switched to milk, eggs, fish, cheese, beans, and rice.

4. Devise a farm system that could be relatively self-sustaining and would minimize the costs of protein conversion.

5. Explain the problems entailed in implementation of the four suggestions at the end of the preceding commentary.

6. What would you recommend doing to stabilize the world population of humans?

Fill-in-the-Blanks

There are now more than (1) _____ billion humans on earth; of these, only
(2) _____ million have adequate diets. About 2 pounds of dried corn can meet the
daily protein needs of an adult male; so will less than 9 ounces of meat. Meat is
considered as (3) _____-_____ protein. About 7-10 pounds of open range
grasses were needed to produce (4) _____ pound(s) of beef, but no humans could
eat grass. Some of the fish under culture in Asian fish ponds grow to about
(5) _____ pounds. Work to incorporate (6) _____-_____ abilities
into the DNA of crop plants is under way. (7) _____-_____ _____
require fertilizers, pesticides, and lots of irrigation water, all of which tend to
raise prices of crops out of reach of the common people. Desert irrigation causes
(8) _____ to accumulate in the soil.

Summary

Primary succession is a gradual process during which vacant land first becomes
populated with pioneer species that later are sequentially replaced by other
communities until a stable, complex climax community is achieved. The process may take
hundreds of years, but sometimes it takes thousands; all the while, the total biomass
slowly increases, offering more possibilities for the partitioning of niches.
   Secondary succession may also occur when an established community is disrupted in
whole or in part. Since most dominant plant species in a climax community may not grow
unless an integrated community already exists, reestablishing a community on the
exposed soil is not simply a matter of seeds from nearby trees drifting over and
germinating. Only when a rich layer of decomposing matter has built up do the trees
characteristic of the climax stage start to germinate. Some communities, such as those
dominated by sequoia trees, depend on intermittent fires for maintaining long-term
stability. Through interlocking food webs, materials and energy are cycled through
communities, and overall community stability is established. If stability in ecosystems
is defined as persistence as conditions change, then some ecosystems are more resilient
than others. Those ecosystems that have few, highly specialized participants in
nutrient cycling, those that have no interchangeable guild species, and those that are

partitioned in delicate interdependencies are particularly vulnerable to disruptions. At the very least, the limits of resilience for our "managed" ecosystems should be determined lest, in the long-term, humans destroy their own basis of sustenance.

The introduction of species that have evolved elsewhere to a region has sometimes been successful (honeybees, ring-necked pheasants), sometimes disastrous. Water hyacinths from South America brought to Louisiana and Florida during the 1880s rapidly displaced the native plants and brought much of the river traffic to a halt. Dutch elm and Chestnut blight fungi have devastated millions of acres of U.S. forests.

Key Terms

climax community (p. 643)
succession (p. 643)
primary
  succession (p. 643)
secondary
  succession (p. 643)
r-selected
  species (p. 643)
opportunistic
  species (p. 643)
pioneer species (p. 643)
equilibrium
  species (p. 643)
K-selected plants (p. 643)
colonization (p. 643)

cyclic replacement (p. 643)
sequoia (p. 643)
reorganization in the
  vegetation cover (p. 644)
meltwater (p. 644)
Dryas, mountain
  avens (p. 644)
glacial till (p. 644)
sitka alders (p. 644)
cottonwood (p. 644)
hemlock (p. 644)
Sitka spruce (p. 644)
peat (p. 644)
water hyacinth (p. 646)
honeybees (p. 647)

mosquitofish (p. 647)
ring-necked
  pheasants (p. 647)
Dutch elm fungus (p. 646)
bark beetle (p. 646)
Chestnut
  blight fungus (p. 646)
Argentine
  fine ant (p. 646)
Japanese beetle (p. 646)
sea lamprey (p. 646)
European
  starling (p. 646)
house sparrow (p. 646)

Objectives

1. Outline the steps by which a climax community on land is shaped. Begin with barren land formed by either volcanic action or deglaciation.

2. Characterize and contrast opportunistic and equilibrium species.

3. Explain why a forest community on the exposed soil created by completely stripping a spruce forest for lumber is not reestablished by seeds from nearby spruces drifting over and germinating.

4. Give an example of a community whose long-term stability depends on intermittent fires.

5. List three introduced species that have been absorbed into existing community structures without creating a great deal of disruption.

6. List eight introduced species that have caused large-scale disruptions to various ecosystems in the U.S.

Self-Quiz Questions

Fill-in-the-Blanks

(1) _____ _____ is a gradual process during which vacant land first

becomes populated with (2) _____ species that later are sequentially replaced by

other communities until a stable (3) _____ community is established.

(4) _____ succession occurs when an established community is disrupted in whole or in part. (5) _____ seeds germinate only on bare mineral soil, so (6) _____ is required to eliminate the litter on the forest floor.

Matching Choose the single best letter for each blank. A letter may be used for more than one blank.

(7) ___ <u>Cerastomella ulmi</u>

(8) ___ Chestnut blight fungus

(9) ___ <u>Dryas</u>

(10) ___ Japanese beetle

(11) ___ mosquito fish

(12) ___ sea lamprey

(13) ___ sequoia

(14) ___ Sitka spruce

(15) ___ water hyacinth

A. a fungus that has killed large numbers of American trees

B. a *K*-selected species in Alaska

C. disrupted the Great Lakes fisheries

D. requires fire to sprout seeds

E. an *r*-selected species in Alaska

F. a very competitive plant that shaded out other vegetation

G. a beneficial insect-eating species

H. an extremely voracious herbivore

INTEGRATING AND APPLYING KEY CONCEPTS

If you were Ruler of All People on Earth, how would you organize industry and human populations in an effort to solve our most pressing pollution problems?

39
THE BIOSPHERE

Summary

The *biosphere* is composed of many ecosystems that are linked through movements of materials and energy that span the globe. *Climate* is the principal factor that determines the specific ecosystem that exists at a specific place on earth. Four factors (amount of incoming solar radiation, Earth's daily rotation and annual migration around the sun, the arrangement of continents and oceans, and the altitude of land masses) predominate and interact to produce the prevailing winds and ocean currents in different regions. These, in turn, affect global weather patterns. Weather and organisms of the region affect the composition and characteristics of soils and sediments (e.g., through erosion), which, in turn, influence the growth rates and distribution of the primary producers. These are the "basis" of all food chains.

Molecules in the atmosphere absorb heat (infrared) and ultraviolet radiation; many of the potentially harsh effects of incoming solar radiation are moderated by the atmosphere. Moist air tends to reduce temperature extremes as day changes to night.

In general, as one goes from the equator to the poles (and from sea level to the high elevations of mountains) specific types of vegetational cover typically occur.

Key Terms

Kalahari Desert (p. 648)	infrared (heat) (p. 649)	temperate (p. 651)
eland (p. 648)	greenhouse	tropical (p. 651)
kudu (p. 648)	effect (p. 649)	shrub (p. 652)
wildebeest (p. 648)	analogous (p. 649)	montane (p. 652)
tsetse flies (p. 648)	westerlies (p. 650)	deciduous (p. 652)
biosphere (p. 649)	tradewinds (p. 650)	coniferous (p. 652)
lithosphere (p. 649)	doldrums (p. 650)	subalpine (p. 652)
hydrosphere (p. 649)	gyres (p. 650)	alpine (p. 652)
atmosphere (p. 649)	countercurrent (p. 651)	sedges (p. 652)
climate (p. 649)	upwelling (p. 650)	rain
ozone (O_3) (p. 649)	California current (p. 651)	shadow (p. 652)

Objectives

1. Describe how the biosphere is related to its three components.

2. Name five abiotic factors that contribute to prevailing weather conditions.

3. Explain how certain components of Earth's atmosphere moderate some of the harsh effects of incoming solar radiation.

4. Define the greenhouse effect and state how it influences ecosystems.

5. State the reason why you think land masses heat and cool more rapidly than oceans do.

6. Explain what causes the prevailing air currents. Compare the prevailing air currents with the prevailing ocean currents and state whether or not they are similar.

7. Explain any differences you may have noted in Question 6 by distinguishing any factors that influence air currents and ocean currents.

Self-Quiz Questions

Choose the single most appropriate letter for each.

Matching I

(1) ____ climate

(2) ____ doldrums

(3) ____ greenhouse effect

(4) ____ gyres

(5) ____ lithosphere

A. absorbs ultraviolet wavelengths of incoming solar radiation
B. located between 0° and ± 30° N and S
C. regions of rising, warmed air spreading northward and southward
D. caused by lack of abundant precipitation and intervening mountains
E. regions of rising, nutrient-laden water
F. prevailing weather conditions that are caused by four principal factors

(6) ___ montane

(7) ___ ozone

(8) ___ rain shadow

(9) ___ tradewinds

(10) ___ upwellings

G. caused by heat radiated from plants and the soil being retained by the atmosphere

H. contains both deciduous and coniferous trees that are tolerant of more moisture and cooler temperatures

I. rocks, soils, sediments, and outer portions of crust

J. circular movements of large water masses

Matching II

(11) ___ Benguela current

(12) ___ Canary current

(13) ___ California current

(14) ___ Gulf Stream

(15) ___ equatorial countercurrent

(16) ___ Humboldt current

(17) ___ Japan current

(18) ___ Labrador current

(19) ___ N. Atlantic Drift

(20) ___ west wind drift

K. a cold water mass that maintains the temperate rain forest associated with Washington and Oregon

L. a southward-flowing current that runs into the Gulf Stream and eddies

M. feeds into the Gulf of Mexico

N. southward off Spain and W. Africa

O. an eastward-flowing warm-water current

P. splits to give rise to a northeastward drift and a southward current

Q. off the west coast of Peru

R. an eastward-flowing *cold* water current of the southern hemisphere

S. flows past the northern coasts of Great Britain and Sweden

T. merges with the California current in the northern Pacific gyre

Fill-in-the-Blanks

From the base of the Rocky Mountains to the peaks, one generally encounters five zones:

(21) _____ and (22) _____ adapted to dry, warm conditions; the

(23) _____ belt with trees that can tolerate moisture and cool temperatures; the

(24) _____ belt with conifers adapted to a still colder climate; the cold

(25) _____ belt where grasses dominate; and on the highest peaks covered with

snow and ice, (26) _____.

When descending dry air warms, it picks up moisture from the land and

(27) _____ are created; most of these occur at about 30° latitudes.

Summary

Material resources flow in cycles. Carbon dioxide, nitrogen, and oxygen flow through all trophic levels in every ecosystem as a gaseous cycle, for the atmosphere is the main storehouse for these three resources. In the hydrologic cycle, water moves through the atmosphere, across the land's surface, to the seas, and back again, largely through evaporation, precipitation, detention, and transportation.

   The cycling of oxygen and carbon dioxide is linked by the processes of photosynthesis and respiration. The turnover rate of carbon depends on environmental conditions. About half of the CO_2 released into the atmosphere from volcanic eruptions and the burning of fossil fuels will be stored either in the oceans or in accumulated plant biomass.

   The earth's crust is the main storehouse for nutrients, such as phosphorus, that flow through sedimentary cycles.

Key Terms

gaseous cycle (p. 653)
hydrologic
  cycle (p. 653)
evaporation (p. 653)

precipitation (p. 653)
detention (p. 653)
transportation (p. 653)
residence time (p. 653)

sedimentary
  cycle (p. 654)
phosphorus cycle (p. 654)
weathering (p. 654)

Objectives

1. Diagram the basic events in (a) the carbon-oxygen cycle, (b) the hydrologic cycle, and (c) the phosphorus cycle. State where the main sinks occur and state the processes that reintroduce the substance into active cycling.

2. State some substances in which carbon, nitrogen, oxygen, and phosphorus occur in living organisms.

3. Explain how atmospheric carbon dioxide increases could submerge New York City and expand existing deserts.

Self-Quiz Questions

   <u>Fill-in-the-Blanks</u>

In (1) _____ cycles, resources move from land, to sediments in the seas, and

back to the land. All living things require (2) _____, which becomes

incorporated into the nucleotides of DNA, RNA, ATP, and the like, which are mined from

deposits in mountain ranges that parallel the coast.

(3) ___ Both photosynthesizers and consumers release CO_2 during respiration.

(4) ___ The *rate* at which CO_2 is fixed during photosynthesis and CO_2's *rate* of release during respiration are approximately equal.

(5) ___ In tropical rain forests, decomposition is rapid and a great deal of carbon is stored in the soil.

Summary

Aquatic ecosystems can be classified according to salinity. *Freshwater ecosystems* with extremely low salinity are either lotic (running-water) or lentic (standing-water) habitats. *Marine ecosystems* include high-salinity oceans, seas, and inland bodies of brackish water, but *estuaries* are regions such as bays, salt marshes, and channels where seawater mixes with freshwater. Productivity in aquatic ecosystems varies with light penetration, temperature, concentrations of nutrients, topography of the basin and sediments, water depth, and pattern of water movement. All of these factors change with daily and seasonal variations in climate, which, in turn, affect the nature and distribution of aquatic life. Deep freshwater lakes have layers of water of differing temperatures, density, nutrients, oxygen, and plankton. These deep lakes generally undergo *spring* and *fall overturns* as they respond to changes in air temperature and are divided into *littoral, limnetic,* and *profundal zones.*

In the oceans, the distribution of producer organisms is governed by such variables as light, temperature, salinity, and available nutrients. Only in the epipelagic (photic) zone, the upper 100 to 400 meters of water, is light intense enough to drive photosynthesis. (The mesopelagic zone [200-1,000 meters below the ocean's surface] has little light but contains most of the inorganic nutrients [nitrates and phosphates].) Temperature and salinity in the bathypelagic zone (between 2,000 and 4,000 meters below the ocean's surface) do not vary much, mainly because major currents tend to circulate water masses on a global scale. At the surface, especially in the shallow intertidal zone (between the high and low tide marks), temperature and salinity vary with degree of latitude and freshwater intrusion. The way materials are cycled back to the producers determines why marine ecosystems are distributed as they are. In shallow nearshore waters and coral reefs, decay occurs in the photic zone, and resources are recycled much as on land. Over open oceans, wastes and decaying organisms generally drift far from the narrow zone of photosynthetic organisms that could thrive on them; thus, essential minerals such as nitrates and phosphates are constantly being removed from the waters of open-water ecosystems, and the productivity is generally low compared to estuaries, nearshore waters, and tropical reefs. Areas of upwelling and the deep-sea geothermal ecosystem of the Galapagos Rift are two notable exceptions to this general rule. Around the world, estuaries, coastal waters, and upwelling areas are being commercially fished at or above levels that are dangerous to their stability.

freshwater
  ecosystems (p. 655)
lentic (p. 655)
lotic (p. 655)
marine (p. 655
estuarine,
  estuary (p. 655)
brackish (p. 655)
stratification (p. 655)
spring overturn (p. 655)

thermocline (p. 655)
fall overturn (p. 655)
littoral (p. 655)
limnetic (p. 655)
profundal (p. 655)
eutrophic (p. 656)
oligotrophic (p. 656)
epipelagic zone (p. 656)
mesopelagic zone (p. 656)
bathypelagic zone (p. 656)

plankton (p. 656)
phytoplankton (p. 656)
zooplankton (p. 656)
nekton (p. 656)
Galapagos Rift (p. 656)
upwelling (p. 657)
El Niño (p. 658)
Spartina (p. 658)
detrital food web (p. 658)

Objectives

1. Define salinity and discuss what you believe to be the reasons why most freshwater organisms cannot adapt to open ocean salinities and why most oceanic species cannot live in freshwater habitats.

2. Describe the causes of temperature stratification in bodies of water and discuss how spring and fall overturns can occur in both freshwater and marine ecosystems.

3. List the variables that determine how producer organisms (and the consumer organisms that depend on them) are distributed in marine environments.

4. Describe the differences in the way materials are cycled (a) in open-ocean communities, and (b) in the intertidal zone.

5. Explain how a hydrothermal vent ecosystem operates.

6. Present a case for protecting the estuaries of the world from being dredged and filled and converted to "reclaimed land."

Self-Quiz Questions

  Fill-in-the-Blanks

(1) _____ ecosystems are inland bodies of standing fresh water; (2) _____

ecosystems include creeks, streams, and rivers. Water is densest at 4°C; at this

temperature it sinks to the bottom of its basin, displacing the nutrient-rich bottom

water upward and giving rise to spring and fall (3) _____. A region where

freshwater mixes with saltwater is a(n) (4) _____. Portions of a lake where

light penetrates to the bottom compose the (5) _____ zone, where rooted

vegetation and decomposers are abundant. The (6) _____ zone includes those areas

of a lake *below* the depth where sufficient light penetration balances the respiration

and photosynthesis of the primary producers; anaerobic bacterial decomposers are the

principal organisms here. (7) _____ lakes are nutrient-poor. In bodies of water that are sufficiently deep, there is a depth where the temperature of the water

decreases rapidly with a small increase in depth; this region is known as the

(8) _____. Organisms that depend on currents and drifts to distribute them are

referred to collectively as (9) _____; larger, more actively-swimming types

compose the (10) _____.

Most marine ecosystems fall within the shallow (11) _____ _____

(between the high and low tide marks) and the (12) _____ _____

(between 2,000 and 4,000 meters beneath the ocean's surface). The upper 100-400 meters

of water, where photosynthetic organisms live, is referred to as the (13) _____

_____. The (14) _____ _____ is a geothermal ecosystem 2,500 meters

beneath the ocean's surface, where sulfur-oxidizing bacteria are the primary producers.

(15) _____ is a marsh grass that is the dominant primary producer in New England

salt marshes.

39–IV   TERRESTRIAL ECOSYSTEMS

Summary

In terrestrial ecosystems, the composition of the underlying soil helps to determine
the kinds of producers that establish the basis of the food webs of the ecosystem.
Soils with large particle sizes (gravel and sand) tend to lose water and minerals
quickly; only plants that can tolerate little water and few minerals can exist in such
places. Soils with a large clay component (which has smaller particle sizes) tend to
hold water and be anaerobic; they can sustain only plants that can tolerate
water-logged, oxygen-poor soils. Loam (clay with a large number of large particles to
prevent packing) is a productive soil.
On land, the distribution of biomes is influenced not only by the amount of
sunlight, but also by temperature and rainfall. The amount of incoming sunlight varies
with latitude, slope exposure, the seasons, and recurring cloud covers or clear skies.
The amount of cloud cover depends on interactions between Earth's topography and
wind and ocean currents. Water droplets in the air moderate extremes in temperature. The
land itself——its elevation, its mineral content, its ability to hold moisture or
encourage runoff, its slope——helps dictate the kinds of life found there.

Desert biomes are arid regions that support little primary production. Because there is little moisture in the air, many deserts undergo large daily temperature fluctuations. About five percent of the world's deserts bake during the day and are extremely cold during the night; the Sahara of Africa and Death Valley in the United States are such deserts. Cacti, yuccas, and a variety of other plants adapted to endure long periods without water constitute the desert producers. When the brief spring rains arrive, these plants depend on extensive root systems just below the soil surface. Occasional oases indicate where rocks saturated with groundwater break through the surfaces at low elevations.

Shrublands such as chaparral, cold deserts, successional shrublands, and thickets generally occur in regions with long, hot, dry summers and cool, moist winters. The plants are highly flammable during the drier seasons and produce abundant seeds, many of which require heat and scarring before they can germinate. In these habitats, periodic fast-spreading fires clear away old growth and speed the recycling of nutrients.

Grasslands prevail where the land is flat, temperatures are moderate, and rainfall is limited by mountains that bar most of the storms moving in from the sea. In the United States, the dominant primary producers were short grasses where rain was sparse and tall grasses where there was a little more rainfall. Today, undisturbed grasslands are rare; tallgrass prairie has been converted into vast fields of corn and wheat, and shortgrass regions sustain herds of imported cattle. Indians and buffalo have been displaced from their ecosystem.

Tropical grasslands include the African savanna with its variable numbers of scattered trees or shrubs as well as the monsoon grasslands of southeast Asia.

Key Terms

soil (p. 658)
silt (p. 658)
clay (p. 658)
A horizon (p. 658)
topsoil (p. 658)
humus (p. 658)
B horizon (p. 658)
subsoil (p. 658)
leached (p. 658)
C horizon (p. 658)

bedrock (p. 658)
loam (p. 658)
biogeographic
    realms (p. 658)
biomes (p. 658)
desert (p. 658)
tundra (p. 658)
Mediterranean-type
    shrublands (p. 661)
cold deserts (p. 662)

successional
    shrublands (p. 662)
tallgrass
    prairie (p. 662)
Dust Bowl (p. 662)
tropical
    grasslands (p. 662)
savanna (p. 662)
monsoon
    grasslands (p. 662-663)

Objectives

1. Describe the typical layered structure of soils and state your understanding of how each layer was formed from its original parent material: solid rock.

2. Arrange according to size (smallest to largest) these particles: sand, silt, and clay. State where loam fits into the series.

3. List the factors that influence the distribution of biomes on land. Then consider the prairie biome and indicate the factors that can bring about more specialized ecosystems within the prairie biome.

4. State the relationship between temperature and rainfall on the abundance and diversity of producers in the different biomes.

5. Contrast the factors that cause cool deserts to form in some regions and warm deserts to form in others.

6. Construct a complex food web that would be characteristic of a warm desert region of the southwestern United States.

7. List the factors that encourage tallgrass prairie and shortgrass prairie to form.

8. Construct a complex food web that would be characteristic of the American prairie as it existed before its conversion to agriculture.

Self-Quiz Questions

Fill-in-the-Blanks

(1) _____ is formed as the products of living and dead organisms (humus) are mixed with weathered and eroded loose rock. As rainwater percolates down through the A horizon, minerals dissolve in it and are carried along with particles to the (2) _____. Tree roots penetrate far below the soil surface, breaking up rocks as they go and absorbing minerals in solution; the region of loose rock that extends to the underlying bedrock is the (3) _____. Arid regions that support little primary production are known as (4) _____. Sagebrush communities are dominant primary producers in (5) _____ _____, and cacti, yuccas, and other drought-resistant plants dominate (6) _____ _____. At one time, (7) _____ extended westward from the Mississippi River region to the Rocky Mountains, from Canada through Texas. (8) _____ _____ was converted into vast fields of corn and wheat, and (9) _____ _____ were converted into ranges for imported cattle. Overgrazing and ill-advised attempts to farm the region led to massive erosion, which crippled the primary productivity of much of this land.

39–V Forests   663–666
Tundra   667
The City as Ecosystem   667–668

Summary

Tropical forests emerge where warm temperatures combine with abundant and fairly uniform rainfall; they are stratified communities dominated by trees spreading their canopies far above the forest floor. At different levels beneath the high canopy in such forests are diverse plant species adapted to ever-diminishing amounts of sun. The complex food webs sustained by all the producer plants are astonishing. In such places, the kinds of organisms living in or on a single tree often exceed the kinds of organisms living in an entire forest to the north. When the forest is cleared of plants and animals for agriculture, most of the nutrients are permanently cleared off with them; as a result, a tropical rain forest is one of the worst places to grow crops. Even with *slash-and-burn agriculture*, most of the nutrients are soon washed away because of the heavy rains and poor soils.

Deciduous forests are those in which most plants lose their leaves during part of the year either as an adaptation to dry seasons (in tropical seasonal forests) or as protection against winter cold (temperate deciduous forests). The tropical seasonal forests of India and Southeast Asia can survive years of drought, but human agriculture in these regions depends entirely on the timely arrival of the monsoons. The best examples of the remains of the temperate deciduous forest are in the Appalachian Mountains in the southern United States.

The *northern coniferous forest*, or *taiga*, exists where there are cold, long winters with fairly constant snow cover and summers warm enough to promote the dense growth of evergreen conifers (spruce, fir, pine, and hemlock), birch, and aspens.

The United States Forest Service manages all wildlife, watersheds, recreation, and lumbering in the national forest system. Part of its program is based on tree farming, which converts a diverse, self-sustaining ecosystem into a monocrop of fast-growing softwood trees whose culture depends upon the widespread use of insecticides and fertilizers and whose harvest is carried out by clear-cutting. Ninety percent of our national forests are potentially open to clear-cutting, which increases the rate of soil erosion and leaves areas scarred and vulnerable for decades.

The *tundra* is a region where it is too cold for large trees to grow but not cold enough to be perpetually frozen over with snow and ice. A permanently frozen layer 610 meters thick (the *permafrost*) underlies the surface soil, and the main limiting factor is the low level of solar radiation. Species diversity is low, and nutrient cycling rates are low. As diverse as all the world's biomes are, as simple or complex as their dominant communities and ecosystems may be, all have a self-perpetuating, self-contained stability. In contrast, an urban region is not self-contained and not stable, even though it may be considered an ecosystem. The foundation of any biome is its characteristic array of producer organisms. Urban centers have few producers. They have mainly consumers and, compared to the human biomass, a negligible number of decomposers. A city endures as an ecosystem only as long as it imports energy and materials from someplace else. The need for imports amplifies pressures on ecosystems far beyond the city's boundaries. Vast transportation networks must be created to ship in essential resources and ship out products and wastes. In return, a city offers employment and manufacturing, entertainment, and a stimulating environment.

Key Terms

tropical rain
  forest (p. 663)
stratified
  forest (p. 663)
canopy (p. 663)
bromeliad (p. 663)
organic debris (p. 664)
decomposition (p. 664)
biomass (p. 664)
slash-and-burn
  agriculture (p. 664)

tropical seasonal
  forests (p. 666)
temperate deciduous
  forests (p. 666)
montane coniferous
  forest (p. 666)
boreal forest (p. 666)
taiga (p. 666)
moose (p. 666)

conifer food web (p. 666)
forage (p. 666)
national forest
  system (p. 666)
tree farming (p. 666)
climax community (p. 666)
simplification (p. 666)
clear-cutting (p. 666)

Objectives

1. List the factors that encourage the development of tropical rain forests.

2. Construct a food web that might typify a tropical rain forest.

3. Describe a tropical seasonal forest and explain what causes a tropical forest to be deciduous rather than a rain forest.

4. Describe a temperate deciduous forest and construct a typical food web for this biome.

5. List the factors that encourage the development of a northern coniferous forest.

6. Construct a food web for a northern coniferous forest.

7. Indicate which of the forest types is least vulnerable to intelligent use by human beings and which is most vulnerable.

8. Explain the benefits and detriments of tree farming in our national forest system.

9. Describe the physical and biotic features of the tundra ecosystem. Distinguish between alpine and arctic tundra.

10. Contrast the benefits of an urban ecosystem with its internal instabilities and the instabilities induced in other ecosystems as a result of urban activities.

11. Explain why an urban "ecosystem" is inherently unstable.

12. Explain why it is difficult, if not impossible, to effectively recycle the by-products of human urban existence.

Self-Quiz Questions

Fill-in-the-Blanks

Warm temperatures and uniform, abundant rainfall combined encourage the growth of

(1) _____ _____ _____, which contain (2) _____ communities,

dominated by trees spreading their (3) _____ far above the forest floor.

Although this type of biome has a very high species (4) _____, it is one of the

(5) ○ best ○ worst places to grow crops. (6) _____-_____-_____

agriculture reduces the forest biomass to ashes, which are then tilled into the soil,

but most of the nutrients are soon washed away because of the heavy rains and poor

soils. In (7) _____ _____ forests, most plants lose their leaves during

part of the year as an adaptation to the dry seasons that alternate with the wet

monsoons. In (8) _____ _____ forests, most of the dominant trees drop

their leaves not in response to dry seasons, but as protection against (9) _____

_____. At lower elevations of the montane coniferous forest (10) _____

_____ and Douglas fir predominate. (11) _____ is synonymous with boreal forest. (12) _____ percent of our national forests are potentially open to

clear-cutting. Just beneath the surface of the (13) _____ is the permafrost. The

main limiting factor in the tundra is the low level of (14) _____ _____.

An urban region may be called a(n) (15) _____, but it is not self-contained and

not stable because urban regions have few producers.

INTEGRATING AND APPLYING KEY CONCEPTS

If, at the end of Chapter 37 you said that the 'cold war' essentially was concerned
with the U.S. and the U.S.S.R. protecting minimal niche conditions (resource
categories) that support living and reproduction by their respective populations, how
do you view the capacity of each nation to bring about a 'nuclear winter' by exploding
just 1000 of their thousands of large warheads? Can you suggest a better global way to
guarantee the minmal resource categories of all nations on our beautiful 'spaceship
Earth'? If so, outline the requirements of such a system and devise a way that it could
be established.

40

HUMAN IMPACT ON THE BIOSPHERE

<table>
<tr><td>Case Study: Solid Wastes</td><td>SOME ENERGY OPTIONS</td></tr>
<tr><td>Case Study: Water Pollution</td><td>Fossil Fuels
Nuclear Energy</td></tr>
<tr><td>Case Study: Air Pollution</td><td>Wind Energy
Solar Energy</td></tr>
<tr><td>Acid Deposition
Industrial and
  Photochemical Smog</td><td>PERSPECTIVE</td></tr>
</table>

Summary

   In natural ecosystems, life's by-products are generally recycled and do not accumulate for long. *Pollutants* are by-products of our existence that are not recycled through natural disposal systems. Humans do not recycle most of their waste products; instead, they reduce, collect, concentrate, bury, and burn wastes, or spread them out through other ecosystems. More and more frequently, as human populations increase, industrial and human wastes that were once put out of sight and out of mind come back in the form of illness and death.

   It is possible to live in the midst of a gradual trend toward unlivable conditions and not even be aware of what is going on. Solid wastes are dumped, burned, or buried at the tremendous rate of 4.5 billion metric tons per year in the United States. Very often hazardous materials are dumped together with paper, glass, aluminum, steel, copper, and plastic objects that could be recycled or burned to generate heat. Garden and food wastes could be used to generate compost useful in rebuilding soil or used as animal feed. Although an ecologically based system for handling solid waste is the most desirable solution, even a recycling system would be preferable to the dump, burn, or bury throwaway system in use now.

   More and more of the water available for use by organisms is becoming polluted because it is used as a dumping ground for by-products of human existence. About 30 percent of the waste water in the United States goes through only *primary* treatment before the liquid effluent is discharged. Mechanical screens and sedimentation tanks are used to force coarse suspended solids out of the water to become a sludge. Chemicals such as aluminum sulfate are added to hasten the sedimentation process, and chlorine is added to kill disease-causing microorganisms. *Secondary* treatment depends on microbial action to degrade the sludge. *Tertiary* treatment involves advanced methods of precipitation of suspended solids and phosphate compounds, adsorption of dissolved organic compounds, reverse osmosis, stripping of ammonia to remove nitrogen from it, and disinfecting the water through chlorination or ultrasonic vibrations. Oxygen-demanding wastes, suspended solids, nitrates, phosphates, and dissolved salts such as heavy metals, pesticides, and radioactive isotopes are the items treated by tertiary treatment.

Key Terms

pollutants (p. 669)
feces (p. 669)
hepatitis (p. 669)
encephalitis (p. 669)
throwaway (p. 670)
recycling (p. 670)
ecologically based
  system (p. 670)
compost (p. 671)
generalized resource
  recovery system (p. 671)
urban "ore" (p. 671)
electromagnets (p. 671)
recovery chambers (p. 671)

mechanical
  screening (p. 671)
flotation (p. 671)
centrifugal
  hurling (p. 671)
smelter (p. 671)
foundry (p. 671)
hydrologic cycle (p. 671)
water pollution (p. 671)
thermal pollution (p. 671)
waste-water
  treatment (p. 671)
primary treatment (p. 671
sedimentation tank (p. 671)

effluent (p. 671)
secondary
  treatment (p. 671)
sludge (p. 671)
suspended solids (p. 671)
heavy metals (p. 672)
settling lagoon (p. 672)
tertiary treatment (p. 672)
precipitation (p. 672)
adsorption (p. 672)
reverse osmosis (p. 672)
stripping of ammonia (p. 672)
ultrasonic
  vibrations (p. 672)

Objectives

1. List the disadvantages and dangers of trying to maintain our present throwaway system for handling solid wastes.

2. Distinguish a recycling system for solid wastes from an ecologically based system. State the conditions under which an ecologically based system for handling solid wastes would benefit the culture more than a recycling system.

3. Describe the basic strategy and layout of a generalized resource recovery system.

4. Define primary, secondary, and tertiary waste-water treatment and list some of the methods used in each of the three types of treatment.

5. Explain what is meant by *oxygen-demanding wastes*. Where do they come from and how do they affect organisms? Then tell how this type of pollutant can be controlled.

6. Identify any products of waste-water treatment that could be of use to humans.

Self-Quiz Questions

Fill-in-the-Blanks

A typical U.S. urban center with a population of 1 million produces (1) _____ tons of trash daily.

In a generalized resource recovery system, (2) _____ could be used to extract steel and iron, and (3) _____ _____ could send plastic and paper to different recovery chambers.

In (4) _____ waste-water treatment, mechanical screens and sedimentation tanks are used to force coarse suspended solids out of the water to become a sludge.

(5) _____ treatment involves advanced methods of precipitation of suspended solids and phosphate compounds. About (6) _____ of the waste water in the United States is not being treated at all.

Summary

Into the thin layer of atmosphere we dump more than 700,000 metric tons of pollutants *each day* in the United States alone. Air pollutants corrode buildings. They ruin agricultural crops and forests. They cause humans to suffer everything from headaches and burning eyes to lung cancer, bronchitis, and emphysema. Pollutants that affect life everywhere are by-products of fossil-fuel burning in transportation vehicles, home heating plants, power plants, and the countless furnaces of industry. Particulates, oxides of sulfur, carbon monoxide, and oxides of nitrogen are routinely pumped into the atmosphere. Under specific conditions of weather and topography, thermal inversions and photochemical and industrial smogs result.

Paralleling the J-shaped curve of human population growth is a steep rise in energy consumption. In the highly industrialized United States, the average individual directly or indirectly uses more than 200,000 kilocalories a day.

Coastal forests and algae were buried and transformed, under the sediments and ooze and compression of time, into fossil fuels: coal, oil, and natural gas. Oil is due to be depleted by 1990 and natural gas by 2010. The coal reserves in the United States represent one-fourth of the world's known coal supply. We must certainly burn more coal over the next two to four decades, but the environmental costs will be great.

Nuclear-powered, electricity-generating stations in the United States produce 8 percent of the nation's electrical energy. The costs of constructing these plants are great, the net energy produced is low, and there is serious concern about short-term safety in and near the plants and about the long-term feasibility of storing nuclear wastes. The most optimistic scientists predict that uranium reserves will be depleted by 2040.

Along the vast plains of the American Midwest, from the Dakotas to Texas, winds are predictable enough and strong enough to be an attractive, non-polluting source of energy. Approximately 300,000 turbine towers could harness the winds of the plains and provide half of our present energy needs. Solar energy ranks with wind energy as the most abundant, the cleanest, and the safest of all potential energy sources now being considered as alternatives to fossil fuels; if 15,000 square miles of desert were set aside in Arizona and California for solar energy collectors that convert sunlight energy directly into electricity, half of the nation's annual energy needs could be satisfied.

Many sources of stress on the biosphere (human overpopulation, overconsumption, ecosystem oversimplification, massive industrial technology, lack of leadership, and me-first behavior) can be traced to indifference to the principles of energy flow and materials reuse. Like all living forms in the past, we are all potentially endangered species, kin to one another in vulnerability because of intricate, often invisible, threads of ecological interdependence stretched through the biosphere.

Key Terms

corrode (p. 672)
emphysema (p. 672)
oxides of sulfur (p. 672)
oxides of nitrogen (p. 672)
dry acid depositions (p. 672)
acid rain (p. 673)
sulfuric acid, H_2SO_4 (p. 672)
nitric acid, HNO_3 (p. 673)
wet acid depositions (p. 673)
methyl mercury (p. 673)
topography (p. 674)
thermal inversion (p. 674)
industrial smog (p. 674)

particulates (p. 674)
photochemical smog (p. 674)
photochemical oxidants (p. 674)
PANS, peroxy-acylnitrates (p. 674)
passive solar heating (p. 675)
net energy (p. 675)
fossil fuels (p. 675)
oil shale (p. 675)
kerogen (p. 675)
coal (p. 675)
black lung disease (p. 675)
strip mining (p. 675)

Operation Plowshare (p. 675)
nuclear enrichment (p. 675)
meltdown (p. 676)
Three Mile Island (p. 676)
"spent" fuel (p. 676)
breeder reactors (p. 676)
liquid metallic sodium (p. 676)
fusion power (p. 677)
net energy yield (p. 677)
turbine towers (p. 677)
feed forward mechanisms (p. 678)

Objectives

1. Identify the principal air pollutants, their sources, their effects, and the possible methods for controlling each pollutant.

2. Define *thermal inversion* and indicate its cause.

3. Distinguish photochemical smog from industrial smog.

4. State when the U.S. petroleum reserves are expected to run out if present consumption rates continue. Natural gas? Uranium?

5. Give the reasons that exploiting oil shale deposits may not be worth doing.

6. Compare the functions of a conventional nuclear power plant reactor and a breeder reactor and assess the principal benefits and risks of each.

7. Explain what a *meltdown* is.

8. Define *fusion power* and assess its prospects of being used.

9. Describe the ways that wind energy could be harnessed and indicate what percentage of U.S. needs could be met by this means.

10. List five ways that you personally could become involved in ensuring that institutions serve the public interest in a long-term, ecologically sound way.

Fill-in-the-Blanks

(1) _____ _____ _____ attack marble, metals, mortar, rubber, and plastic; they also form droplets of (2) _____ _____ that create holes in nylon stockings. When a layer of dense, cool air gets trapped beneath a layer of warm air, the situation is known as a(n) (3) _____ _____. When (4) _____ and nitrogen dioxide are exposed to sunlight, they become converted to poisonous substances collectively called (5) _____ smog. Oil shale is buried rock that contains (6) _____, a hydrocarbon compound. Nuclear-powered electricity-generating plants produce (7) _____ percent of the U.S. electrical energy. Only (8) _____ has good long-term, intermediate, and short-term availability as an energy option for the United States. Unlike conventional nuclear-fission reactors, (9) _____ reactors could potentially explode like atomic bombs. If (10) _____ square miles of desert were set aside in Arizona and California for solar energy collectors that convert sunlight energy directly into electricity, we could develop a system for providing (11) _____ percent of our nation's annual energy needs even decades from today.

INTEGRATING AND APPLYING KEY CONCEPTS

If you were Ruler of All People on Earth, how would you encourage people to depopulate the cities and adopt a life where they could supply their own resources from the land and dispose of their own waste products safely on their own land?

Explain why some biologists believe that the endangered species list now includes all species.

41
ANIMAL BEHAVIOR

Summary

A *motor score* is a sequence of motor output that is determined by the central nervous system. Innate behavior is represented by genetically determined motor scores that predictably run their course once some stimulus sets them in motion. Any stimulus that activates a motor score, or even some part of it, is called a *releaser*. Perception of what constitutes the correct releaser for a particular motor score also has a genetic basis; this and the motor score may undergo mutation. In any case, any behavior is subject to selection pressures; if advantageous to survival and reproduction, the motor scores will probably be transmitted to the next generation; disadvantageous motor scores tend to be eliminated.

As long as a motor score contributes to adaptability, selection would tend to favor individuals in which that behavior is rigidly locked into the genetic program. But if some innate behavior could be modified and if the change made it more successful than other members of the population in finding food, escaping from predators, or producing offspring, then its flexibility of response would tend to be perpetuated. If such changes are more than one-time motor responses to a new cluster of stimuli and if there is an enduring potential for adapting future responses as a result of past experience, the modification is called *learning*. In constant or highly predictable environments, rigid genetically determined motor patterns are common; in changing and unpredictable environments, the capacity for behavioral flexibility and learning tends to prevail.

Classical *conditioning*, whereby a connection is made between a new stimulus and a familiar one, is a form of *associative learning*. It is adaptive because it enables animals to anticipate and recognize potentially dangerous situations in time to deal with them adequately and thus survive. Another form of associative learning is *instrumental conditioning*, in which a reinforcing stimulus (reward or punishment) appears after a behavioral response is given; the animal learns by trial and error. In the absence of the reinforcing stimulus, the learned behavior may soon become extinguished; this is a learning process called *extinction*. Exploration (curiosity) encourages *latent learning*. Some primates demonstrate *insight learning* (alternative responses to a situation are first evaluated mentally with reference to stored information in the brain, and an enlightened response is made). The limited time spans during which learning certain forms of behavior occurs are the sensitive periods of development; *imprinting*, in which an enduring preference for a particular object (or other animal or place) is formed occurs during an animal's critical period. Stimulus enrichment during the first two years of human life is critically important in developing intelligence in humans.

Key Terms

adaptive
  significance (p. 679)
territorial
  display (p. 679)
nudibranch (p. 679)
Willows (p. 679)
behavior (p. 679)
neuromotor
  response (p. 679)
flight
  response (p. 679)
innate releasing
  mechanisms (p. 680)
releaser (p. 680)

motor score (p. 680)
instinctive (p. 680)
innate
  behavior (p. 680)
learning (p. 680)
associative
  learning (p. 680)
Pavlov (p. 680)
conditioned
  reflex (p. 680)
conditioned
  stimulus (p. 680)
reinforcing
  stimulus (p. 680)

instrumental
  conditioning (p. 680)
T-maze (p. 680)
extinction (p. 680)
latent
  learning (p. 680)
insight
  learning (p. 680)
imprinting (p. 681)
sensitive period (p. 681)
Lorenz (p. 682)
Tryon (p. 682)
artificial
  selection (p. 682)

Objectives

1. Define *behavior* and name four factors that produce it.

2. Explain what a motor score is and state its relationship to innate behavior.

3. Explain how a researcher can determine whether a behavior is innate or learned.

4. Define *releaser* and give one example of how a releaser works.

5. Provide an example of selection pressures operating to eliminate a disadvantageous behavior.

6. Give one example of a motor score that has been modified in ways that are adaptive to changing times and circumstances.

7. Define *learning* and distinguish it from innate behavior.

8. State the generalization that expresses the relationship between predictability of environmental events and the capacity for behavioral change and learning.

9. Explain what is meant by classical Pavlovian conditioning and distinguish it from instrumental conditioning by describing an example of each process. Explain why both forms of conditioning are examples of associative learning.

10. Name some of the simplest animals in which learning has been demonstrated. Describe the experiment that demonstrated the the associative learning process

11. Define *insight* and give an example.

12. Explain whether or not you think that humans have any critical periods for establishing the ability to learn. Also state whether or not you think humans undergo imprinting.

Self-Quiz Questions

Fill-in-the-Blanks

A (1) _____ _____ is a sequence of motor outputs determined by the central nervous system that results in a coordinated movement; it is genetically determined, that is, (2) _____, and predictably runs its course once some stimulus sets it in motion. Any stimulus that activates a motor score is known as a (3) _____.

(4) _____ is a change in behavior that involves an enduring potential for adapting future responses as a result of past experience. In (5) _____ environments, rigid, genetically determined motor patterns are common. The work of Ivan Pavlov centered on a form of (6) _____ learning, classical conditioning, in which new responses called (7) _____ _____ were established in the behavior of dogs. Another form of associative learning, (8) _____ conditioning, uses a reinforcing stimulus (a reward or punishment) that appears *after* a response is given, and learning occurs by (9) _____-_____-_____. If the reinforcing stimulus is stopped, the learned behavior is discontinued—a learning process called (10) _____. (11) _____ _____ has been demonstrated

to occur only in some primates; it is a trial-and-error process that goes on in the brain before an appropriate response is made. For many animals, learning certain forms of behavior occurs only during limited time spans called (12) _____ _____ of development; preferential behavior toward a stimulus acquired during such a time is called (13) _____.

41–II   ECOLOGICAL ASPECTS OF BEHAVIOR

Summary

Much of animal behavior is correlated with rhythms of the planet. Circadian rhythms are based on the daily rotation of the earth about its long axis. Photoperiodism in animals is based on changes in day length that occur as the earth makes its annual journey around the sun. In some vertebrates, the pineal gland secretes melatonin synthesized from serotonin; the "timekeeping" enzyme helps catalyze the synthesis of melatonin. When melatonin levels are lowest, motor and metabolic activities are at their peak. A migration is a cyclic surge of a whole population away from one region toward another that beckons it with compelling force: *homing behavior* returns them to their home base. Homing behavior may involve having a well-developed *map sense, compass sense,* or *magnetic sense.*

Many prey organisms have developed *avoidance behavior* to a fine art: sophisticated escape responses based on ultrasensitive receptors, "freeze" behavior in combination with environmental cryptic coloration, and behavior that threatens the predator in some way. Predators, on the other hand, have developed "lures" and camouflage with productive results, and some have employed the use of tools.

Key Terms

circadian
  rhythm (p. 683)
biological
  clocks (p. 683)
pineal gland (p. 683)
melatonin (p. 683)
migration (p. 684)

homing
  behavior (p. 684)
albatross (p. 684)
map sense (p. 684)
compass sense (p. 684)
magnetic sense (p. 684)
aggressive mimicry (p. 685)

predator avoidance
  behavior (p. 685)
echolocating
  pulses (p. 685)
ultrasonic
  frequencies (p. 685)

Objectives

1. Define *migration* and give examples of a fish and a bird that migrate.

2. Explain why it would be advantageous for some birds to migrate. Speculate about why the redwinged blackbird doesn't migrate.

3. List the physiological changes that happen in birds that do migrate.

4. Describe how starlings are guided during their migrations.

5. Explain how you think geese are guided to their breeding grounds.

6. Give two components of human behavior that are affected by circadian factors.

7. Define *photoperiodism* and describe the basic biochemical events that underlie photoperiodic behavior in some vertebrates.

Self-Quiz Questions

Fill-in-the-Blanks

The cyclic surging of whole populations away from one region toward another that beckons with compelling force is called a (1) _____. (2) _____ activate the quest in response to the change in length of (3) _____. Many birds learn to recognize familiar landmarks in the area around their nests; the ability to locate one's nest is (4) _____ _____. (5) _____ rhythms occur in twenty-four-hour cycles, and (6) _____ events are cued to the seasonal changes of day length that occur because the earth journeys around the sun.

41–III BEHAVIORAL ADAPTATION: WHO BENEFITS?

Summary

In the unconscious competition between bearers of "selfish" alleles and others with self-sacrificing tendencies, the selfish types would become more common as time progresses. More behavioral traits are convincingly explained using Darwinian individual selection than good-of-the-group selection.

Natural selection favors species with reproductively selfish behavior, as exemplified by siblicide in egrets. Prolonged courtship rituals protect the reproductive concerns of both parents and help to ensure successful rearing of the young. Male territorial displays and mate-guarding behavior help to ensure that the fittest males, their mates, and offspring will have access to the best environmental resources.

Submission finds expression in two ways: avoidance behavior and appeasement behavior (exaggerated displays of submission). In a natural environment, a dominant animal rarely takes advantage of such total vulnerability, so selection pressures must favor the behaviors helping to keep the group together rather than ones that would promote mutual destruction.

Key Terms

individual
  selection (p. 686)
good-of-the-group
  selection (p. 686)
Williams (p. 686)
selfish gene (p. 686)
altruism (p. 686)
behavioral
  adaptation (p. 686)
siblicide (p. 686)
species-survival
  promotions (p. 686)

synchronous
  incubation (p. 686)
sibling
  aggression (p. 686)
courtship
  behavior (p. 686)
rituals (p. 686)
territory (p. 687)
vocalizations (p. 687)
visual displays (p. 687)

aggressive
  display (p. 687)
courtship
  display (p. 687)
pair bond (p. 687)
harem (p. 688)
combative
  structure (p. 688)
mate-guarding
  behavior (p. 689)

Objectives

1. Explain why reproductively selfish types increase their numbers of offspring, whereas altruistic types decrease their numbers.

2. Contrast individual and group selection.

3. Explain how siblicidal tendencies could benefit a population.

4. Describe a complex courtship ritual and suggest a reason why it could be linked with territoriality. List two ways in which prolonged courtship can protect the interests of both potential parents.

5. Explain why mate-guarding behavior by males can be viewed as a selfish behavior to protect interests, rather than an altruistic behavior on behalf of the female.

Self-Quiz Questions

Fill-in-the-Blanks

(1) _____ traits as well as structural traits are selected for by natural selection mechanisms. (2) _____ selection occurs through the differential reproduction of genetically varied individuals in a population. An area defended rigorously against intruders is called a(n) (3) _____; ownership is announced through (4) _____ and (5) _____ _____. Sneaky tactics of defeated males apparently have fostered (6) _____-_____ behavior.

Summary

Because of the genetic continuity between a parent and its offspring, it is possible for a parent to die in the defense of its young and thus give up some of its own chances at future reproduction and still have its genes spread through the population.

In natural settings, *aggregations* of animals that are nothing more than a response to some conditions in the local environment can occur; such groupings are not considered to be true social groups. But other groups in which there is a division of labor that promotes the chances of survival also occur: *pair formation, family groupings, troops, schools, herds, flocks,* and *societies.* In pair formation, a male and female mate and share such tasks as food getting, nest building, brooding, and defense. A family group consists of one or more sets of offspring remaining with one or more parents; the young of a family group thus acquire a set of learned behavior patterns that increase their chances for survival and reproduction. Larger social groups generally form in open environments where food resources are adequate for supporting large populations but where there aren't many places to hide; extremely complex group-oriented behavior patterns tend to evolve in such groups. Selection favors the dominant animal whose competitive show of strength ensures it will get the choicest food and its pick of potential mates. A dominant animal advertises its higher status in formal displays, or ritualized behavior, which are exaggerated ordinary functional movements that send distinct signals to conspecifics. Initially, fighting establishes dominance by one individual and submission by others. A dominance hierarchy is established and actual fighting dwindles as a result. Because less energy is wasted on aggression, more is available for the business of survival.

Among the social insects such as bees and termites, division of labor reaches the level of a complex society; individuals have become so specialized that they cannot survive on their own. When some members become specialized in food gathering, others in defense, and others in reproduction, there is little energy wasted in the group as a whole, and selective agents must act upon the instinctive cooperative behavior of all the individuals.

If natural selection is the basis for evolutionary change, then behavioral as well as structural traits that are the most adaptive—in other words, provide the edge in competition for resources—will be selected for. Social groups are based, to varying degrees, on mutual cooperation. Aggression is dispersive; cooperation is cohesive. Many writers have attempted to decipher what is the inherent, "natural" behavior of humans, and many tend to overestimate the role of heredity in determining behavior and status in human societies.

Key Terms

"society" (p. 690)
school (p. 690)
herd (p. 690)
dilution effect (p. 690)
dominance
   hierarchy (p. 691)
"alpha" position (p. 691)
subordinate
   members (p. 692)
appeasement
   gestures (p. 692)

individual selection
   theory (p. 692)
true altruism (p. 693)
pheromone (p. 694)
bee dance (p. 694)
guard bees (p. 694)
queen (p. 694)
stingless
   drones (p. 694)
worker bees (p. 694)

nest parasite (p. 696)
suicidal
   behavior (p. 696)
sterility (p. 696)
sexual jealousy (p. 696)
biologically
   determined (p. 697)
adaptive (p. 697)

Objectives

1. Explain the factors that encourage large social groups such as schools, herds, and flocks to form.

2. Define *aggressive behavior* and state what it accomplishes for an animal.

3. Explain why an animal that became established at a low position in a dominance hierarchy wouldn't simply leave the group and seek a more dominant position elsewhere.

4. State any long-term values to an animal population that might accrue from assigning dominant and submissive statuses to its members.

5. Define *ritualized behavior* and indicate when it is used.

6. Contrast avoidance behavior and appeasement behavior.

7. Explain how a pheromone differs from a hormone and state some of the uses of pheromones.

8. Contrast the structure of a society with that of a family group.

Self-Quiz Questions

Fill-in-the-Blanks

Usually, larger social units such as herds and flocks form in (1) _____ environments—the arctic tundra, savanna plains, open seas—where food is adequate but there aren't many places to (2) _____. Fighting results in establishing (3) _____ for one individual and submission by others. The (4) _____ animals end up being separated by a greater distance from the rest of the group, and a ranking of its members—a (5) _____ _____ —serves to lessen actual fighting. Dominant animals advertise their status in formal displays, or (6) _____ _____. Submission finds expression in avoidance behavior and (7) _____ behavior as a further show of deference. Among the (8) _____ _____, division of labor reaches the level of a complex society; the (9) _____ becomes the reproductive "organ" for all, and the workers forage for food and repel invaders. (10) _____ are male members of the colony that have no sting; some mate with the queen. People who believe in the notion of (11) _____ _____ tend to overestimate the role of heredity in determining behavior and status in human societies.

INTEGRATING AND APPLYING KEY CONCEPTS

If you were Ruler of All Human Populations on Earth, what mechanisms would you employ to distribute essential resources so as to minimize deadly competition and wars? Would a person who believed in biological determinism be likely to advocate the equal distribution of essential resources among all members of a social group?

CROSSWORD NUMBER SIX

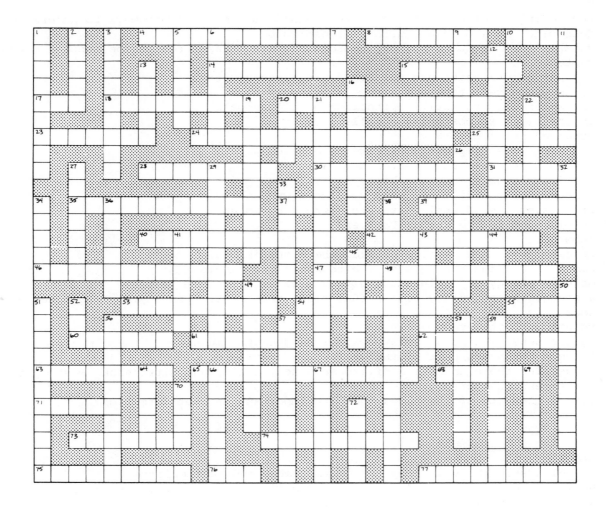

Across

4. ___ canals detect acceleration or a change in the direction of movement
8. leads eggs from the ovary generally to a uterus
10. the ear ___ separates the outer ear from the middle ear
14. ___ eye
15. one of the major membranes that forms around the embryos of reptiles, birds, and mammals
17. nerve ___
18. has small clusters of living cells suspended in a firm, rubbery matrix, which they themselves secreted
20. an increasing concentration of nervous structures and coordinative functions in the "front" end, or head
23. fatty

24. detects energy associated with changes in pressure, position, or acceleration
25. the tough outer layer of the eyeball
28. hair ___
30. the adding together of graded potentials from different pathways to determine determine the course of further transmission
31. stimuli that activate sensory neurons and are transmitted to the central nervous system
35. ___ nerves dominate internal events in times of stress
37. rapid eye movements
39. the outermost protective epithelium of all animals
40. in the small spaces between cells
42. arranged in layers
46. ___ disks are components of cardiac muscle
47. having both sexes in one organism
53. capillary clusters in the kidneys
54. a receptor that distinguishes up from down
55. stimulates the adrenal cortex to secrete hormones
60. an organism in the phylum Porifera
61. inborn
62. release eggs that develop outside the mother's body
63. embryonic tissue that gives rise to bones and muscles
65. detects energy of visible and ultraviolet light
68. a spongelike and fatty tissue that fills the interior cavity of long bones or bone ends
71. relating to the eye
73. length-sensitive organ located within a skeletal muscle
74. the most anterior part of the brain
75. concerning the elimination of excess water, salts, and nitrogenous waste products
76. electroencephalogram
77. formed elements of the blood that are important in clot formation

Down

1. medulla ___
2. ___ glands in the skin help to control water and salt balance
3. having male individuals separate and distinct from female individuals
5. a fatty derivative that coats certain neurons
6. adenyl ___ is an enzyme important in producing cAMP
7. a photoreceptor cell concerned with perceiving gray and black shading in dim light, and with perceiving movement
9. the neck of the uterus
11. the adrenal ___ secretes epinephrine
12. ___ tissues bind together and support other animal tissues
13. a group of fibers that connects the central nervous system and the organs or parts of the body
16. fundamental unit of muscle contraction
19. emit high-frequency sound waves followed by detection of waves that bounce back to their ears from objects in the surroundings
20. a photoreceptor cell that detects specific colors
21. ___ nerves generally dominate internal events when environmental conditions permit normal body functioning
22. a periodically recurring physiological state of rest characterized by lessened responsiveness to external stimuli
26. an agent that slows or stops an activity

27. alternative responses to a situation are first evaluated mentally so that accumulated experiences may suggest an appropriate response
29. neuron that connects sensory neurons with motor neurons
32. like cells united in form and function
33. connects the urinary bladder with the external environment
34. sound receptors in the human ear are hair cells in the organ of ___
36. the storage of individual bits of information somewhere in the brain
38. alimentary tract
41. shock
43. sticking to
44. contains circular and radial muscles in a contractile ring
45. connects a kidney to the urinary bladder
48. lens adjustments that bring about precise focusing onto the retina
49. relating to the sense of smell
50. the study of the function of a structure
51. synonym for platelet
52. reticular activating system
56. capable of long-range transmission of unaltered messages
57. ___ period: a time of insensitivity to stimulation
58. tubal ___, a form of sterilization
59. large white blood cell that phagocytizes
64. ___ window through which any vibrational energy remaining is dissipated
66. endocrine cell product
67. a movement of electrical energy
69. ___ blood cell
70. osseous tissue
72. an immature motile form of an organism that next becomes a juvenile

ANSWERS

1 On the Unity and Diversity of Life

1-I   ORIGINS AND ORGANIZATION

1.	T	7.	G
2.	T	8.	D
3.	T	9.	B
4.	F	10.	H
5.	C	11.	A
.6.	E	12.	I

1-II   UNITY IN BASIC LIFE PROCESSES
Metabolism
Growth, Development, and Reproduction

1.	T	8.	fat or carbon-
2.	T		containing
3.	T		molecules
4.	T	9.	ATP
5.	F	10.	metabolism
6.	T	11.	egg
7.	F	12.	larva
		13.	pupa
		14.	adult

1-III   Homeostasis
DNA: Storehouse of Constancy and Change

1.	T	5.	T
2.	F	6.	homeostasis
3.	F	7.	mutations
4.	F		

1-IV   DIVERSITY IN FORM AND FUNCTION
The Tropical Reef
The Savanna
A Definition of Diversity

1.	F	6.	H
2.	T	7.	F
3.	T	8.	G
4.	A	9.	B
5.	E	10.	C
		11.	D

1-V   ENERGY FLOW AND THE CYCLING OF
     MATERIAL RESOURCES
PERSPECTIVE

1.	F	8.	A
2.	F	9.	B
3.	T	10.	producers
4.	F	11.	herbivores
5.	C	12.	carnivores
6.	B	13.	decomposers
7.	D	14.	nutrients
			(raw materials)

2 Methods and Organizing Concepts in Biology

2-I   WHAT DO YOU DO WITH AN OBSERVATION?
ON THE "SCIENTIFIC METHOD"
COMMENTARY: Process of Scientific
  Inquiry

1.	F	6.	d	11.	O
2.	T	7.	f	12.	O
3.	T	8.	e	13.	C
4.	F	9.	a	14.	O
5.	b	10.	c	15.	C

2-II   EMERGENCE OF EVOLUTIONARY THOUGHT
Linnean System of Classification
Challenges to the Theory of Unchanging Life
Lamarck's Theory of Evolution

1.	Lamarck	7.	B	13.	F
2.	Buffon	8.	C	14.	F
3.	D	9.	G	15.	F
4.	F	10.	F	16.	T
5.	A	11.	F	17.	F
6.	E	12.	T	18.	F
				19.	d

2-III EMERGENCE OF THE PRINCIPLE
  OF EVOLUTION
Naturalist Inclinations of the
  Young Darwin
Voyage of the *Beagle*
Darwin and Wallace: The Theory
  Takes Form

1. T
2. F
3. T

2-IV AN EVOLUTIONARY VIEW OF DIVERSITY
PERSPECTIVE

1. F
2. T
3. T

3 Atoms, Molecules, and Cell Substances

3-I ORGANIZATION OF MATTER
The Nature of Atoms

1.	T	8.	C	15.	J
2.	T	9.	N	16.	E
3.	T	10.	K	17.	F
4.	F	11.	L	18.	B
5.	F	12.	G	19.	H
6.	T	13.	A	20.	I
7.	D	14.	M		

3-II How Electrons are Arranged in
  Atoms
COMMENTARY: Keeping Track of Electrons:
  Some Options
Electron Excitation

1.	F	3.	F
2.	T	4.	T

3-III INTERACTION AMONG ATOMS IN THE
  CELLULAR WORLD
Ionic Bonding
Covalent Bonding
Hydrogen Bonding

1.	T	6.	T	11.	F
2.	F	7.	F	12.	F
3.	T	8.	T	13.	T
4.	F	9.	T		
5.	T	10.	F		

  14-17 See pages 43-44 of the text
         for the answers.

18.	hydroxyl	20.	phosphate
19.	carboxyl	21.	amino

3-IV Hydrophobic Interactions
Bond Energies
ACIDS, BASES, AND SALTS
MAKING AND BREAKING BONDS IN
  BIOLOGICAL MOLECULES

1.	F	11.	bicarbonate
2.	T	12.	salts
3.	T	13.	dissociate
4.	T	14.	F
5.	F	15.	E
6.	T	16.	B
7.	T	17.	C
8.	K^+, Na^+, Ca^{++},	18.	A
	Mg^{++} or Cl^-	19.	G
9.	same as above	20.	D
10.	HCO_3^-		

3-V CARBOHYDRATES
Monosaccharides
Disaccharides and Polysaccharides

1.	F	10.	CD
2.	T	11.	AJ
3.	F	12.	AJ
4.	T	13.	AJ
5.	T	14.	CE
6.	T	15.	BG
7.	T	16.	BH
8.	CE	17.	AI
9.	CD	18.	BF

3-VI LIPIDS
True Fats and Waxes
Steroids and Related Substances
Phospholipids and Glycolipids

1.	F	8.	E
2.	T	9.	BF
3.	T	10.	DG
4.	T	11.	H
5.	T	12.	A
6.	F	13.	B
7.	E	14.	C

3-VII PROTEINS
Primary Structure: A String of
  Amino Acids
Spatial Patterns of Protein Structure
Protein Denaturation
Protein-Based Substances

1.	F	6.	F
2.	T	7.	D
3.	T	8.	B
4.	F	9.	A
5.	T	10.	-

4 Cell Structure and Function: An Overview

4-I GENERALIZED PICTURE OF THE CELL
Emergence of the Cell Theory
Basic Aspects of Cell Structure
  and Function
Cell Size

1.	T	9.	light
2.	F	10.	light
3.	T	11.	Resolution
4.	E	12.	nanometer
5.	A	13.	transmission
6.	D	14.	scanning
7.	C	15.	nucleus
8.	B		

4-II PROKARYOTIC CELLS
EUKARYOTIC CELLS

1.	D	6.	E
2.	D	7.	Eukaryotes
3.	B	8.	prokaryotes
4.	F	9.	Animal
5.	C	10.	nucleoid

3-VIII NUCLEIC ACIDS AND OTHER NUCLEOTIDES

1.	T	6.	T
2.	F	7.	T
3.	T	8.	BD
4.	F	9.	CE
5.	T	10.	AF

4-III THE NUCLEUS
Nuclear Envelope
Chromosomes
Nucleolus

1.	T	4.	nucleolus
2.	T	5.	chromatin
3.	F	6.	chromosomes

4-IV ORGANELLES OF AN ENDOMEMBRANE SYSTEM
Endoplasmic Reticulum and Ribosomes
Golgi Bodies
Lysosomes
Microbodies

1.	T	6.	F
2.	T	7.	C
3.	T	8.	A
4.	F	9.	B
5.	E	10.	D

4-V   OTHER CYTOPLASMIC ORGANELLES
Mitochondria
Chloroplasts and Other Plastids
Central Vacuoles in Plant Cells

1. F
2. T
3. F
4. F
5. T
6. T
7. C
8. A
9. D
10. B
11. G
12. infolded plasma membrane
13. chloroplast
14. NADPH$_2$ (ATP)
15. ATP (NADPH$_2$)
16. ATP

4-VI   INTERNAL FRAMEWORK OF EUKARYOTIC CELLS
The Cytoplasmic Lattice
Microfilaments and Microtubules

1. microtubules
2. Microfilaments
3. cytoplasmic lattice
4. Microtubules
5. tubulins

4-VII   STRUCTURES AT THE CELL SURFACE
Cell Walls and Other Surface Deposits
Microvilli
Cilia and Flagella
Cell-to-Cell Junctions

1.	H	6.	E
2.	A	7.	E
3.	C	8.	F
4.	D	9.	B
5.	A	10.	G

4-VIII   SUMMARY OF MAJOR CELL STRUCTURES AND THEIR FUNCTIONS

Figs. 4.8 and 4.9 (page 46)

1. cell wall
2. plasma membrane
3. ribosomes
4. Gogli complex
5. rough endoplasmic reticulum
6. nucleolus
7. nuclear pore
8. mitochondrion
9. vacuole
10. microfilaments
11. flagellum
12. ribosmomes
13. nuclear envelope
14. nucleus
15. centriole
16. lysosome
17. mitochondrion
18. Golgi complex
19. microfilaments
20. cytoplasm

Chart (p. 47)

1.	+	2.	+	3.	+	4.	√	5.		
6.	√	7.	√	8.	√	9.	√	10.	√	
11.	+	12.	+	13.		14.	√	15.		
16.		17.	+	18.		19.	√	20.		
21.		22.	√	23.	√	24.	√	25.	√	
26.	√	27.	√	28.	√	29.	√	30.	√	
31.		32.	√	33.	√	34.	√	35.	√	
36.		37.	√	38.	√	39.	√	40.	√	
41.		42.	+	43.	+	44.	+	45.	√	56.	T
46.	√	47.	√	48.	√	49.	√	50.	√	57.	T
										58.	T
51.		52.	√	53.	√	54.	√	55.	√	59.	F

5 Water, Membranes, and Cell Functioning

5-I ON CELLULAR ENVIRONMENTS
WATER AND CELL FUNCTIONING
The Importance of Hydrogen Bonds
Solvent Properties of Water

1. A calorie
2. hydrogen bonds
3. heat of vaporization
4. evaporation
5. absorbed
6. motion
7. cohesion
8. solvent
9. solutes

5-II CYTOPLASMIC ORGANIZATION OF WATER, SOLUTES, AND LARGER PARTICLES

1. hydrophobic
2. Phospholipids
3. phosphate
4. sol
5. T
6. T
7. F
8. F
9. T

5-III CELL MEMBRANES: FLUID STRUCTURES IN A LARGELY FLUID WORLD
MOVEMENT OF WATER AND SOLUTES ACROSS MEMBRANES
Overview of Membrane Transport Systems

1. Freeze-fracturing
2. freeze-etching
3. lipid bilayer
4. proteins
5. "mosaic"
6. differentially permeable
7. water
8. Active transport systems

5-IV Diffusion
Osmosis

1. Water (O_2, CO_2)
2. oxygen (H_2O, CO_2)
3. carbon dioxide (O_2, H_2O)
4. gases
5. energy of motion
6. greater
7. lesser (lower)
8. bulk flow
9. T
10. F
11. T

5-V Active Transport
Endocytosis and Exocytosis

1. T
2. F
3. F
4. T

5-VI MEMBRANE SURFACE RECEPTORS
PERSPECTIVE

1. carbohydrate
2. proteins
3. metabolism
4. tissue formation
5. microfilaments

6 Energy Transformations in the Cell

6-I ON THE AVAILABILITY OF USEFUL
   ENERGY
Two Energy Laws

1.	F	3.	T
2.	T	4.	T

6-II Metabolic Reactions: Energy
   Changes in Cells
Equilibrium and the Cell

1.	F	3.	F
2.	F	4.	T

6-III ENZYME FUNCTION
Activation Energy and Enzymes
Enzyme Structure and Functioning

1.	T	4.	T
2.	F	5.	F
3.	T	6.	T

6-IV Cofactors

Effects of pH and Temperature on
   Enzyme Activity
Controls Over Enzymes

1. Temperature (pH)
2. pH (Temperature)
3. hydrogen
4. hydrophobic
5. denaturation
6. active
7. product
8. active
9. active

6-V TYPES OF ENERGY TRANSFERS IN CELLS
Formation and Use of ATP

1.	Biosynthesis	7.	water
2.	active transport	8.	enzyme
3.	movement	9.	ADP
4.	adenine	10.	phosphate
5.	ribose	11.	energy
6.	phosphate		

6-VI Electron-Transfer Reactions

1. F
2. T
3. T
4. electron transport
5. electron
6. oxidation-reduction
7. electron
8. carrier
9. energy
10. phosphate
11. oxidized
12. $NADP^+$
13. NAD^+
14. FAD
15. cytochromes

7 Energy-Acquiring Pathways

7-I FROM SUNLIGHT TO CELLULAR WORK:
PREVIEW OF THE MAIN PATHWAYS

1. Photosynthetic autotrophs
2. Chemosynthetic autotrophs
3. monerans (bacteria)
4. monerans (bacteria)
5. Heterotrophic (consumer)
6. glycolysis
7. respiration

7-II PHOTOSYNTHESIS
Simplified Picture of the Two Stages
  of Photosynthesis
Chloroplast Structure and Function
THE LIGHT-DEPENDENT REACTIONS
Light Absorption in Photosystems
Two Pathways of Electron Transfer
Mechanism of ATP Synthesis
Summary of Light Reactions

1. Autotrophic
2. heterotrophic
3. carbon dioxide
4. water
5. oxygen
6. energy-rich food molecules
7. sunlight
8. pigment molecules
9. ATP
10. NADPH
11. Cyclic
12. ATP
13. non-cyclic photophosphorylation
14. NADPH
15. hydrogen
16. electrons
17. NADP$^+$
18. electrons

7-III THE LIGHT-INDEPENDENT REACTIONS
Summary of Light-Independent Reactions
How Autotrophs Use Intermediates and
  Products of Photosynthesis
Environmental Effects on Rates of
  Photosynthesis

1. ATP
2. NADPH
3. carbon dioxide
4. ribulose biphosphate
5. PGA
6. carbon dioxide fixation
7. six
8. PGA
9. ATP
10. NADPH
11. light-dependent reactions
12. PGAL
13. glucose
14. ribulose biphosphate

7-IV CHEMOSYNTHESIS

1. autotrophic
2. inorganic
3. oxidation
4. ammonium
5. sulfur
6. nitrifying
7. nitrogen
8. nitrogen

8 Energy-Releasing Pathways

8-I OVERVIEW OF THE MAIN ENERGY-
   RELEASING PATHWAYS
Glycolysis
Aerobic Respiration
Anaerobic Respiration
Fermentation Pathways

1. Autotrophic
2. Glucose
3. ATP
4. ATP (NADH)
5. NADH (ATP)
6. pyruvic acid (pyruvate)
7. O_2, oxygen
8. fermentative (anaerobic)
9. lactate
10. ethanol (CO_2)
11. carbon dioxide (ethyl alcohol)
12. aerobic
13. krebs
14. electron transport
    phosphorylation
15. 36
16. ATP
17. aerobic respiration
    (cellular respiration)
18. carbon dioxide
19. electrons
20. NAD^+ (FAD)
21. FAD (NAD^+)
22. transport system (transport
    chain)
23. ATP
24. oxygen
25. mitochondria

8-II A CLOSER LOOK AT AEROBIC RESPIRATION
First Stage: Glycolysis
Second Stage: The Krebs Cycle
Third Stage: Electron Transport Phosphorylation
Glucose Energy Yield: Summing Up

1. Phosphorylating the sugars makes them
   unstable and quite reactive.
2. It is converted to PGAL.
3. It is used in the second part.
4. It is used to do work in the cell.
5. Its energy is used to convert pyruvic
   acid to fermentative products or
   transferred to ATP.
6. They act as temporary acceptors for
   hydrogen ions and electrons released
   during the oxidation of pyruvic acid
   to CO_2 and H_2O.
7. NADH and $FADH_2$
8. electron transport phosphorylation
9. 0, 2, 2
10. 2, 6, 2
11. 0, 2
12. 3, 2
13. 4
14. $10 \times 3 = 30$
15. $2 \times 2 = 4$
16. Pyruvic acid is fed into one of several
    possible fermentative pathways.
17. Nineteen times as many.

8-III FUELS OR BUILDING BLOCKS?
CONTROLS OVER EBERGY METABOLISM

1. F   4. T
2. T   5. F
3. T

8-IV PERSPECTIVE

1. F   3. T
2. F   4. T

9 Cell Reproduction

9-I SOME GENERAL ASPECTS OF CELL REPRODUCTION
The Molecular Basis of Cell Reproduction
Kinds of Division Mechanisms

1.	life cycle	9.	lipids (fats)
2.	reproduction	10.	replication
3.	DNA	11.	D
4.	Genes	12.	C
5.	RNA molecules	13.	A
6.	RNA	14.	B
7.	enzymes	15.	T
8.	carbohydrates (lipids)	16.	T

9-II COMPONENTS OF EUKARYOTIC CELL DIVISION
Chromosomes: Eukaryotic Packaging of DNA
Microtubular Spindles
THE CELL CYCLE

1.	T	5.	DNA
2.	F	6.	microtubules
3.	T	7.	chromosomes
4.	F		

9-III MITOSIS: MAINTAINING THE NUMBER OF CHROMOSOME SETS
Phrophase: Mitosis Begins
Metaphase
Anaphase
Telophase
CYTOKINESIS: DIVIDING THE CYTOPLASM

1.	I	9.	D
2.	F	10.	E
3.	H	11.	F
4.	B	12.	F
5.	A	13.	T
6.	G	14.	F
7.	J	15.	T
8.	C		

9-IV MEIOSIS: HALVING THE NUMBER OF CHROMOSOME SETS
Prophase I Activities
Separating the Homologues
Separating the Sister Chromatids

1.	B	5.	D	9.	F
2.	E	6.	G	10.	T
3.	C	7.	A	11.	F
4.	H	8.	F	12.	T

9-V WHERE CELL DIVISIONS OCCUR IN PLANT AND ANIMAL LIFE CYCLES
Animal Life Cycles
Plant Life Cycles
PERSPECTIVE

1.	2	6.	T
2.	5	7.	F
3.	4	8.	T
4.	1	9.	T
5.	3	10.	F

10 Observable Patterns of Inheritance

10-I MENDEL AND THE FIRST FORMULATION OF THE PRINCIPLES OF INHERITANCE
Mendel's Experimental Approach
The Concept of Segregation

1. traits
2. strains
3. peas
4. cross-fertilization
5. true-breeding
6. hybrids
7. recessive
8. blending
9. monohybrid
10. segregation

10-II Testcrosses: Support for the Concept
Current Terms for Mendelian Units of Inheritance
Probability: Predicting the Outcome of Crosses

1. F
2. T
3.

N?n	nn	
normal × woman	albino man	children's genotypes

	N	Nn
n		
n	nn	nn

Half of the children will be normally pigmented; half will be albinos.

4. c

10-III The Concept of Independent Assortment

1. d
2. a
3. d

10-IV VARIATIONS ON MENDEL'S THEMES
Dominance Relations and Multiple Alleles
Variable Expression of Single Genes
Interactions Between Gene Pairs
Multiple Effects of Single Genes

1. F
2. F
3. F
4. F
5. T
6. F

10-V Environmental Effects on Phenotype
Continuous and Discontinuous Variation
PERSPECTIVE

1. T
2. F

11 Emergence of the Chromosomal Theory of Inheritance

11-I RETURN OF THE PEA PLANT
THE CHROMOSOMAL THEORY

1. Flemming
2. Weismann
3. Flemming
4. meiosis
5. E
6. A
7. D
8. G
9. F
10. B
11. C

11-II CLUES FROM THE INHERITANCE OF SEX
Enter the Fruit Fly
Sex Determination
Sex-Linked Traits
LINKAGE GROUPS AND THE NUMBER OF CHROMOSOMES

1. c
2. a
3. b
4. T
5. F
6. T

11-III RECOMBINATION AND LINKAGE
  MAPPING OF CHROMOSOMES
VISUALIZATION OF CHROMOSOMES

1.	c	5.	b
2.	d	6.	a
3.	b	7.	b
4.	a		

11-IV CHANGES IN CHROMOSOME STRUCTURE
CHANGES IN CHROMOSOME NUMBER
Missing or Extra Chromosomes
Three or More Chromosome Sets
Chromosome Fission and Fusion
COMMENTARY: Cytogenetics, Mr. Sears, and
  a Lot More Bread

1.	deletion	7.	trisomy
2.	translocation	8.	crossing-over
3.	inversion	9.	tetraploid
4.	polyploidy	10.	bridging
5.	Colchicine	11.	chromosome fusion
6.	Nondisjunction		

12 The Rise of Molecular Genetics

12-I THE SEARCH FOR THE HEREDITARY
  MOLECULE
The Puzzle of Bacterial Transformation
Bacteriophage Studies

1.	F	5.	F
2.	F	6.	F
3.	F	7.	T
4.	F		

12-II THE RIDDLE OF THE DOUBLE HELIX
Patterns of Base Pairing

1.	F	5.	T
2.	T	6.	F
3.	T	7.	T
4.	F		

12-III Unwinding the Double Helix:
  The Secret of Self-Replication
DNA REPLICATION
Meanwhile, Back at the Forks . . .
Enzymes of DNA Replication and Repair
Summary of DNA Replication

1.	F	4.	T
2.	F	5.	T
3.	T	6.	c

12-IV DNA RECOMBINATION
Molecular Basis of Recombination
Recombinant DNA Research

1.	T	3.	T
2.	F	4.	F

12-V CHANGE IN THE HEREDITARY MATERIAL
Mutation at the Molecular Level
PERSPECTIVE

1. Ultra-violet light
   (High energy radiation)
2. mutagenic chemicals
3. Insertions (Additions)
4. deletions
5. Frameshift

13 From DNA to Protein: How Genes Function

13-I PREVIEW OF THE STEPS AND PARTICIPANTS IN PROTEIN ASSEMBLY
EARLY IDEAS ABOUT GENE FUNCTION

1. phenylketonuria
2. alkaptonuria
3. Garrod
4. enzyme
5. unit of heredity
6. enzyme
7. Beadle
8. Tatum
9. pink bread mold
10. haploid
11. Asexually
12. sugar
13. gene
14. enzyme
15. RNA molecules
16. RNA
17. RNA
18. ribose
19. uracil
20. adenine

13-III TRANSCRIPTION OF DNA INTO RNA
MESSENGER RNA TRANSLATION
SUMMARY OF PROTEIN SYNTHESIS
Meanwhile, Back at the Mitochondrion . . .

1. Messenger
2. translation
3. Ribosomal (r)
4. transfer (t)
5. amino acid
6. protein
7. codon
8. anticodon
9. transcription
10. DNA
11. RNA polymerase
12. one of the strands
13. complementary
14. mRNA
15. cytoplasm

13-II The Genetic Code

1. primary
2. mRNA
3. sixteen
4. mRna
5. codon
6. 3rd

14 Controls Over Gene Expression

14-I TYPES OF GENE CONTROLS
GENE REGULATION IN PROKARYOTES
The Operon: Coordinated Unit of Transcription
Induction-Repression
Corepression
Other Prokaryotic Control Mechanisms

1. controls
2. Feedback inhibition
3. allosteric
4. short
5. development
6. differentiation
7. Repression
8. repressor
9. RNA polymerase
10. operon
11. repressor
12. corepressor
13. operator
14. regulator
15. promoter
16. translational

14-II GENE REGULATION IN EUKARYOTES
Comparison of Prokaryotic and Eukaryotic DNA
Transcript Processing
Chromosome Organization and Gene Activity

1. single-copy
2. Moderately repetitive
3. rRNA (tRNA)
4. tRNA (rRNA)
5. Introns
6. exon
7. Histones
8. nucleoprotein
9. nucleosome
10. Barr body
11. Lyonization

14-III   VARIABLE GENE ACTIVITY IN
   EUKARYOTIC DEVELOPMENT
Case Study: The Not-So-Simple
   Slime Molds
Case Study: *Acetabularia*
Observations of Variable Gene Activity

1. differentiation   6. T
2. slug   7. T
3. sporocarp   8. T
4. cyclic adenosine   9. T
   monophosphate   10. T
5. environmental
   change

14-IV   CANCER: WHEN CONTROLS BREAK DOWN
PERSPECTIVE

1. F   5. DNA (genes or chromosomes)
2. T   6. expressing (transcribing
3. F   and translating)
4. F   7. nucleus (environment)
           8. cells

15   Human Genetics

15-I   AUTOSOMAL RECESSIVE INHERITANCE
SEX-LINKED RECESSIVE INHERITANCE
Hemophilia
Green Color Blindness
CHANGES IN CHROMOSOME NUMBER
Down's Syndrome
Turner's Syndrome
Trisomy XXY and XYY

1. C   4. B
2. A   5. C
3. B   6. C

15-II   PROSPECTS AND PROBLEMS IN
   HUMAN GENETICS
Treatment for Phenotypic Defects
Genetic Screening
Genetic Counseling and Prenatal Diagnosis
Gene Replacement
COMMENTARY: A Holistic View of Human
   Genetic Disorders

1. H   5. D
2. G   6. B
3. C   7. F
4. E   8. A

16   Plant Cells, Tissues, and Systems

16-I   THE PLANT BODY: AN OVERVIEW
On "Typical" Plants
How Plant Tissues Arise

1. apical   7. mesophyll
2. protoderm   (or parenchyma)
3. procambium   8. epidermis
4. ground meristem   9. xylem
5. Pith   10. phloem
6. Cortex

16-II   PLANT TISSUES AND THEIR
   COMPONENT CELLS
Parenchyma: Photosynthesis, Food Storage
Collenchyma: Supporting Young Plant Parts
Sclerenchyma: Strengthening Mature
   Plant Parts

1. Parenchyma   6. Collenchyma
2. pith   7. pectin
3. epidermis   8. Schlerenchyma
4. mesophyll   9. sclereids
5. Parenchyma

16-III Xylem: Water Transport,
  Storage, and Support
Phloem: Food Transport, Storage,
  and Support

1. salts (ions)
2. vessel members
3. tracheids
4. pits
5. food
6. sieve-tube
    members
7. sieve plates
8. Companion cells

16-IV Epidermis and Periderm:
  Interface with the Environment

1. epidermis
2. water
    absorption
3. water loss
4. microbial
    attack
5. Stomata
6. guard cells
7. Periderm

16-V THE ROOT SYSTEM
Root Primary Structure
Secondary Growth in Roots

1. Taproot
2. fibrous
3. cap
4. apical meristem
5. Epidermis
6. cortex
7. vascular column
8. endodermis
9. vascular cambium
10. cork cambium

16-VI THE SHOOT SYSTEM
Stem Primary Structure
Leaf Primary Structure
Secondary Growth in Stems

1. nodes
2. Monocot
3. vascular bundles
4. sunlight
5. carbon dioxide
6. water
7. photosynthesis
8. vertical (ray)
9. ray (vertical)
10. ray
11. parenchyma
12. vascular cambium

17 Water, Solutes, and Plant Functioning

17-I INTRODUCTION

1. Water
2. vacuole
3. less expensive

17-II ESSENTIAL ELEMENTS AND THEIR
  FUNCTIONS
Oxygen, Carbon, and Hydrogen
Mineral Elements

1. nitrogen bases
2. nucleic acids
3. nitrate (NO_3^-)
4. ammonium (NH_4^+)
5. Oxygen
6. Hydrogen
7. Turgor pressure
8. unwilted
9. B
10. A, I
11. B
12. B
13. B, H
14. A, D
15. B
16. B, F
17. A, G
18. A, C
19. A, E
20. A
21. B

17-III WATER UPTAKE, TRANSPORT, AND LOSS
Water Absorption by Roots
Transpiration
Cohesion Theory of Water Transport
The Paradox in Water and Carbon
  Dioxide Movements
Stomatal Action

1. Transpiration
2. cohesion
3. hydrogen bonding
4. potassium
5. active transport
6. heat
7. abscisic acid
8. potassium

17-IV UPTAKE AND ACCUMULATION OF MINERALS
Active Transport of Mineral Ions
Feedback Controls Over Ion Absorption
TRANSPORT OF ORGANIC SUBSTANCES
  THROUGH PHLOEM
Translocation
Pressure Flow Theory

1. active
    transport
2. xylem
3. Starch
4. sucrose
5. translocation
6. pressure flow
7. sieve tube
8. water
9. pressure
10. roots

18   Plant Reproduction and Embryonic Development

1. sepals
2. petals
3. Stamens
4. anther
5. microspores

6. Carpels
7. ovary
8. ovule
9. megaspores

1. C
2. F
3. A
4. E

5. B
6. G
7. D

1. anthers
2. stigmas
3. pollen tube
4. sperm nuclei
5. zygote
6. endosperm
7. seeds

8. fruit
9. 125
10. Cenozoic
11. red
12. yellow
13. ultraviolet

1. angiosperm
2. monocots
3. dicots
4. cotyledons
5. digestive
   enzymes
6. ovary

7. Endosperm
8. ovule
9. wind
10. petals (flowers)
11. Polarization

1. parthenogenesis
2. Nodes
3. bulbs
4. runners
5. clones

19   Plant Growth and Development

1. polarization
2. Germination
3. Microtubles

4. hormones
5. target cells

1. adventitious
2. radicle
3. enlarge (expand)
4. meristematic

5. water
6. turgor pressure
7. polysaccharides
8. Microtubular

19-III PLANT HORMONES
Types of Plant Hormones
Examples of Hormonal Action

1. hormones
2. Auxins
   Gibberellins)
3. gibberellins
   (auxins)
4. Cytokinins

5. Abscisic acid
6. gravitropism
7. Ethylene
8. IAA (auxin)

19-IV HORMONES, THE ENVIRONMENT, AND
  PLANT DEVELOPMENT
Effects of Temperature
Effects of Sunlight
The Many and Puzzling Tropisms

1. promotes
2. inhibits
3. Phytochrome
4. germination
5. inactive
6. red

7. active
8. Pfr is
   converted
   to Pr
9. A
10. D
11. C
12. A

19-V THE FLOWERING PROCESS
SENESCENCE
DORMANCY

1. biennials
2. perennials
3. Daylength
4. Reproductive

5. sieve tubes
6. parenchyma
7. Ethylene
8. Abscisic acid

19-VI BIOLOGICAL CLOCKS IN PLANTS
Case Study: From Embryogenesis to the
  Mature Oak

1. Endogenous
2. exogenous
3. Phytochrome
4. Mycorrhizae

20   Systems of Cells and Homeostasis

20-I SOME CHARACTERISTICS OF ANIMAL
  CELLS AND TISSUE

1. tissue
2. organ
3. tissue
4. Somatic

5. epithelial
6. connective
7. muscular
8. nervous

20-II KINDS OF ANIMAL TISSUES
Epithelial Tissues

1. T
2. F
3. F

4. F
5. F

20-III Connective Tissue

1. matrix (ground substance)
2. Loose
3. collagen
4. elastin
5. Tendons (ligaments)
6. ligaments (tendons)
7. Bones (Cartilage)
8. cartilage (bone)
9. Haversian
10. Red marrow
11. plasma
12. red blood cells
13. white blood cells
14. platelets
15. Adipose

20-IV Nerve Tissues
Muscle Tissues

1. T   4. T
2. T   5. F
3. F   6. T

20-V OVERVIEW OF ORGAN SYSTEMS AND
  THEIR FUNCTION

1. epidermis   6. physiology
2. dermis   7. metabolic
3. subcutaneous   8. chemical
4. adipose   (hormonal)
5. Anatomy   9. nervous

21 Integration and Control: Nervous Systems

21-I THE NEURON

1. neurons   6. terminals
2. Neuroglial   7. Sensory
3. dendrites   8. interneurons
4. conducting   9. motor
    zone
5. axon

21-II MESSAGE CONDUCTION
The Neuron "At Rest"
Changes in Membrane Potential

1. differentially   8. resting
2. concentration   membrane
3. against   potential
4. potassium   9. sodium-
5. sodium   potassium pump
6. negative   10. message
7. millivolts

21-III Action Potentials
Summary of Message Conduction Mechanisms

1. action   7. node of
    potential   Ranvier
2. nerve impulse   8. salatory
3. all-or-nothing   9. graded
4. threshold   potentials
5. refractory   10. summed
6. myelin sheaths

20-VI HOMEOSTASIS AND SYSTEMS CONTROL
The Internal Environment
Homeostatic Control Mechanisms

1. Interstitial   5. nutrients
2. liver   6. F
3. contraction   7. T
4. relaxation   8. F

21-IV MESSAGE TRANSFERS
Synaptic Transmission Between Neurons

1. synapse
2. transmitter substances
3. Acetylcholine (or dopamine or
    epinephrine or norepinephrine)
4. Integration
5. summed
6. integration
7. excitatory
8. inhibitory

21-V Junctional Transmission Between
  Neurons and Muscle Cells
THE REFLEX ARC: FROM STIMULUS TO RESPONSE

1. reflex   3. muscle spindles
2. stretch reflex 4. spinal cord

21-VI NERVOUS SYSTEMS: INCREASING THE
  OPTIONS FOR RESPONSE
Evolution of Nervous Systems

1. radial   6. muscles (nerves)
2. ganglia   7. brain
3. Cephalization   8. central
4. bilateral   9. peripheral
5. nerves (muscles)

21-VII Central Nervous Systems
Peripheral Nervous Systems

1. spinal cord
2. white matter
3. gray matter
4. synaptic
5. medulla oblongata
6. cerebellum
7. midbrain
8. brain
9. hypothalamus
10. endocrine
11. thalamus
12. somatic
13. autonomic
14. sympathetic
15. parasympathetic

21-VIII THE HUMAN BRAIN
Regions of the Cortex
Information Processing in the Cortex
Conscious Experience

1. cerebral cortex
2. motor cortex
3. somatosensory
4. visual cortex
5. occipital
6. auditory cortex
7. association cortex
8. stimulus type
9. Stimulus intensity (strength)
10. neuron
11. nerve
12. receptive fields
13. Stimulus location
14. corpus callosum
15. conscious
16. Broca's area

21-IX Memory
Sleeping and Dreaming
DRUG ACTION ON INTEGRATION AND CONTROL

1. memory
2. memory trace
3. short term formative
4. long term storage
5. electroencephalogram
6. alpha rhythm
7. slow-wave sleep
8. EEG arousal
9. REM sleep
10. reticular formation (sleep centers)
11. serotonin
12. Endorphins (enkephalins)
13. pain
14. transmitter substances
15. tranquilizers

22 Integration and Control: Endocrine Systems

22-I HORMONE FUNCTION
Comparison of Neural and Endocrine
  Cell Secretions
Mechanisms of Hormone Action
Factors Influencing Hormone Levels
  in the Bloodstream
COMPONENTS OF ENDOCRINE SYSTEMS

1. Neuro-transmitters
2. Neurosecretory
3. neurohormones
4. hormones
5. target
6. exocrine
7. same
8. cAMP
9. ovaries
10. testes

22-II NEUROENDOCRINE CONTROL CENTER
On Neural and Endocrine Links
Interactions Between the Hypothalamus
  and Pituitary

1. hypothalamus
2. neurohormones
3. pituitary
4. glandular (secretory)
5. nervous
6. portal
7. hypothalamus
8. ACTH
9. TSH
10. prolactin

22-III OTHER ENDOCRINE ELEMENTS
Adrenal Glands
Thyroid and Parathyroid Glands
Gonads
Glandular Tissues of the Pancreas, Stomach,
   and Small Intestine
Hormone Functions of Kidneys
Pineal Gland and Thymus Gland
Prostaglandins

1.	kidneys	8.	parathyroid glands
2.	angiotension	9.	promotes
3.	high	10.	Glucagon
4.	adrenal cortex	11.	insulin
5.	reproductive	12.	epinephrine (adrenalin)
6.	thyroid	13.	norepinephrine (noradrenalin)
7.	inhibits	14.	thymus

23 Reception and Motor Response

23-I SENSING ENVIRONMENTAL CHANGE
Primary Receptors
Chemical Reception
Mechanical Reception

1. primary receptors
2. stimulus
3. Chemoreceptors
4. mechanoreceptors
5. photoreceptors
6. thermoreceptors
7. nocireceptors
8. electroreceptors
9. Olfactory
10. Stretch
11. amplitude
12. frequency
13. higher
14. mechanoreceptors
15. middle ear
16. cochlea
17. organ of Corti
18. semicircular canals

23-II Light Reception

1.	photons	7.	retina
2.	Photoreception	8.	ommatidia
3.	Vision	9.	focal point
4.	Eyespots	10.	Accommodation
5.	Eyes	11.	Far
6.	cornea	12.	Cone
		13.	fovea

23-III MOVEMENT IN RESPONSE TO CHANGE
Motor Systems
Muscle Contraction
Summary of Muscle Contraction

1. contract
2. relax
3. contractile
4. antagonistic
5. hydraulic
6. exoskeleton
7. reciprocal innervation
8. myosin
9. actin
10. sarcomere
11. calcium

23-IV Types of Muscular Responses
Neuromotor Basis of Behavior

1. Twitches
2. tetanic
3. muscle tone

24 Circulation

24-I   INTERNAL DISTRIBUTION STRATEGIES
COMPONENTS OF BLOOD
Blood Composition and Volume
Types of Blood Cells and Their Function

1. connective
2. pH
3. closed
4. lymph
5. vascular system
6. polypeptide (protein) chains
7. iron
8. oxyhemoglobin
9. erythropoietin
10. bone marrow
11. Stem
12. Antibodies
13. plasma cells
14. Neutrophils
15. Platelets

24-II   STRUCTURE AND FUNCTION OF BLOOD
CIRCULATION SYSTEMS
Systemic and Pulmonary Circulation
The Human Heart
Blood Vessels
Blood Pressure and the Distribution
 of Flow

1. atrium
2. ventricle
3. systole
4. diastole
5. aorta
6. left pulmonary veins
7. semilunar valve
8. left ventricle
9. inferior vena cava
10. atrioventricular valve
11. right pulmonary artery
12. superior vena cava
13. artery
14. capillary
15. endothelial
16. capillary bed
17. interstitial
18. Valves
19. veins (venules)
20. venules (veins)
21. pulmonary
22. systemic
23. F
24. F

24-III   HEMOSTASIS
COMMENTARY: On Cardiovascular Disorders

1. hemostasis
2. platelet plug formation
3. coagulation
4. collagen
5. calcium
6. thrombin
7. fibrin
8. stroke
9. coronary occlusion
10. Plaque
11. low

24-IV   LYMPHATIC SYSTEM
Lymph Vascular System
Lymphoid Organs
NONSPECIFIC DEFENSE RESPONSES
External Barriers to Invasion
Inflammatory Response

1. right lymphatic duct
2. tonsils
3. left subclavian vein
4. right subclavian vein
5. thymus
6. thoracic duct
7. spleen
8. bone marrow
9. Lymph
10. fats
11. small intestine
12. bone marrow
13. spleen
14. hemorrhaging
15. Complement

24-V SPECIFIC DEFENSE RESPONSES:
  THE IMMUNE SYSTEM
Evolution of Vertebrate Immune Systems
Killer T-Cells and the B-Cell Ways of
  Chemical Warfare

1. phagocytosis
2. immune surveillance
3. lymphocyte
4. T-cells (B-cells)
5. B-cells (T-cells)
6. stem cells
7. receptors
8. precommitted
9. B-cell (plasma cell)
10. primary immune pathway
11. immunization
12. nonspecific reactions
13. primary immune reactions
14. specific lymphocytes
15. T-cells
16. B-cells
17. memory lymphocytes
18. antigens
19. memory lymphocytes
20. plasma cells
21. antibodies

24-VI COMMENTARY: Cancer, Allergy,
  and Attacks on Self
The Basis of Self-Recognition
Case Study: The Silent, Unseen Struggles

1. major histocompatibility complex
2. Autoimmune disease
3. Allergy
4. immune surveillance theory
5. killer T-cells
6. lymphokines
7. Interferons
8. monoclonal (pure) antibodies
9. Targeted
10. complement

25 Respiration

25-I OXYGEN ACQUISITION: SOME
  ENVIRONMENTAL CONSIDERATIONS
RESPIRATION: AN OVERVIEW

1. gill
2. countercurrent
    exchange mechanism
3. tracheae
4. pleura
5. alveoli

25-II HUMAN RESPIRATORY SYSTEM
Respiratory Organs
Relation Between Lungs and the Pleural Sac
VENTILATION
Inhalation and Exhalation
Controls Over Breathing

1. diaphragm
2. rib cage
3. increases
4. drops
5. brainstem
    (medulla
    oblongata)
6. pleural sac
7. larynx
8. glottis
9. bronchi
10. bronchioles
11. alveoli

25-III GAS EXCHANGE AND TRANSPORT
Gas Exchange in Alveoli
Gas Transport Between Lungs and Tissues
CILIARY ACTION IN RESPIRATORY TRACTS
COMMENTARY: When the Lungs Break Down

1. partial pressure
2. Passive diffusion
3. carbon dioxide
4. Hypoxia
5. bicarbonate
6. oxyhemoglobin (hemoglobin)
7. systemic
8. carbonic anhydrase
9. Emphysema
10. lung cancer

26 Digestion and Organic Metabolism

26-I   FEEDING STRATEGIES

1. Nutrition
2. Ingestion
3. digestion
4. absorption
5. assimilated

26-II   HUMAN DIGESTIVE SYSTEM:
AN OVERVIEW
Components of the Digestive System
General Structure of the
  Gastrointestinal Tract
Gastrointestinal Motility
CONTROL OF GASTROINTESTINAL MOVEMENTS
  AND SECRETIONS
STRUCTURE AND FUNCTION OF
  GASTROINTESTINAL ORGANS
Mouth and Salivary Glands
Pharynx and Esophagus
Stomach
Small Intestine
Large Intestine
Enzymes of Digestion: A Summary

1. amylase
2. epiglottis
3. esophagus
4. peristalsis
5. stomach
6. small
   intestine
7. Pepsin
8. Bile
9. small
   intestine
10. Amylase,
    lipase,
    trypsin or
    chymotrypsin
11. salivary glands
12. oral cavity
13. liver
14. stomach
15. gall bladder
16. small intestine
17. large intestine
18. rectum
19. pancreas
20. esophagus
21. pharynx
22. T
23. F
24. F
25. T
26. F

26-III   HUMAN NUTRITIONAL REQUIREMENTS
Energy Needs
Carbohydrates and Lipids
Proteins
COMMENTARY: Human Nutrition and
  Gastrointestinal Disorders
Vitamins and Minerals

1. D
2. calcium
3. A
4. D
5. K
6. niacin (vitamin B)
7. C
8. phosphorus
9. Iodine
10. Iron

26-IV   ORGANIC METABOLISM
The Vertebrate Liver
Absorptive and Post-Absorptive States
Case Study: Feasting, Fasting, and Systems
  Integration

1. general systemic
2. hepatic portal vein
3. ammonia
4. urea
5. insulin
6. glucagon
7. glycogen
8. epinephrine (norepinephrine)
9. norepinephrine (epinephrine)
10. glycogen
11. glucose
12. glucose

27 Regulation of Body Temperature and Body Fluids

27-I CONTROL OF BODY TEMPERATURE
Temperature Range Suitable for Life
Heat Production and Heat Loss
Thermal Regulation in Ectotherms
Thermal Regulation in Birds and Mammals
COMMENTARY: Falling Overboard and the
  Odds for Survival

1. hypothalamus   4. hypothermia
2. core   5. T
3. epinephrine   6. T

27-II CONTROL OF EXTRACELLULAR FLUIDS
Water and Solute Balance
Mammalian Urinary System
Kidney Function

1. T   13. Ureters
2. F   14. urinary bladder
3. T   15. urethra
4. F   16. Bowman's capsule
5. T   17. glomerulus
6. Filtration   18. proximal
7. permeability   convoluted tubule
8. pressure   19. loop of Henle
9. filter   20. distal convoluted
10. reabsorption   tubule
11. solutes   21. collecting duct
12. tubular   22. peritubular
    secretion   capillaries

27-III CONTROL OF EXTRACELLULAR FLUIDS
Controls Over Fluid Volume and Composition
COMMENTARY: Kidney Failure, Bypass Measures,
  and Transplants

1. Antidiuretic hormone
2. Diuresis
3. hypothalamus
4. thirst
5. dialysis

28 Principles of Reproduction and Development

28-I THE BEGINNING: REPRODUCTIVE MODES
Asexual Reproduction
Sexual Reproduction
Some Strategic Problems in Having
  Separate Sexes

1. zygote   6. Parthenogenesis
2. Cleavage   7. hermaphrodites
3. larva   8. ovaries
4. regeneration   9. testes
5. budding

28-II STAGES OF EARLY EMBRYONIC DEVELOPMENT
Egg Formation and the Onset of Gene Control
Visible Changes in Amphibian Eggs
  at Fertilization

1. F   9. Gastrulation
2. F   10. oviviviparity
3. T   11. oviparity
4. gametogenesis 12. RNA transcripts
5. egg   (maternal messages)
6. fertilization 13. animal pole
7. Cleavage   14. gray crescent
8. blastula   15. body axis

28-III Cleavage: Carving Up Cytoplasmic
    Controls
Gastrulation: When Embryonic Controls
    Take Over
Organ Formation

1. F
2. cleavage
3. blastula
4. gastrulation
5. germ
6. gastrula
7. nervous
8. gut
9. skeleton
10. incomplete
11. yolk
12. five

28-IV MORPHOGENESIS AND GROWTH
Growth Patterns
Cell Movements and Changes in Cell Shapes
Pattern Formation
Controlled Cell Death
CELL DIFFERENTIATION

1. T
2. T
3. embryonic
    regulation
4. identical
    twins
5. selective
6. Morphogenesis
7. chemical gradients
8. epithelial sheets
9. embryonic
    induction
10. host tissues
11. ooplasmic
    localization
12. controlled cell
    death

28-V CELL DIFFERENTIATION

1. T
2. F

28-VI METAMORPHOSIS
AGING AND DEATH
COMMENTARY: Death in the Open

1. T
2. metamorphosis
3. indirect

29 Human Reproduction and Development

29-I DEFINITION OF PRIMARY
    REPRODUCTIVE ORGANS
MALE REPRODUCTIVE SYSTEM
Male Reproductive Tract
Spermatogenesis
Sperm Movement Through the Reproductive
    Tract
Hormonal Control in the Male

1. seminiferous tubules
2. epididymis
3. testosterone
4. vas deferens
5. prostate
6. sphincter
7. lytic (digestive) enzymes
8. midpiece
9. Prostaglandins
10. 24, 72

29-II FEMALE REPRODUCTIVE SYSTEM
Female Reproductive Tract
Menstrual Cycle: An Overview
Ovarian Function
Uterine Function
SEXUAL UNION AND FERTILIZATION
COMMENTARY: In Vitro Fertilization

1. estrogen (progesterone)
2. progesterone (estrogen)
3. oviducts
4. uterus
5. estrus
6. follicle
7. ovulation
8. oviduct
9. corpus luteum
10. endometrium
11. Progesterone
12. LH
13. FSH
14. menstruation

29-III FROM FERTILIZATION TO BIRTH
Early Embroyonic Development
Case Study: Mother as Protector,
 Provider, Potential Threat

1. oviduct   4. second
2. implantation   5. fetus
3. placenta

29-IV CONTROL OF HUMAN FERTILITY
Some Ethical Considerations
Possible Means of Birth Control

1. 1,330,560   7. estrogens
2. 200,000   (progesterones)
3. 1 million   8. progesterones
4. abstention   (estrogens)
5. Condoms   9. gonadotropins
6. diaphragm   10. tubal ligation

30 Individuals, Populations, and Evolution

30-I VARIATION: WHAT IT IS, HOW IT ARISES
THE HARDY-WEINBERG BASELINE FOR MEASURING CHANGE
FACTORS BRINGING ABOUT CHANGE
Effects of Mutations
Genetic Drift
Gene Flow
Natural Selection

1. typological
2. population
3. Hardy-Weinberg rule
4. allele frequencies
5. genetic equilibrium
6. mutations
7. genetic drift
8. Gene flow
9. polymorphism
10. natural selection
11. $p^2 = 0.81$

   $p = \sqrt{0.81} = 0.9 =$ the frequency of the dominant allele

  $p + q = 1$

   $q = 1 - 0.9 = 0.1 =$ the frequency of the recessive allele

  $2pq = 2 \cdot (0.9) \cdot (0.1) = 2 \cdot (0.09) = 0.18$

     $= 18\%$ which is the percentage of heterozygotes

12. homozygous dominant $= p^2 \times 200 = (0.8)^2 \times 200 = 0.64 \times 200 = \underline{128}$ individuals

   homozygous recessive $= q^2 \times 200 = (0.2)^2 \times 200 = (0.04)(200) = \underline{8}$ individuals

   heterozygotes $= 2pg \times 200 = 2 \times .8 \times .2 \times 200 = .32 \times 200 = \underline{64}$ individuals

   Check: $128 + 8 + 64 = 200$

30-II EXAMPLES OF NATURAL SELECTION
Stabilizing Selection
Directional Selection
Disruptive Selection
Sexual Selection
Selection and Balanced Polymorphism

1. Directional selection
2. Disruptive selection
3. Horseshoe crabs
4. balanced polymorphism

30-III EVOLUTION OF SPECIES
Defining a Species
Reproductive Isolating Mechanisms
Modes of Speciation

1. geographic isolation
2. genetic drift
3. divergence
4. Speciation
5. reproductive
6. Allopatric
7. polyploidy
8. polyploid

30-IV MACROEVOLUTION
A Time Scale for Macroevolution
Rates and Patterns of Change

1. Proterozoic
2. prokaryotes
3. eukaryotes
4. Paleozoic
5. selection
6. speciation
7. adaptive radiation

30-V EVOLUTIONARY TRENDS
Phyletic Evolution
Quantum Evolution
Mosaic Evolution
Extinction

1. gradualistic
2. phyletic
3. punctuational
4. quantum speciation
5. *Archaeopteryx*
6. "mosaic"
7. key characters
8. feathers
9. Extinction

31 Origins and the Evolution of Life

31-I ORIGIN OF LIFE
The Early Earth and Its Atmosphere
Spontaneous Assembly of Organic Compounds
Speculations on the First Self-
  Replicating Systems

1. plates   5. clay crystals
2. mantle   6. amino acids
3. 3   7. Microspheres
4. 3.7   8. Liposomes

31-II THE AGE OF PROKARYOTES
THE RISE OF EUKARYOTES
Divergence Into Three Primordial Lineages
Symbiosis in the Evolution of Eukaryotes
Origin of the Nucleus
EUKARYOTES OF THE PRECAMBRIAN

1. 2   6. stromatolites
2. oxygen   7. aerobic
3. anaerobic   respiration
4. 1.4   8. symbiotic
5. aerobic
   (cellular)

31-III LIFE DURING THE PALEOZOIC

1. Gondwana   5. amphibians
2. trilobites   6. Carboniferous
3. invertebrate   7. Pangea
4. Jawless

31-IV MESOZOIC: AGE OF THE DINOSAURS
CENOZOIC: THRESHOLD TO THE PRESENT
PERSPECTIVE

1. volcanic   10. flowering
2. Fragmentation   11. Gymnosperms
3. uplifting   (Reptiles)
4. rifts   12. reptiles
5. mountain   (gymnosperms)
   building   13. mammals
6. sea level
7. temperature
8. grasslands
9. mammals
   (birds)

32 Viruses, Bacteria, and Protistans

32-I BACTERIA
Characteristics of Bacteria
Archaebacteria
Eubacteria
Fifteen Thousand Species of Bacteria
  Can't Be All Bad

1. Chemosynthetic
2. decomposers
3. parasites
4. obligate anaerobes
5. peptidoglycan
6. Archaebacteria
7. methanogens
8. Cyanobacteria
9. nitrogen fixation
10. endospores
11. Mycoplasmas
12. Rickettsiae
13. cyanobacteria
14. bacterial (binary) fission
15. spores

32-II VIRUSES
Characteristics of Viruses
COMMENTARY: Bacteria, Viruses, and
  Sexually Transmitted Diseases
Viroids

1. virions
2. nucleic acid
3. viral capsid (protein coat)
4. Viruses
5. pandemic
6. herpesvirus
7. Viroids
8. spirochaete (treponeme)
9. chancre

32-III PROTISTANS
Euglenids
Chrysophytes and Diatoms
Dinoflagellates

1. Symbiosis
2. Dinoflagellates
3. bioluminescence
4. golden algae
5. diatoms
6. heterotrophic

32-IV Protozoa

1. pseudopodia
2. actin
3. ATP
4. Foraminiferans
5. *Plasmodium*
6. sporozoan (parasite)
7. gametes
8. sporozoites
9. freshwater
10. contractile vacuoles
11. gullet
12. food vacuoles
13. Trichocysts

32-V ON THE ROAD TO MULTICELLULARITY

1. *Prochloron*
2. *Chlamydomonas*
3. volvocines
4. plants
5. *Trichoplax adharens*
6. colony
7. multicellular

33 Fungi and Plants

33-I   KINGDOM OF FUNGI
On the Fungal Way of Life
Fungal Body Plans
Overview of Reproductive Modes
Major Groups of Fungi

1.	plants	7.	asci
2.	spore	8.	ascocarp
3.	parasitic	9.	sac
4.	mycelium	10.	club
5.	hyphae	11.	ten
6.	zygote		

33-II   Mycorrhizal Mats and the Lichens
Species of Unknown Affiliations

1.	Mycorrhizae	3.	pseudoplasmodium
2.	Lichens	4.	fruiting body

33-III   KINGDOM OF PLANTS
Evolutionary Trends Among Plants
Algae: Plants that Never Left the Water

1.	400	7.	pigments
2.	passive	8.	Brown
3.	gametophyte	9.	kelps
4.	meiosis	10.	holdfast
5.	spores	11.	green
6.	Red	12.	land

33-IV   The Bryophytes
Ferns and Their Allies: First of the
   Vascular Plants

1.	vascular	9.	diploid
2.	sexually	10.	sporophyte
3.	mosses	11.	sporophyte
4.	gametophyte	12.	gametophyte
5.	sporophyte	13.	vascular
6.	homospores	14.	ferns
7.	gametophyte	15.	rhizomes
8.	gametes	16.	sporophyte

**33-V   Gymnosperms: The Conifers and
   Their Kin**
Angiosperms: The Flowering Plants
PERSPECTIVE

1.	heterosporous	7.	seed coat
2.	female	8.	gymnosperms
3.	pollen grains	9.	Cycads
4.	wind	10.	Ginkgos
5.	angiosperms	11.	Conifers
6.	seeds	12.	Mesozoic

34 Animal Diversity

34-I   GENERAL CHARACTERISTICS OF ANIMALS
Body Plans
Major Groups of Animals
SPONGES
CNIDARIANS

1.	invertebrates	6.	nerve
2.	metazoans	7.	radially
3.	Sponges	8.	nematocysts
4.	collar cells	9.	planula
5.	Communication	10.	medusa

34-II   FLATWORMS
Turbellarians
Trematodes
Cestodes
Flatworm Origins
ROUNDWORMS

1.	planula
2.	mesoderm
3.	salt (ion)
4.	paired organs
5.	respiratory (circulatory)
6.	digestive
7.	regional specialization
8.	Nematodes (sac worms)
9.	cuticle
10.	anaerobic

34-III TWO MAIN LINES OF DIVERGENCE: PROTOSTOMES AND DEUTEROSTOMES

1. Deuterostomes
2. anus
3. mouth
4. protostomes
5. earthworms
6. insects (spiders)
7. snails (clams)
8. parallel

34-IV MOLLUSKS
Gastropods
Bivalves
Cephalopods

1. Turbellarians
2. mollusks
3. cephalopods
4. brains
5. vision
6. motor
7. mantle
8. mantle cavity

34-V ANNELIDS

1. Annelids
2. polychaetes (marine worms)
3. anus
4. coelom
5. blood
6. hemoglobin
7. Segmentation
8. ventral nerve cord
9. nephridia
10. parapodia

34-VI ARTHROPODS
Arthropod Adaptations
Chelicerates
Crustaceans
Insects and Their Kin

1. cuticle
2. exoskeletons
3. crustaceans
4. arachnids
5. spiders
6. Centipedes
7. Insects
8. gills
9. tracheal
10. metabolic
11. annelids

34-VII ECHINODERMS
CHORDATES
From Notochords to Backbones
Evolutionary Potential of Bones and Jaws
Lungs and the Vertebrate Heart
On the Vertebrate Nervous System

1. deuterostomes
2. echinoderms
3. radial
4. bilaterally
5. water-vascular
6. tube feet
7. Sea urchins
8. Crinoids
9. Tunicates (sea squirts)
10. tadpoles
11. notochord
12. endoskeleton
13. lancelets
14. filter feeding
15. gill slits
16. jawless fishes
17. jaws
18. swim bladder
19. two-
20. atria
21. pulmonary

35 Human Origins and Evolution

35-I THE PRIMATE FAMILY
Primate Origins
General Characteristics of the Primates

1. Cenozoic
2. Insectivora
3. shrews
4. primates
5. depth
6. grasping

35-II EARLY HOMINIDS (PERHAPS) AND THEIR PREDECESSORS (MAYBE)
Ancient Apelike Forms
Australopithecines
The First Humans
COMMENTARY: A Biological Perspective on Human Organs

1. *Aegyptopithecus*
2. quadrupedal
3. 25
4. rainfall
5. grasslands
6. savannas
7. dryopithecines
8. adaptive radiation
9. *Ramapithecus*
10. six
11. australopithecines
12. bipedal
13. two
14. Neanderthalers
15. *Homosapiens*

36 Population Ecology

36-I ECOLOGY DEFINED
POPULATION DENSITY AND DISTRIBUTION
Ecological Density
Distribution in Space
Distribution Over Time
POPULATION DYNAMICS
Parameters Affecting Population Size
Life Expectancy and Survivor Curves

1. population
2. biosphere
3. Fitness
4. ecological density
5. crude density
6. clumped
7. immigration
8. emigration
9. Suvivorship curves
10. III

36-II POPULATION GROWTH
Biotic Potential
Exponential Growth
Environmental Resistance to Growth
Tolerance Limits
Carrying Capacity and Logistic Growth
REPRODUCTIVE RESPONSES TO GROWTH LIMITS

1. exponential
2. reproductive
3. J
4. limiting
5. plateau
6. tolerance range
7. Carrying capacity
8. biotic potential
9. logistic growth
10. r-selected

36-III FACTORS THAT REGULATE POPULATION GROWTH
Density-Dependent and Density-Independent Influences
Extrinsic Influences
Intrinsic Influences

1. dependent
2. carrying capacity
3. independent
4. intrinsic limiting
5. extrinsic limiting
6. D
7. A
8. B, C, D
9. A, B, C

36-IV HUMAN POPULATION GROWTH
Doubling Time for the Human Population
Where We Began Sidestepping Controls
Age Structure and Fertility Rates

1. age structure
2. fertility rate
3. 2.5
4. 70-100
5. an immense number of children are still in the pre-reproductive base
7. 20
8. 70

37 Community Interactions

37-I HABITAT AND NICHE
TYPES OF COMMUNITY INTERACTIONS

1. ecosystem
2. community
3. habitat
4. niche
5. realized niche
6. resource gradient
7. commensal
8. mutualistic

37-II INTERSPECIFIC COMPETITION
The Concept of Competitive Exclusion
Field Evidence of Competitive Exclusion
Coexistence in Resource Categories
PREDATION

1. competition
2. competitive exclusion
3. niche
4. interference competition
5. resource partitioning
6. cyclic oscillations
7. territories
8. plants
9. animals
10. anything

37-III COEVOLUTION
Warning Coloration and Mimicry
Camouflage
Moment-of-Truth Defenses
Chemical Defenses
Predation and Seed Dispersal
Energy Outlays for Predation
Mutualism
Parasitism

1.	Coevolution	7.	C	13.	D
2.	disguise	8.	C	14.	C
3.	Camouflage	9.	A	15.	D
4.	Müllerian	10.	C	16.	E
5.	D	11.	B	17.	D
6.	A	12.	A	18.	B

38 Ecosystems

38-I ORGANIZATION OF ECOSYSTEMS
Trophic Levels
Food Webs
Case Study: Biological Concentration
  of DDT

1.	protistans (or some bacteria)	5.	Herbivores
		6.	Decomposers
		7.	mosquitos
2.	energy	8.	malaria
3.	nutrients	9.	house cats
4.	trophic level	10.	biological concentration

38-II ENERGY FLOW THROUGH ECOSYSTEMS
Primary Productivity and Energy Storage
Major Pathways of Energy Flow
Energy Budgets

1.	net primary production	3.	dentrital food web
2.	Biomass	4.	detritus
		5.	ten

38-III NUTRIENT CYCLING
A Model of Nutrient Flow
Case Study: Nutrient Recovery in the
  Arctic Tundra
Nitrogen Cycling in Ecosystems

1.	brown lemming	6.	nitrous oxide
2.	Bacteria	7.	nutrient recovery hypothesis
3.	nitrogen fixation		
4.	nitrification		
5.	nitrite (NO_2^-)	8.	nutrient availability

38-IV COMMENTARY: Resources and the Human Condition

1.	4.7	5.	35
2.	700	6.	nitrogen-fixing
3.	high-grade	7.	High-yield crops
4.	one	8.	salt

38-V SUCCESSION
Succession Defined
Opportunistic and Equilibrium Species
Disturbances in Succession
SPECIES INTRODUCTIONS

1.	Primary succession	8.	A
		9.	E
2.	pioneer	10.	H
3.	climax	11.	G
4.	Secondary	12.	C
5.	Sequoia	13.	D
6.	fire	14.	B
7.	A	15.	F

39 The Biosphere

39-I COMPONENTS OF THE BIOSPHERE
GLOBAL PATTERNS OF CLIMATE
Moderating Effects of the Atmosphere
Air Currents
Ocean Currents
Seasonal Variations in Climate
Regional Climates

1.	F	15.	O
2.	C	16.	Q
3.	G	17.	T
4.	J	18.	L
5.	I	19.	S
6.	H	20.	R
7.	A	21.	grasslands
8.	D	22.	shrubs
9.	B	23.	montane
10.	E	24.	subalpine
11.	M	25.	alpine
12.	N	26.	lack of
13.	K		vegetation
14.	P		cover
		27.	deserts

39-II BIOGEOCHEMICAL MOVEMENTS
Hydrologic Cycle
Gaseous Cycles
Sedimentary Cycles

1.	sedimentary	4.	T
2.	phosphorus	5.	F
3.	T		

39-III AQUATIC ECOSYSTEMS
Freshwater Ecosystems
Marine and Estuarine Ecosystems

1.	Lentic	11.	intertidal zone
2.	lotic	12.	bathpelagic
3.	overturns		(abyssal) zone
4.	estuary	13.	epipelagic
5.	littoral		(photic) zone
6.	profundal	14.	hydrothermal vent
7.	Oligotrophic	15.	*Spartina*
8.	thermocline		
9.	plankton		
10.	nekton		

39-IV TERRESTRIAL ECOSYSTEMS
Effect of Soils
The Concept of Biomes
Deserts
Shrublands
Grasslands

1. Topsoil (or, the A horizon)
2. subsoil (or, B horizon)
3. loose rock (or C horizon)
4. desert
5. cold (cool) deserts
6. hot (warm) deserts
7. grasslands (prairie)
8. Tallgrass prairie
9. shortgrass prairie

39-V Forests
Tundra
THE CITY AS ECOSYSTEM

1.	tropical rain forests	6.	Slash-and-burn	11.	Taiga
2.	complex	7.	tropical seasonal	12.	Ninety
3.	canopy	8.	temperate deciduous	13.	tundra
4.	diversity	9.	winter cold	14.	solar radiation
5.	worst	10.	ponderosa pine	15.	ecosystem

40   Human Ecology

40-I   Case Study: Solid Wastes
Case Study: Water Pollution

1. 2000
2. electromagnets
3. air blowers
4. primary
5. Tertiary
6. half

40-II   Case Study: Air Pollution
Acid Deposition
Industrial and Photochemical Smog
SOME ENERGY OPTIONS
Fossil Fuels
Nuclear Energy
Wind Energy
Solar Energy
PERSPECTIVE

1. Dry acid
   depositions
2. sulfuric acid
3. thermal
   inversion
4. hydrocarbons
5. photochemical
6. kerogen
7. 8
8. coal
9. breeder
10. 15,000
11. 50

41   Animal Behavior

41-I   THE DIVERSITY OF BEHAVIOR
Innate Behavior
Categories of Learning
AN EVOLUTIONARY APPROACH TO BEHAVIOR
Genes, Environment, and the Development
  of Behavior
Evolution of Behavior

1. motor score
2. innate
3. releaser
4. Learning
5. constant
6. associative
7. conditioned
   reflexes
8. instrumental
9. trial-and-error
10. extinction
11. Insight
    learning
12. sensitive
    periods
13. imprinting

41-II   ECOLOGICAL ASPECTS OF BEHAVIOR
Biological Clocks, Compasses, and Maps
Predator and Prey Behavior

1. migration
2. Hormones
3. days
4. homing behavior
5. Circadian
6. photoperiodic

41-III   BEHAVIORAL ADAPTATION: WHO BENEFITS?
On Individual Selection
Case Study: Siblicide Among the Egrets
Case Study: Courtship Behavior
Case Study: Competition for Females
Case Study: Tactics of Defeated Males

1. Behavioral
2. Individual
3. territory
4. vocalizations
5. visual displays
6. mate-guarding

41-IV   ON "SELFISH BEHAVIOR AND SOCIAL LIFE"
Parenting as Genetically Selfish Behavior
Individual Advantages to Group Living
Cooperative Societies of Birds and Mammals
Suicide, Sterility, and Social Insects
SELECTION THEORY AND THE EVOLUTION OF
  HUMAN BEHAVIOR

1. open
2. hide
3. dominance
4. dominant
   (stronger)
5. dominance
   hierarchy
   (social
   hierarchy)
6. ritualized
   behavior
7. appeasement
8. social insects
9. queen
10. Drones
11. biological
    determinism

CROSSWORD NUMBER ONE

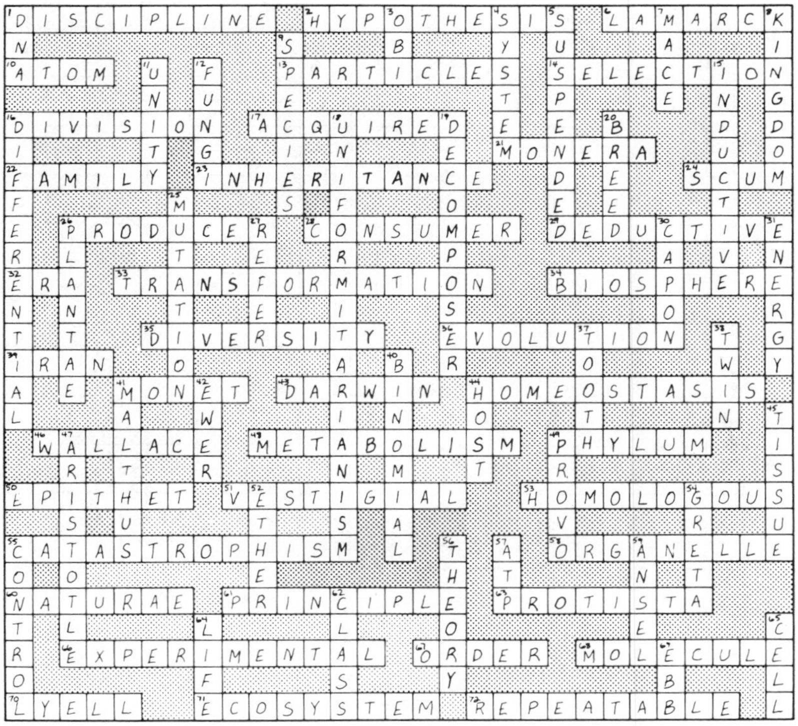

CROSSWORD NUMBER TWO

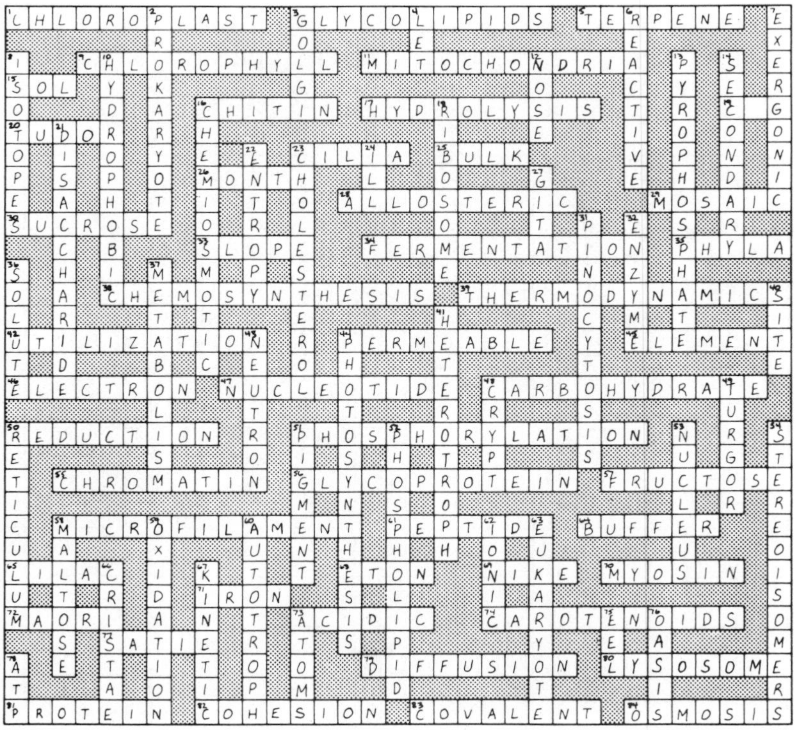

CROSSWORD NUMBER THREE

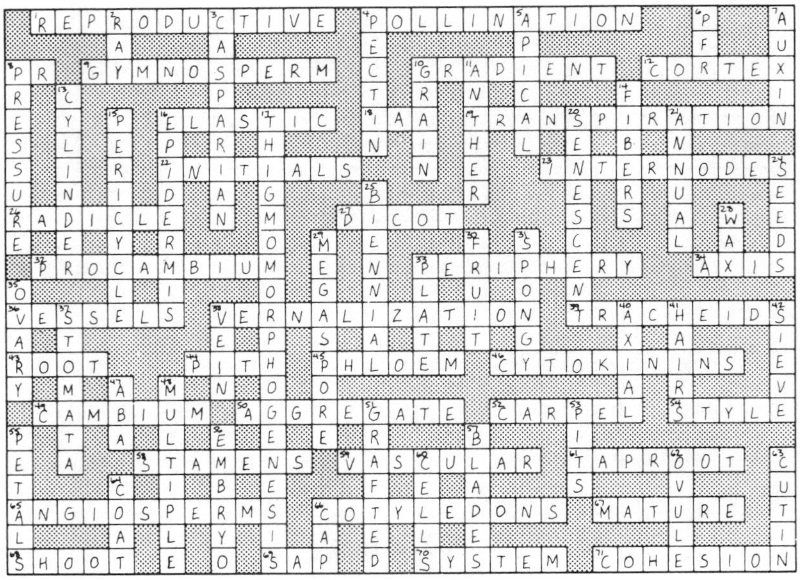

CROSSWORD NUMBER FOUR

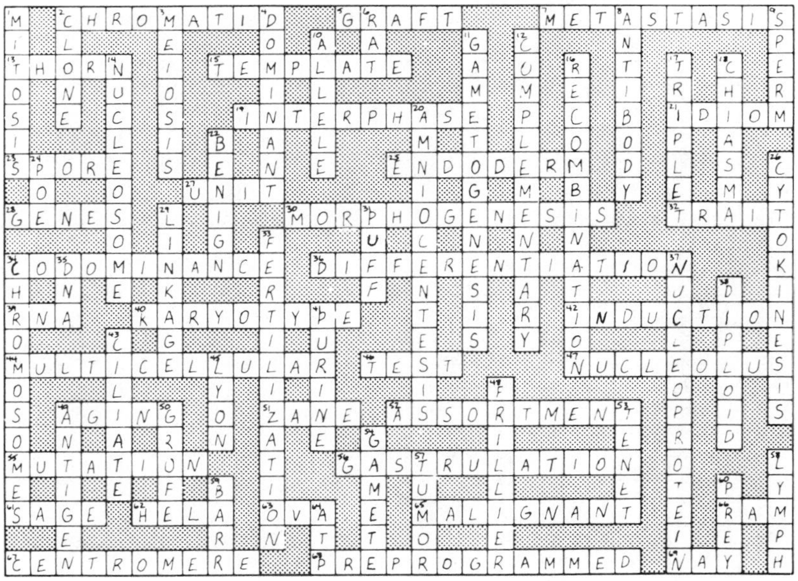

CROSSWORD NUMBER FIVE

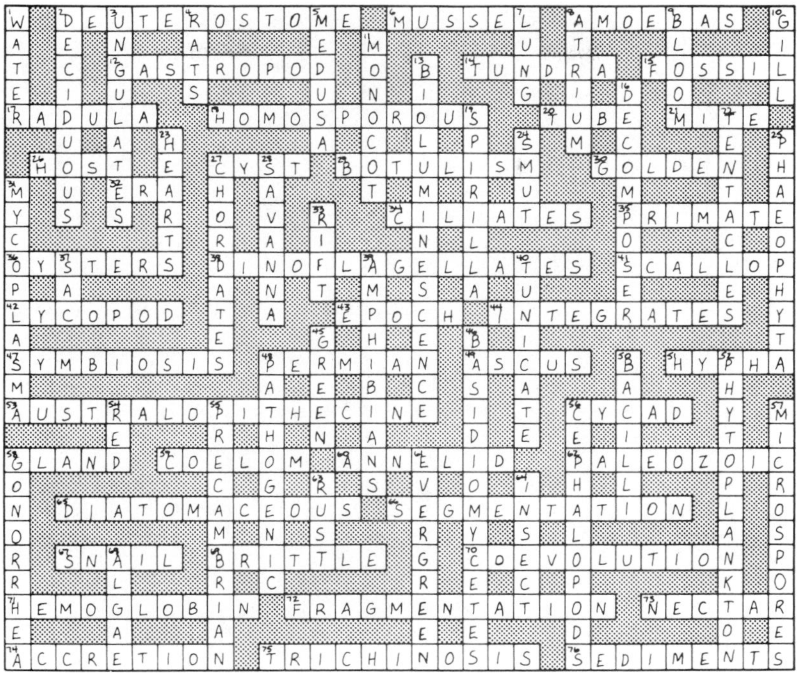

CROSSWORD NUMBER SIX

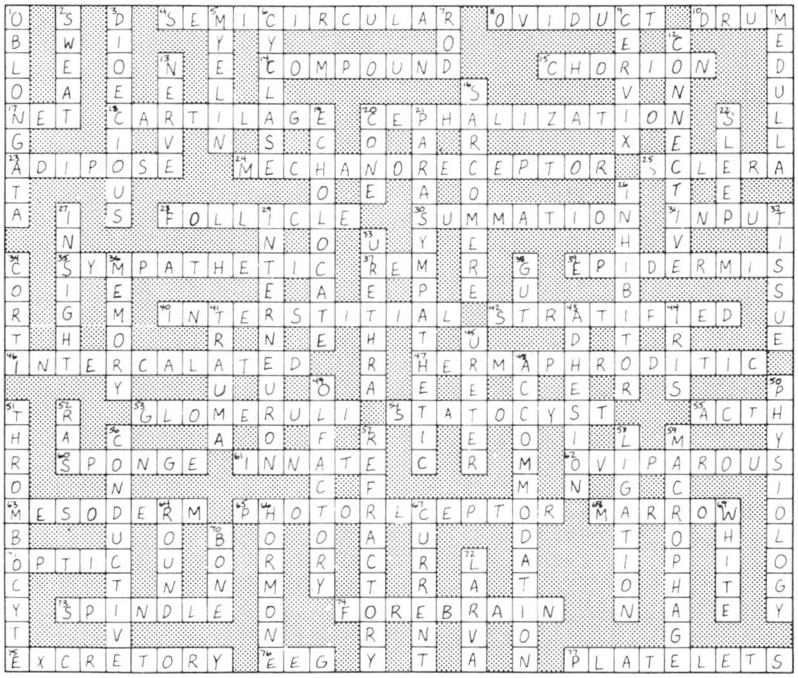